Epigenetic Drug Discovery

This book reviews the evolving field of epigenetics and its implications for drug discovery and precision medicine. It also focuses on the intricate mechanisms governing gene regulation and the impacts of epigenetics on health and disease. The book encompasses the complexities of epigenetic mechanisms and their role in diseases such as cancer, autoimmune disorders, inflammatory conditions, and neurodegenerative diseases. Additionally, it examines the role of miRNA as an epigenetic drug in treating cancer and in silico approaches to epigenetic drug discovery. The book uncovers the potential of epigenetic drug discovery, including insights into traditional medicine, marine sources, and omics technologies. Moreover, it explores the interplay between gut microbiota, probiotics, and prebiotics in epigenetic therapeutic approaches. With a focus on personalized medicine and recent advancements in drug delivery systems, this book is intended for researchers, clinicians, and pharmaceutical professionals seeking to explore epigenetic drug discovery and development.

- Discusses epigenetic mechanisms in different diseases and epigenetics drug discovery.
- Reviews targeted therapies developed based on epigenetic mechanisms for cancer, autoimmune diseases, neurodegenerative conditions, and inflammatory diseases.
- Emphasizes the importance of precision medicine and personalized therapies based on epigenetic profiling.
- Explores applications of artificial intelligence in designing novel epigenetic drug delivery systems.
- Presents role of advance in-silico and drug discovery tools in the development of epigenetic drugs.

Epigenetic Drug Discovery

Advancements, Challenges, and Applications in Precision Medicine

Edited by
M V N L Chaitanya, Trudi Collet, Jon Adams,
Kamal Dua, and Sachin Kumar Singh

CRC Press
Taylor & Francis Group
Boca Raton London New York

CRC Press is an imprint of the
Taylor & Francis Group, an **informa** business

Designed cover image: M V N L Chaitanya, Trudi Collet, Jon Adams, Kamal Dua and Sachin Kumar Singh

First edition published 2026
by CRC Press
2385 NW Executive Center Drive, Suite 320, Boca Raton FL 33431

and by CRC Press
4 Park Square, Milton Park, Abingdon, Oxon, OX14 4RN

CRC Press is an imprint of Taylor & Francis Group, LLC

ISBN: 978-1-032-70649-8 (hbk)
ISBN: 978-1-032-70650-4 (pbk)
ISBN: 978-1-032-70651-1 (ebk)

DOI: 10.1201/9781032706511

Typeset in Times
by Apex CoVantage, LLC

Contents

About the Editors vii
List of Abbreviations ix

Chapter 1 Epigenetics Challenges and Prospects 1

Kanaka Raju A, Vijay Srinivas P, Rushikesh B Katkar, and Gana Manjusha Kondepudi

Chapter 2 Epigenetic Disease Mechanisms 14

Aman Khandelwal, Suguna Kotte, and S. Ponnusankar

Chapter 3 Epigenetic Drug Discovery and Its Importance in Precision Medicine 36

Mohammad Aleemuddin and Ayesha Sultana

Chapter 4 Traditional Medicine and Its Importance in Epigenetic Healing 46

Venkateswara Rao Amara, Ajay Sharma, Abhishek Tiwari, Varsha Tiwari, Vineet Mittal, Mohan Gandhi Bonthu, Bhanu Sirisha Kagitapalli, and Arya Lakshmi Marisetti

Chapter 5 Epigenetic Drug Discovery in Multiresistant Cancers and Autoimmune Diseases 58

Stuti Biswal, Rajeev Kumar Sahoo, Sambit Kumar Patra, Shikshya Swarupa Panda, and Bijesh K Biswal

Chapter 6 Epigenetic Drug Discovery in Inflammatory Diseases 85

Silpa R S, Akey Krishna Swaroop, Esakkimuthukumar M, Saranya R, Prabha T, Suresh S, and Jubie S

Chapter 7 Epigenetic Drug Discovery in Neurodegenerative Diseases 99

Madhavi Patel, Priya Shah, and M V N L Chaitanya

Chapter 8 *In Silico* Tools, Molecular Docking Knowledge, and Epigenetic Drug Discovery 115

Prabha Thangavelu, Pradeep Kumar MR, Jawahar Natarajan, Veerasamy Ravichandran, and Muthu Kumaradoss Kathiravan

Chapter 9 The Role of Indian Traditional Knowledge in Epigenetic Drug Discovery 134

Rajan Logesh, Maruveettill Arjun, Saladula Hima, Prabitha Prabhakaran, and Kumarappan Chidambaram

Chapter 10 The Importance of Omics in Epigenetic Drug Discovery 153

Kella Alekhya, Anuciya R, Jawahar Rohith R, and Kaviyarasan D

Chapter 11 The Role and Mechanisms of Gut Microbiota in Epigenetic Healing 159

Arshia Razdan, Romi Shreshtha, M V N L Chaitanya, and Smriti Arora

Chapter 12 Prebiotics and Probiotics in Epigenetics Therapies 165

Saranya Punniyakotti, Krishna Prabha Naduchamy, and Nithya Varadarajan

Chapter 13 Epigenetic Drug Discovery and Development from Marine Sources 180

Rashmi Saxena Pal, Yogendra Pal, Shubham Kumar, Sharma Vedika Vijaypal, Nisha Kumari, and Satya Kumar Avula

Chapter 14 Mapping Novel Epigenetic Drug Delivery Systems with Artificial Intelligence..... 196

Gana Manjusha K, Chiranjeevi P, and Aruna Kumari D

Chapter 15 Recent Developments in Epigenetics-Based Personalized Medicine 208

Deepti Katiyar, Shreya Dixit, Rajnandani Singh, Rashmi Saxena Pal, and Pradeep Singh

Index 241

About the Editors

Dr M V N L Chaitanya, BPharm, M Pharm (Phytopharmacy and Phytomedicine), PhD (Pharmacognosy), Educationist, Researcher, was born on November 10, 1982, in Rajahmundry (A.P.), India. He has 14 years of experience in industry, academia, and herbal research. He is currently working as a professor in the School of Pharmaceutical Sciences at the Lovely Professional University, Phagwara, Punjab, India. He is a well-recognized conceptualist in Indian and African herbal medicine, anticancer drug discovery and development, crystal healing, and psychic reading. He has keen skills in laboratory drug discovery and development from herbs, weeds, algae, and mushrooms, with a proven ability in quality documentation. He has received awards including the Indian National Science Academy Visiting Scientist Fellowship (2019), the Artificial Intelligence Media Festival Award (2019), the National Forum of Pharmacy Students Best Professor Award (2022), and the DK International Best Researcher Award for 2023. He is a lifetime member of the Association of Pharmaceutical Teachers of India, the Indian Pharmaceutical Association, the Institute of Scholars, and the International Natural Product Sciences Task Force. He visited three African countries as part of academic research to explore the phytochemicals in Indo–African plants against resistant diseases such as cancer, malaria, and tuberculosis. Dr Chaitanya's extensive body of work includes numerous research papers, peer-reviewed articles, books, and book chapters, and he holds multiple patents for his innovations.

Dr Trudi Collet, a biochemist, was awarded her PhD from the Queensland University of Technology in 2007. She worked as a postdoctoral fellow at the Queensland Institute of Medical Research and investigated novel serological markers for the differential diagnosis of inflammatory bowel disease. Since 2009, she has returned to Queensland University of Technology and is now the leader of the Innovative Medicines Group (IMG), which is based at the Kelvin Grove campus. The IMG conducts research on the medicinal properties of Australian native plants. Using plant-derived bioactive compounds and extracts, the IMG evaluates the biological potential including bactericidal effects, the deadly ESKAPE pathogens, viricidal effects against members of the *Flaviviridae* family (dengue, Zika and West Nile viruses); wound healing properties; antimicrobial effects against sexually transmitted diseases and novel pharmacotherapeutic targets for common neurological disorders.

Prof Jon Adams is Distinguished Professor of Public Health, Director of the Australian Research Centre in Complementary and Integrative Medicine and Deputy Head of School (Research) at the University of Technology Sydney (UTS) School of Public Health. Professor Adams was recently awarded a prestigious Fulbright Senior Scholarship (2019–2020). He is the world-leading public health and health services researcher focused upon self-care, informal healthcare and traditional, complementary, integrative, and indigenous medicine use and practice. Jon's research program has attracted over $20.7M in competitive external funding to date including 23 Category 1 grants (with 17 NHMRC and ARC grants). The professor's extensive research program, conducted with many industry partners (>$10M in partnership funding), includes focus upon practice-based research networks, implementation science, translation research, and mixed-methods design. Professor Adams has authored over 560 peer-reviewed academic publications and is Chief Editor of eight international research books and author of over 70 book chapters with leading publishers.

Dr Kamal Dua is Associate Professor in the Discipline of Pharmacy at the University of Technology Sydney (UTS). Dr Dua has been recognized and named one of Australia's Top 250 researchers in 2025 and 2024 in *Research Magazine* by "The Australian" in the research discipline of health and medical sciences and no. 1 research leader in the area of pharmacology and pharmacy (2025) and toxicology (2024), respectively. With over 17 years of research experience, he has specialized in drug delivery targeting inflammatory diseases. Additionally, he also leads drug delivery research at the Woolcock Institute of Medical Research. In this role, he focuses on advancing targets identified in research projects to develop innovative formulations, taking the first steps toward clinical translation. His research encompasses two complementary areas: drug delivery and immunology. His work explores how these disciplines can mutually benefit, contributing to the goal of promoting longer and healthier lives for the community. His commitment to synergy is reflected in his extensive publication record in reputable journals. His research interests are centered on harnessing the pharmaceutical potential of modulating critical regulators, such as interleukins and microRNAs. He also specializes in developing new and effective drug delivery formulations for managing inflammation in chronic airway diseases and cancer.

Dr Sachin Kumar Singh, a professor at School of Pharmaceutical Sciences, Lovely Professional University, India, holds a master of pharmacy in pharmaceutical analysis from J.S.S. College of Pharmacy, Ooty, Tamil Nadua and a PhD in pharmaceutical sciences from Karpagam University, Coimbatore, India. He has completed one DST-SERB-funded project on the development of SNEDDS (self-nano-emulsifying drug-delivery systems) of polypeptide-k for treatment of diabetes mellitus. Dr. Singh has published more than 200 research papers and has a cumulative impact of 400 and co-authored 10 book chapters. He was granted two patents in India and two in Australia. In January 2020, he received the Dr. B C Deb memorial award for popularization of science by the Indian Science Congress Association at the 107th Indian Science Congress held at University of Agricultural Services, Bengaluru. He is the first recipient from pharmaceutical sciences. He also received the 2014 Bharat Shiksha Ratan award from the Global Society for Health and Educational Growth, Delhi. His core area is improving the oral bioavailability of poorly water-soluble drugs using formulation approaches including nanotechnology, complexation, solid dispersion, and liquisolid technology.

Abbreviations

5-FU	5-fluorouracil
Aβ	Amyloid beta protein
A-SAA	Acute phase serum amylase-A
A1AT	Alpha 1 antitrypsin
ACS	Acute coronary syndrome
ACS	American Cancer Society
AITDs	Thyroid autoimmune diseases
AKAP12	A-kinase anchoring protein 12
AKT/PKB	Protein kinase B
AML	Acute myeloid lymphoma
APEX1	Apurinic/apyrimidinic endodeoxyribonuclease 1
APP	Amyloid precursor protein
BMP3	Bone Morphogenetic Protein 3
BNP	Brain-type natriuretic peptide
BRCA1	Breast Cancer gene 1
BRCA2	Breast Cancer gene 2
CA19-9	Cancer antigen 19-9
CALCA	Calcitonin-related polypeptide alpha
CBP	CREB binding protein
ccfDNA	Cell free tumor DNA
ccRCC	Clear cell renal cell carcinoma
CD 4+	Cluster of differentiation 4
CD 40 LG	Cluster of differentiation 40 ligand
CD40	Cluster of differentiation 40
CD 8+	Cluster of differentiation 8
CD70	Cluster of differentiation 70
CD247	Cluster of differentiation 247
CDH13	Cadherin 13
CDKN2A	Cyclin-dependent kinase inhibitor 2A
CDO1	Cysteine dioxygenase 1
CEA	Carcinoembryonic antigen
CGRP	Calcitonin gene related peptic
CLIP4	CAP-Gly Domain Containing Linker Protein Family Member 4
CMML	Chronic myelomonocytic leukemia
CRLR	Calcitonin receptor like receptor
CRP	C-reactive protein
CSCs	Cancer stem cells
CTCL	Cutaneous T-cell lymphoma
CTLA 4	Cytotoxic T-lymphocyte–associated antigen 4
cTn	Cardiac troponin
cTnI	Cardiac troponin I
cTnT	Cardiac troponin T
DDR1	Discoidin domain receptor tyrosine kinase 1
DJ1	Protein DJ-1
DLB	Dementia Lewy bodies
DMARDs	Disease modifying Antirheumatic drugs

DNMT	DNA methyltransferase
DOT1L	Treg, T regulatory cells
EAC	Esophageal Cancer
EBV	Epstein Barr Virus
EGCG	Epigallocatechin gallate
ELSI	Ethical, Legal, and Social Issues
EMT	Epithelial-to-mesenchymal transition
ENS	enteric nervous system
ERα	Estrogen Receptors Alpha
ESCC	Esophageal squamous cell carcinoma
ESR	Erythrocyte sedimentation rate
FOBT	Fecal occult blood test
FOXP	Forkhead box protein
GD	Graves' Disease
GNAT	GCN5-related N-acetyltransferases
GPX 7	Glutathione Peroxidase
GSTP1	Glutathione S-transferase P1
HAT	Histone acetyltransferase
HATs	Histone acetyltransferases
HDAC	Histone deacetylase
HDACi	Histone deacetylase inhibitors
HDACs	Histone deacetylases
HDM	Histone de-methylase
HLA	Human leucocyte antigen
HMGB1	High-mobility group box 1 protein
HMTs	Histone methyltransferases
HT	Hashimoto's Thyroiditis
IARC	International Agency for research on cancer
IBD	Inflammatory bowel disease
ICAM1	intercellular adhesion molecule
IDHs	Isocitrate dehydrogenases
IFN-γ	Interferon γ
JNK	Jun N-terminal kinase
KEAP1	Kelch-like ECH-associated protein 1
KRAS	Kirsten rat sarcoma
LncRNAs	Long Noncoding RNAs
LRRK2	Leucine rich repeat serine/threonine protein kinase 2
MDR	Multidrug resistance
MSA	Multiple system atrophy
NAD	Nicotinamide adenine dinucleotide
ncRNA	Non-coding ribonucleic acid
NDRG4	N-Myc downstream-regulated gene 4
NELL1	Nerve Epidermal Growth Factor-like 1
NF-κβ	Nuclear factor κβ
NT proBNP	N-terminal pro b-type natriuretic peptide
PARK2	Parkin
PCR	Polymerase chain reaction
PD	Parkinson's Disease
PDD	Parkinson's disease dementia

PET	Positron emission tomography
PI3K	Phosphoinositide 3 kinase
PINK1	PTEN-induced kinase protein 1
PIWI	P Element induced wimpy testis
PIWIL1	P-element induced wimpy testis-like protein 1
PMCA	Protein misfolding cyclic amplification
POLE3	Pole holoenzyme 3
PRF1	Platelet-rich fibrin
PSP	Progressive supranuclear palsy
PTCL	Peripheral T-cell lymphoma
RA	Rheumatoid arthritis
RAMP 1	Receptor Activity Modifying Protein 1
RARβ2	Retinoic acid receptor beta 2
RF	Rheumatoid factor
ROS	Reactive oxygen species
RT-QuIC	Real time quaking induced conversion
RUNX3	RUNX Family Transcription Factor 3
SAM	S-adenosyl methionine
SARDs	Systemic autoimmune rheumatic diseases
sCD40	Soluble form of CD40
SCLC	Small cell lung cancer
SDC2	Syndecan 2
SHIP1	SRC homology2 domain containing inositol 5 phosphate
SLE	Systemic lupus erythematosus
SN	Substantia nigra
SNCA	Synuclein Alpha
SNP	Single Nucleotide Polymorphism
SNPs	Single Nucleotide Polymorphisms
SNX19	Sorting Nexin 19
SOCS1	Suppressor of cytokine signaling 1
SPECT	Single photon emission CT
SS	Sjogren's syndrome
SSc	Systemic sclerosis
T1D	Diabetes type 1
TET	Ten-eleven translocation
TGF-β	Transforming growth factor β
Th17	T helper 17
TNF	Tumor necrosis factor
TNFRSF5	Tumor necrosis factor receptor superfamily member 5
TSG	Tumor suppressor gene
TWIST 1	Twist-related protein 1
UCHL1	Ubiquitin C-terminal hydrolase L1
Wnt	Wingless related integration site

1 Epigenetics Challenges and Prospects

Kanaka Raju A,[1] Vijay Srinivas P,[2] Rushikesh B Katkar,[3] and Gana Manjusha Kondepudi[1]

[1]Vignan Institute of Pharmaceutical Technology, Visakhapatnam, Andhra Pradesh, India

[2]Stira Pharma LLC, Fairfield, New Jersey, USA

[3]Ashokrao Mane Institute of Pharmaceutical Sciences and Research, Save, Kolhapur, Maharashtra, India

1.1 INTRODUCTION

1.1.1 Definition

The prefix *epi* comes from Greek and means "upon, on, over, or beside." Epigenetics involves the extra layer of instructions that lie over or on top of DNA and regulate how genes are expressed. All cells in the body share the same genetic code, but each cell has a unique structure and unique functions; a heart cell, for example, is quite distinct from a nerve cell [1], and epigenetics aims to explain how this occurs. It investigates how chemical tags linked to various regions of DNA and associated proteins can activate or mute genes. Such chemical alterations are referred to as epigenetic modifications.

This procedure does not change the underlying genetic code. Instead, it controls which genes are activated or deactivated in individual cells, determining the cell's distinct shape and function. Epigenetic changes can also help explain discrepancies between identical twins who have the same genetic code. The epigenetic system is vital for cell formation and maintaining the body's equilibrium or homeostasis. Any disruption of the epigenetic process can have serious health and behavioral effects [1]. Internal imbalances inside the body, as well as environmental and lifestyle variables, can all have an impact on this. Most epigenetic alterations occur just once in the lifespan of an individual organism. These changes, however, can have long-term consequences and be passed down to offspring and future generations.

1.1.2 Discovery and Importance

Epigenetics, the analysis of variations in gene activity that require no amendments to the underlying genome array, has garnered much attention since its inception in the mid-twentieth century. The word "epigenetics" was initially used in 1942 by British biologist Conrad H. Waddington to describe the processes that allow a fertilized egg to evolve into a fully formed creature[1]. Over time, the word "inherited" came to refer to heritable alterations in gene expression that did not affect the DNA itself.

The pace of epigenetic research accelerated significantly in the early 2000s. Between 2000 and 2010, the number of articles on epigenetics in scientific databases increased dramatically. This growth was fueled by increasing facts that epigenetic mechanisms are involved in a wide array of ailments, including cancers, cardiovascular disorders, autoimmune diseases, and neurodegenerative

DOI: 10.1201/9781032706511-1

conditions [2]. Lifestyle aspects like diet, exercise, and exposure to environmental toxins like heavy metals and tobacco smoke are also known to induce epigenetic changes, further emphasizing the importance of understanding this field [3].

The Human Epigenome Project was established in 2010, marking a significant milestone in the research. The initiative sought to map epigenetic markers, specifically genetic methylation patterns, spanning the entire human genome in order to better understand gene regulation in health and illness [4]. By 2015, the experiment had yielded vital information about how these markers affect gene expression in different human tissues [5].

Epigenetics is already making a tangible impact in medicine. Several drugs approved by the US Food and Drug Administration (FDA) are designed to reverse aberrant epigenetic changes linked to cancer, and ongoing research suggests that these drugs may enhance the effectiveness of immune checkpoint inhibitors [6]. Epigenetic biomarkers are also being explored as diagnostic tools for diseases like cancer, with companies developing tests to detect these markers.

The field of epigenetics is poised for further breakthroughs, with the potential to revolutionize our understanding of gene regulation, disease mechanisms, and therapeutic interventions [7]. As research continues, epigenetics promises to unlock new ways to treat diseases and improve health outcomes, potentially transforming medicine in the coming years [8].

1.2 BASIC PRINCIPLES OF EPIGENETIC MECHANISMS

One of the most important processes discovered in this research was DNA methylation, which involves adding a methyl group to a DNA strand to affect gene function without changing the DNA sequence. The first identification of methylated cytosine occurred in 1948, but it was not until the 1960s and 1970s that the relationship between DNA methylation and gene regulation was established [9]. Since the mid-1980s, the field of epigenetics has gained clarity. Key studies have revealed that DNA methylation could possess an essential role in turning genes on and off during biological development. By the 1990s, breakthroughs in DNA sequencing and cloning techniques enabled larger-scale studies of DNA methylation that demonstrated its importance in controlling gene expression. Researchers revealed that DNA methylation can silence genes by altering certain DNA sequences known as cytosine-phosphodiester-guanine (CpG) islands [9]. This was a significant advance since it illustrated how epigenetic markers might affect gene function.

In addition to DNA methylation, histone modification has emerged as a significant epigenetic process. Histones are proteins that wrap around DNA and help package it into the cell's nucleus [10]. Chemical alterations to histones, such as acetylation, phosphorylation, and ubiquitination, can tighten or loosen DNA coils. These alterations control how accessible a gene is to the cell's transcription machinery. Acetylation, for example, loosens the DNA coil, which boosts gene expression [11]. Figure 1.1 presents other mechanisms of epigenetics.

1.3 DNA METHYLATION AND GENE REGULATION

1.3.1 The Fundamentals of DNA Methylation

DNA methylation is a crucial epigenetic mechanism, involving the addition of a methyl group to cytosine residues, particularly in CpG dinucleotides, orchestrated by DNA methyltransferases (DNMTs). Notably, genomic regions with dense CpG sequences, called CpG islands, are abundant in gene promoter regions, influencing gene expression through methylation patterns [12].

In cancer, DNA methylation dynamics are complex and play a dual role in oncogenesis:

- **Hypomethylation**: Global loss of methylation, often seen in coding regions, disrupts genomic stability. This leads to chromosomal rearrangements, activation of oncogenes

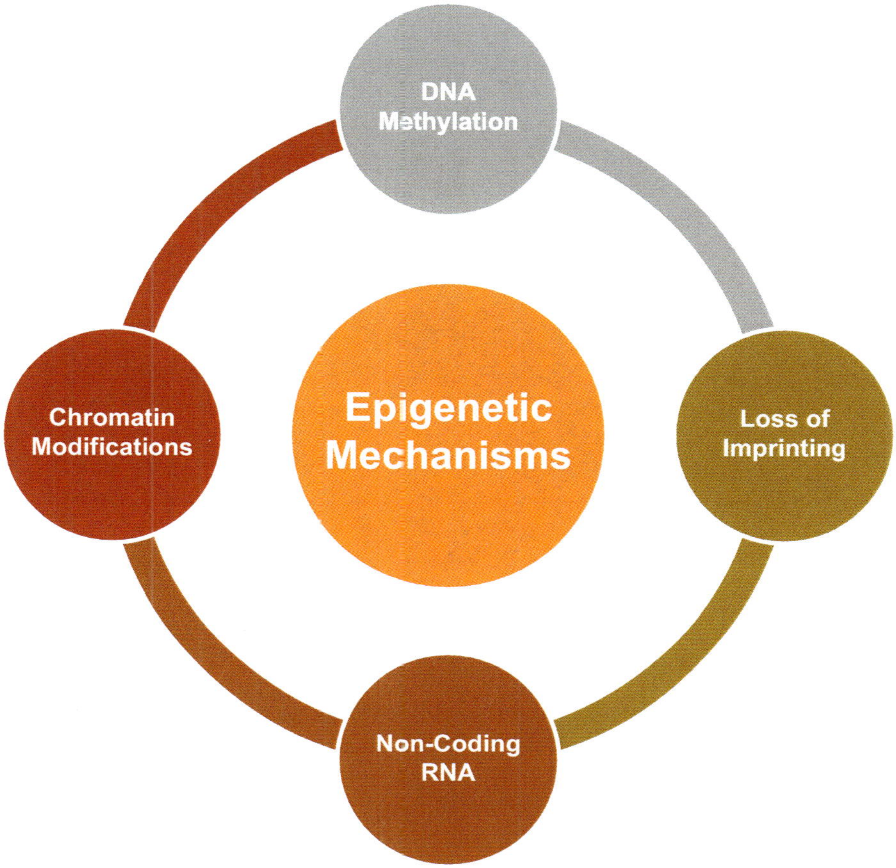

FIGURE 1.1 Epigenetic mechanisms.

(e.g., *c-Jun*, *c-Myc*), and progressive tumorigenesis. Hypomethylation in repetitive DNA elements (e.g., SAT2, LINE1) is associated with poor cancer prognoses [13,14].

- **Hypermethylation**: Selective hypermethylation silences tumor suppressor genes by altering promoter regions. This repression impairs critical pathways in cell cycle regulation, DNA repair, and apoptosis. While physiologically normal in processes like X-chromosome inactivation and aging, hypermethylation in cancer contexts leads to aberrant gene silencing [15,16].

1.3.2 DNA Methylation Dysregulation

The juxtaposition of hypermethylation in promoter regions and hypomethylation in coding regions creates a paradoxical landscape. Methyl groups influence DNA-protein interactions, directly impacting transcriptional machinery. DNMT1's fidelity in replicating methylation patterns during mitosis underscores the heritability of these modifications, with errors potentially cascading into oncogenic events [17].

In the conventional approaches to detecting DNA methylation, early methods employed isoschizomers to differentiate methylated sites but were limited in scope. Bisulfite conversion revolutionized detection by selectively altering unmethylated cytosines to uracils, facilitating methylation-specific PCR; this method provides gene-specific methylation profiles [18]. More advanced techniques include pyrosequencing, which combines real-time sequencing with high-throughput capabilities, enabling precise CpG analysis[19]; MeDIP, in which immunoprecipitation using antibodies against 5-methylcytidine enriches methylated DNA for global analysis[20]; and MCA enrichment of methylated DNA fragments using selective enzyme digestion and PCR amplification [21].

Researchers are now exploring epigenetic reversibility. Gene expression microarrays combined with demethylating agents identify genes reactivated by reversing methylation-induced silencing. This method unveils potential therapeutic targets in cancer.

1.3.3 Histone Acetylation: Gatekeeper of Gene Transcription

The structural organization of DNA is tightly regulated by histones, which form nucleosome complexes with DNA wrapped around them. This configuration controls access to transcription machinery. A crucial epigenetic regulator is histone acetylation, which occurs on histone tails extending from the nucleosome. When histone tails remain unacetylated, the DNA stays condensed and inaccessible to transcription factors. Acetylation, however, reduces the positive charge on lysine residues within histone tails, weakening the interactions between DNA and histones. This relaxation of chromatin structure facilitates the binding of transcription factors, effectively enabling gene expression [22–24].

1.3.4 Detecting Chromatin Modifications

The dynamic nature of histone modifications demands robust detection methods. Mass spectrometry is a gold standard for identifying these modifications, offering precision but requiring specialized skills and instruments. Alternatively, chromatin immunoprecipitation (ChIP) serves as a versatile approach by combining immunoprecipitation with DNA sequencing to identify histone modifications across the genome. Advanced adaptations like "ChIP on chip" integrate microarray platforms to enable high-throughput analysis. Nonetheless, the accuracy of ChIP depends heavily on the quality of the antibodies used for targeting specific histone modifications [25].

1.3.5 The Complexity of Chromatin Modifications

Post-translational modifications of histones, including methylation, acetylation, phosphorylation, and ubiquitination, dictate chromatin behavior. These modifications can either enhance or repress transcription, influence DNA repair mechanisms, and impact chromosomal organization. For instance, histone acetylation generally promotes gene expression, while methylation exerts varying effects based on the site and context of modification. These interactions often occur in tandem with DNA methylation, underscoring the intricate network of chromatin modifications that regulate gene function [25].

1.3.6 Loss of Imprinting: An Epigenetic Vulnerability

Imprinted genes exhibit mono-allelic expression, which means that only one parental copy is active. This regulation is assisted by DNA methylation and histone changes. Disruptions in imprinting mechanisms, referred to as loss of imprinting, can cause unnatural gene expression, either by stimulating both alleles or by silencing genes entirely. This functional haploidy makes imprinted genes especially sensitive to epigenetic disturbances and mutations, linking imprinting loss to developmental diseases such Angelman, Prader–Willi, and Beckwith–Wiedemann syndromes. Furthermore, this mechanism has been implicated in complex diseases like bipolar disorder, autism, and schizophrenia, indicating that epigenetic inheritance plays a role in these ailments [26–28].

1.3.7 Noncoding RNAs: Regulators of Gene Translation

Noncoding RNAs (ncRNAs), particularly microRNAs (miRNAs), represent another layer of epigenetic regulation. These small RNA molecules, approximately 22 nucleotides in length, are transcribed from long non-coding RNA or gene introns. MiRNAs influence protein synthesis by binding to messenger RNA (mRNA). If the miRNA sequence fully complements the mRNA, it facilitates degradation via the RNA-induced silencing complex. Partial complementarity, on the other hand, inhibits mRNA translation by blocking transfer RNA action. Erroneous expression of miRNAs has been associated with several gastrointestinal disorders, including inflammatory bowel disease, cholangiocarcinoma, esophageal adenocarcinoma, and hepatocellular carcinoma, underscoring its clinical significance [29–32].

1.4 EPIGENETIC REGULATION AND GENE EXPRESSION

Epigenetic changes can be described as "heritable modifications in gene function, inherited through mitotic and/or meiotic cell divisions that do not involve alterations to the underlying DNA sequence." These changes contribute to heritable phenotypic variance [33]. Epigenetic regulation can be relatively stable, as seen in X-chromosome inactivation, genomic imprinting, gene silencing, and chromatin border creation, or more dynamic [34–36]. When dynamic, this control is referred to as epigenetic memory, which is a phenomenon in which gene expression or cellular behavior changes and is kept in response to prior events. These stimuli may be the product of developmental processes or environmental impacts.

Epigenetic memory acts through a multitude of mechanisms, encompassing chromatin-level changes such as DNA methylation, histone modifications, and the amalgamation of histone variations [37]. DNA methylation is notable for its capacity to sustain heredity, as methylation patterns are kept by maintenance DNA methyltransferases that work on hemimethylated DNA after replication [38]. In contrast, the heredity of histone modifications is unclear, and whether any histone alterations may self-propagate is dubious.

Three types of epigenetic memory are identified, each using similar mechanisms but occurring over distinct timescales: (1) cellular memory, which refers to mitotically inherited transcriptional conditions created during development in response to developmental signals; (2) transcriptional memory, which entails mitotically heritable changes in an organism's response to environmental stimuli based on prior experiences; and (3) transgenerational memory, which includes meiotically inherited changes in gene expression and physiological traits influenced by previous generations' experiences [39–43].

1.5 THE ROLE OF EPIGENETICS IN DISEASE DEVELOPMENT

Examples of roles of epigenetics in disease developments are outlined in Tables 1.1–1.4.

1.6 ENVIRONMENTAL AND LIFESTYLE INFLUENCES ON THE EPIGENOME

Early life experiences can have lasting epigenetic effects, influencing stress responses through DNA methylation (Figure 1.2; Table 1.2) [48–50].

1.7 CURRENT EPIGENETIC RESEARCH TOOLS AND TECHNOLOGIES

Current epigenetic research tools and technologies provide powerful insights into the dynamic regulation of gene expression without altering DNA sequences (see Table 1.5). These advancements enable precise analysis of chromatin modifications, DNA methylation patterns, and non-coding RNA, facilitating breakthroughs in understanding complex biological processes and disease mechanisms [50].

TABLE 1.1
Epigenetics in Disease Mechanisms

Disease Category	Epigenetic Mechanisms	Key Findings	Therapeutic Implications	Challenges
Neurodegenerative disorders[44]	DNA methylation, histone modifications, non-coding RNAs (ncRNAs)	- Dysregulated DNA methylation and histone acetylation in Alzheimer's and Parkinson's diseases - ncRNAs modulate neuroinflammation and synaptic function - HDAC inhibitors show promise for treatment	- HDAC inhibitors offer therapeutic potential - Biomarkers like lncRNA and miRNA are being explored for diagnostics and treatment - DNA methylation profiling holds promise	- Side effects of HDAC inhibitors including oxidative stress and apoptosis - Complex interplay of epigenetic mechanisms and brain recovery pathways
Metabolic and cardiovascular diseases[45]	DNA methylation, histone modifications, chromatin remodeling, ncRNAs	- Aberrant DNA methylation linked to diabetes and obesity - Histone modifications affect lipid metabolism - ncRNAs regulate pathways in atherosclerosis and NAFLD	- Epigenetic biomarkers for early diagnosis - Epigenetic drugs targeting DNA methylation and histone acetylation - Potential for gene-editing therapies	- Lack of epigenetic drugs approved for metabolic diseases - Off-target effects of epigenetic therapies
Autoimmune diseases[46]	DNA methylation, histone modifications, ncRNAs	- Hypomethylation in SLE and RA leads to overactivation of immune genes - ncRNAs regulate inflammatory cytokines in SS and AITD - Histone modifications drive fibrosis in SSc	- HDAC inhibitors and DNMT inhibitors under investigation - Biomarkers like miRNAs for diagnostics - Potential for CRISPR-based epigenetic editing	- Few clinical studies on epigenetic therapies - Complex tissue-specific epigenetic changes require further research
Acute brain injuries[44]	DNA methylation, histone modifications, ncRNAs	- Epigenetic changes influence recovery and neuronal regeneration - Dysregulated DNA methylation impacts memory formation and cognition	- Epigenetic drugs show promise in enhancing neurological recovery - Biomarkers for early diagnosis of brain injury	- Limited understanding of the interplay between injury, epigenetics, and brain environment
Cancer[47]	DNA methylation, histone modifications, chromatin remodeling, ncRNAs	- Aberrant hypermethylation silences tumor suppressor genes - Dysregulated histone modifications promote cancer progression	- Epigenetic drugs like HDAC inhibitors and DNMT inhibitors approved for some cancers - Potential for ncRNA-based therapies	- Off-target effects of epigenetic drugs - Need for personalized therapies based on individual epigenetic profiles

TABLE 1.2
Nutritional Components and Their Role in Epigenetic Modulation

Nutritional Component	Key Effects	Mechanism of Action	Implications
Polyunsaturated fatty acids	- Generate free radicals and oxidative stress - Modulate gene methylation - Suppress tumorigenesis	- Promote pro-angiogenic pathways (e.g., arachidonic acid) - Inhibit NF-kappaB pathway and inflammation	- Protective role in reducing cancer risk - Influence endothelial function and angiogenesis
Folate and vitamin B12	- Essential for SAM and DNA methylation - Prevent DNA hypomethylation - Regulate miRNA expression	- Maintain SAM ratio for methylation capacity - Support metionine synthesis from homocysteine	- Reduce colorectal cancer risk - Reverse miRNA dysregulation in carcinogenesis
Polyphenols	- Modulate DNMT, HAT, and HDAC activity - Reactivate silenced genes - Reverse CpG island methylation	- EGCG from green tea inhibits DNMT activity - Soy polyphenols (e.g., genistein) downregulate DNMTs and HDACs	- Reduce risk of hormone-related and other cancers - Reactivate tumor suppressor genes
Selenium	- Inhibits DNMT and HDAC - Reactivates hypermethylated tumor suppressor genes	- Downregulates DNMTs and HDACs - Restores methylation in genes like *GSTP1*, *APC*, and *CSR1*	- Protects against oxidative damage - Reduces cancer risk, especially prostate cancer

TABLE 1.3
Environmental Pollutants and Their Epigenetic Effects

Pollutant	Exposure Source	Epigenetic Effects	Mechanisms and Observations	Health Implications
Arsenic	Drinking water (India, Bangladesh)	- DNA hypermethylation of *p53* and *p16* promoter regions - Global DNA hypomethylation and dose-dependent hypermethylation	- Subtracts methyl donors from DNA methylation reactions - Folate modifies the methylation effects	- Increased cancer risk - Chronic exposure linked to skin lesions
Air pollution (PM, BC)	Particulate matter (PM10, black carbon)	- Decreased *iNOS* promoter methylation - LINE-1 and Alu hypomethylation	- *iNOS* demethylation increases inflammation and oxidative stress - LINE-1 hypomethylation linked to traffic particles	- Linked to cardiorespiratory diseases and cancer - miRNA changes (e.g., miR-222, miR-21) linked to inflammation

(Continued)

TABLE 1.3 (*Continued*)
Environmental Pollutants and Their Epigenetic Effects

Pollutant	Exposure Source	Epigenetic Effects	Mechanisms and Observations	Health Implications
Aromatic hydrocarbons	Benzene, PAHs, gasoline emissions	- LINE-1 and Alu hypomethylation - Gene-specific hypermethylation (e.g., *p15*, *p53*, *p16*).	- Benzene metabolites induce hypomethylation - PAHs alter methylation of gene promoters and repetitive elements	- Increased risk of AML, multiple myeloma, and other cancers - Fetal exposure linked to altered methylation patterns
Persistent organic pollutants (POPs)	Pesticides (DDT, DDE), industrial chemicals (PCBs, chlordane)	- Inverse relationship with global DNA methylation (e.g., Alu elements)	- Observed in Inuit populations with high POP levels - Methylation changes linked to persistent pollutant exposure	- High prevalence of cancers and immune system alterations

TABLE 1.4
The Influence of External Factors on Epigenetic Changes

Factor	Epigenetic Mechanisms	Key Findings
Psychological stress	DNA methylation	- Childhood abuse is linked to hypermethylation of the glucocorticoid receptor gene promoter in the hippocampus in suicide victims. - Positive early parental care (e.g., maternal grooming in animals) leads to hypomethylation of the same gene, reducing stress responses later in life.
Shift work	DNA methylation, histone acetylation	- Disruption of circadian rhythms due to shift work alters methylation patterns of genes including CLOCK and inflammatory markers (e.g., IFN-γ, TNF-α). - Chromatin remodeling regulated by circadian genes impacts health outcomes.
Obesity and physical activity	DNA methylation, histone acetylation, miRNA regulation	- Genes involved in obesity pathways (e.g., SOCS1, PTEN, CDKN1A) show methylation changes. - Physical activity increases LINE-1 methylation, reducing inflammation and stabilizing chromosomes. - Exercise induces global histone acetylation and miRNA alterations in muscle and neutrophils.
Tobacco smoke	DNA methylation, histone modifications, miRNA regulation	- Cigarette smoke reduces histone acetylation and trimethylation levels in respiratory cells. - Hypermethylation of tumor suppressor genes (e.g., p16, MGMT) occurs early in lung carcinogenesis. - Maternal smoking alters LINE-1 methylation and miRNA levels in placental tissue.
Alcohol consumption	DNA methylation	- Alcohol acts as a co-carcinogen, altering methylation in genes related to colorectal cancer (e.g., APC-1A, p16INK4A). - Fetal alcohol exposure causes demethylation of neural genes (e.g., NR2B) and hypermethylation of cell cycle genes, impacting neural development and metabolism.

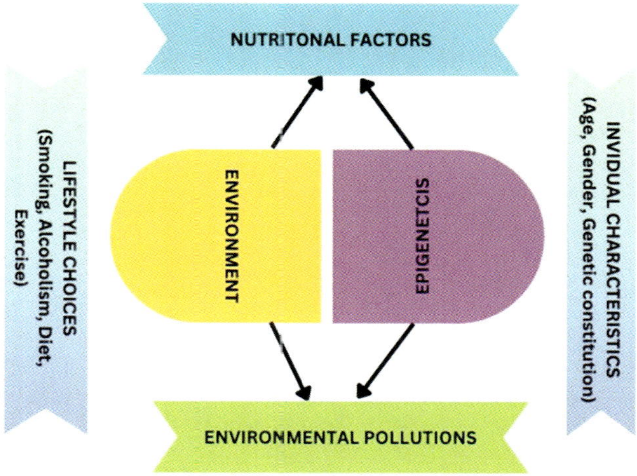

FIGURE 1.2 Environmental and lifestyle influences on the epigenome.

TABLE 1.5
Advanced DNA Methylation Technologies

Technology/Method	Description	Advantages	Limitations
Reduced representation bisulfite sequencing	Combines restriction enzyme digestion, bisulfite conversion, and NGS to target CpG-rich regions like promoters, covering 85% of CpG islands	Cost-effective for large-scale or exploratory studies; targets regulatory regions effectively	DNA degradation due to bisulfite treatment; limited to CpG-enriched regions
Whole genome bisulfite sequencing	Provides comprehensive methylation profiling across the entire genome, including low-density CpG regions	Unbiased and comprehensive; covers distal regulatory elements and intergenic regions	High cost and data processing requirements; requires significant computational resources
Methylation-sensitive restriction enzymes	Uses enzymes to differentiate methylated and unmethylated DNA regions	Cost-effective and avoids harsh chemical treatments	Low resolution and genome coverage; cannot distinguish individual methylation contexts
Methylated DNA immunoprecipitation (MeDIP)	Employs antibodies to capture methylated DNA, followed by high-throughput sequencing	Useful for broad methylation studies; avoids bisulfite treatment	Low resolution; cannot provide single-base methylation context
Enzyme-linked immunosorbent assay	Detects global DNA methylation levels using commercial kits for 5mC quantification	Quick and easy for rough estimation of global methylation	High variability; low quantitative accuracy; limited to global assessments

(Continued)

TABLE 1.5 (*Continued*)
Advanced DNA Methylation Technologies

Technology/Method	Description	Advantages	Limitations
Single-cell bisulfite sequencing	Enables analysis of individual cell methylation heterogeneity	High resolution; suitable for understanding tumorigenesis	Low-input DNA requirements; suffers from DNA degradation and restricted methylation heterogeneity assessment
Post-bisulfite adapter tagging	Performs bisulfite treatment before adapter tagging, allowing for low DNA input in single-cell methylation studies	Overcomes DNA degradation challenges; improves single-cell methylation analysis	Limited throughput compared with bulk sequencing methods
Single-molecule real-time sequencing	Detects base modifications directly via polymerase kinetics during DNA synthesis using fluorescent nucleotides	Avoids DNA degradation; simultaneous detection of multiple methylation types (5mC, 5hmC, 6mA, 4mC)	Sensitivity differences for modifications; requires significant computational support; best suited for bacterial genomes
Nanopore sequencing	Detects methylation as single-stranded DNA passes through nanopores, with ion current deviations signaling modifications	Can distinguish multiple cytosine and adenine variants (e.g., 5mC, 5hmC, 6mA)	Accuracy depends on algorithms; improved with machine learning techniques like DeepSignal
Oxidative bisulfite sequencing	Uses chemical oxidation to distinguish 5hmC from 5mC by converting 5hmC to 5fC while 5mC remains intact	Identifies oxidized methylation forms; highlights the role of 5hmC in regulation	DNA degradation during multiple bisulfite treatments; high sequencing depth requirements
Hmtop-seq	Bisulfite-free method to map 5hmC using tethered oligonucleotide-primed sequencing	High resolution; cost-efficient; avoids bisulfite-induced degradation	Requires specialized equipment and expertise

1.8 CHALLENGES IN EPIGENETIC RESEARCH

Epigenetic networks are complex; the intricate relationships between DNA, histones, and non-coding RNA have rendered it harder to figure out accurate mechanisms. Additionally, high-throughput strategies such as chromatin immunoprecipitation sequencing and mass spectrometry require particular expertise, expensive equipment, and high-quality antibodies, minimizing their widespread use. Linking epigenetic modifications to functional consequences is also challenging because of heterogeneity in tissue type, developmental stage, and environmental circumstances. Another challenge is that integrating enormous epigenetic datasets with genomic and transcriptome information requires sophisticated computational techniques and models. Finally, therapeutic translation is difficult; developing medications for specific epigenetic pathways is challenging due to potential off-target effects and the reversibility of many epigenetic modifications.

1.9 FUTURE DIRECTIONS FOR EPIGENETIC RESEARCH

Future research promises to advance single-cell epigenomics; analyzing epigenetic modifications in individual cells may uncover regulatory mechanisms and tissue heterogeneity. Improving diagnostic tools will involve designing less invasive technologies such as liquid biopsies to identify epigenetic biomarkers for early identification of diseases and prognoses. Epigenome editing will also advance; CRISPR-based technologies offer therapeutic potential for diseases such as cancer and neurological disorders. Ongoing research on understanding the different roles of long non-coding RNAs and microRNAs in epigenetic control could lead to new therapeutic targets. Also in terms of research, exploring how environmental factors like nutrition, pollution, and stress affect epigenetic patterns will improve knowledge of gene–environment interactions.

1.10 CONCLUSION

Epigenetics has revolutionized our understanding of gene control, highlighting how heritable but reversible modifications affect health and illness. Despite great progress, there are still obstacles in understanding the intricate nature of epigenetic networks and transferring results into therapeutic applications. However, emerging technologies such as single-cell analysis, epigenome editing, and integrative omics techniques seem to hold possibilities for the future of epigenetic research. These advances will not only improve our understanding of gene–environment interactions but also pave the way for precision medicine, which will provide novel diagnostic and therapeutic options for a wide spectrum of disorders.

REFERENCES

[1] "What Is Biotechnology—Epigenetics." Accessed November 22, 2024. https://www.whatisbiotechnology.org/index.php/science/summary/epigenetics#:~:text=In%20the%20late%201970s%20Holliday,developed%20to%20inhibit%20DNA%20methylation

[2] Biswas, S, and CM Rao. "Epigenetic Tools (The Writers, The Readers and The Erasers) and Their Implications in Cancer Therapy." *European Journal of Pharmacology* 837 (2018): 8–24. https://doi.org/10.1016/j.ejphar.2018.08.021

[3] Baylin, SB, and PA Jones. "A Decade of Exploring the Cancer Epigenome: Biological and Translational Implications." *Nature Reviews Cancer* 11 (2011): 726–734. Accessed November 22, 2024. http://www.ncbi.nlm.nih.gov/pubmed/21941284

[4] *Nature Reviews Genetics*. "Insights on Epigenetics." Accessed November 22, 2024. http://www.ncbi.nlm.nih.gov/pubmed/30479381

[5] Lamb, YN, and S Dhillon. "Epi proColon® 2.0 CE: A Blood-Based Screening Test for Colorectal Cancer." *Molecular Diagnosis & Therapy* 21 (2017) 225–232. https://doi.org/10.1007/s40291-017-0260-1

[6] Ned, RM, S Melillo, and M Marrone. "Fecal DNA Testing for Colorectal Cancer Screening: The ColoSure™ Test." *PLoS Currents* 3 (2011): RRN1220. Accessed November 22, 2024. http://currents.plos.org/genomictests/article/fecal-dna-testing-for-colorectal-cancer-od1hzthyodr3-1

[7] Delgado-Morales, R, RC Agís-Balboa, M Esteller, and M Berdasco. "Epigenetic Mechanisms During Ageing and Neurogenesis as Novel Therapeutic Avenues in Human Brain Disorders." *Clinical Epigenetics* 9 (2017): 67. Accessed November 22, 2024 http://www.ncbi.nlm.nih.gov/pubmed/28670349

[8] Mithraprabhu, S, A Kalff, A Chow, T Khong, and A Spencer. "Dysregulated Class I Histone Deacetylases Are Indicators of Poor Prognosis in Multiple Myeloma." *Epigenetics* 9 (2014): 1511–1520. https://doi.org/10.4161/15592294.2014.972903

[9] Han, M, L Jia, W Lv, L Wang, and W Cui. "Epigenetic Enzyme Mutations: Role in Tumorigenesis and Molecular Inhibitors." *Frontiers in Oncology* 9 (2019): 194. https://doi.org/10.3389/fonc.2019.00194

[10] Yang, L, R Rau, and MA Goodell. "DNMT3A in Haematological Malignancies." *Nature Reviews Cancer* 15 (2015): 152–165. Accessed November 22, 2024.

[11] Daigle, SR, EJ Olhava, CA Therkelsen, A Basavapathruni, L Jin, PA Boriack-Sjodin, et al. "Potent Inhibition of DOT1L as Treatment of MLL-Fusion Leukemia." *Blood* 122 (2013): 1017–1025. Accessed November 22, 2024. http://www.ncbi.nlm.nih.gov/pubmed/23801631

[12] Gibney, ER, and CM Nolan. "Epigenetics and Gene Expression." *Heredity* 105 (2010): 4–13. https://doi.org/10.1038/hdy.2010.54

[13] Feinberg, AP, and B Vogelstein. "Hypomethylation Distinguishes Genes of Some Human Cancers from Their Normal Counterparts." *Nature* 301 (1983): 89–92. https://doi.org/10.1038/301089a0

[14] Calvisi, DF, S Ladu, A Gorden, M Farina, JS Lee, EA Conner, et al. "Mechanistic and Prognostic Significance of Aberrant Methylation in the Molecular Pathogenesis of Human Hepatocellular Carcinoma." *Journal of Clinical Investigation* 117 (2007): 2713–2722. https://doi.org/10.1172/JCI31457

[15] Roberts, LR, and GJ Gores. "Hepatocellular Carcinoma: Molecular Pathways and New Therapeutic Targets." *Seminars in Liver Disease* 25 (2005): 212–225. https://doi.org/10.1055/s-2005-871200

[16] Lee, HS, BH Kim, NY Cho, EJ Yoo, M Choi, SH Shin, et al. "Prognostic Implications of and Relationship between CpG Island Hypermethylation and Repetitive DNA Hypomethylation in Hepatocellular Carcinoma." *Clinical Cancer Research* 15 (2009): 812–820. https://doi.org/10.1158/1078-0432.CCR-08-0266

[17] Herman, JG, and SB Baylin. "Gene Silencing in Cancer in Association with Promoter Hypermethylation." *New England Journal of Medicine* 349 (2003): 2042–2054. https://doi.org/10.1056/NEJMra023075

[18] Herman, JG, JR Graff, S Myohanen, BD Nelkin, and SB Baylin. "Methylation-Specific PCR: A Novel PCR Assay for Methylation Status of CpG Islands." *Proceedings of the National Academy of Sciences* 93 (1996): 9821–9826. https://doi.org/10.1073/pnas.93.18.9821

[19] Dupont, JM, J Tost, H Jammes, and IG Gut. "De Novo Quantitative Bisulfite Sequencing Using the Pyrosequencing Technology." *Analytical Biochemistry* 333 (2004): 119–127. https://doi.org/10.1016/j.ab.2004.05.007

[20] Weber, M, JJ Davies, D Wittig, EJ Oakeley, M Haase, WL Lam, et al. "Chromosome-Wide and Promoter-Specific Analyses Identify Sites of Differential DNA Methylation in Normal and Transformed Human Cells." *Nature Genetics* 37 (2005): 853–862. https://doi.org/10.1038/ng1598

[21] Toyota, M, C Ho, N Ahuja, KW Jair, Q Li, M Ohe-Toyota, et al. "Identification of Differentially Methylated Sequences in Colorectal Cancer by Methylated CpG Island Amplification." *Cancer Research* 59 (1999): 2307–2312.

[22] Esteller, M. "Epigenetics in Cancer." *New England Journal of Medicine* 358 (2008): 1148–1159. https://doi.org/10.1056/NEJMra072067

[23] Gibney, ER, and CM Nolan. "Epigenetics and Gene Expression." *Heredity* 105 (2010): 4–13. https://doi.org/10.1038/hdy.2010.54

[24] Esteller, M. "Cancer Epigenomics: DNA Methylomes and Histone-Modification Maps." *Nature Reviews Genetics* 8 (2007): 286–298. https://doi.org/10.1038/nrg2005

[25] Hamilton, JP. "Epigenetic Mechanisms Involved in the Pathogenesis of Hepatobiliary Malignancies." *Epigenomics* 2 (2010): 233–243. https://doi.org/10.2217/epi.10.9

[26] Wu, F, S Zhang, T Dassopoulos, ML Harris, TM Bayless, SJ Meltzer, et al. "Identification of MicroRNAs Associated with Ileal and Colonic Crohn's Disease." *Inflammatory Bowel Diseases* 16 (2010): 1729–1738. https://doi.org/10.1002/ibd.21267

[27] Jirtle, RL, and MK Skinner. "Environmental Epigenomics and Disease Susceptibility." *Nature Reviews Genetics* 8 (2007): 253–262. https://doi.org/10.1038/nrg2045

[28] Feinberg, AP. "Epigenetics at the Epicenter of Modern Medicine." *JAMA* 299 (2008): 1345–1350. https://doi.org/10.1001/jama.299.11.1345

[29] Olaru, AV, FM Selaru, Y Mori, C Vazquez, S David, B Paun, et al. "Dynamic Changes in the Expression of MicroRNA-31 During Inflammatory Bowel Disease-Associated Neoplastic Transformation." *Inflammatory Bowel Diseases* 17 (2011): 221–231. https://doi.org/10.1002/ibd.21359

[30] Selaru, FM, AV Olaru, T Kan, S David, Y Cheng, Y Mori, et al. "MicroRNA-21 Is Overexpressed in Human Cholangiocarcinoma and Regulates Programmed Cell Death 4 and Tissue Inhibitor of Metalloproteinase 3." *Hepatology* 49 (2009): 1595–1601. https://doi.org/10.1002/hep.22838

[31] Kan, T, F Sato, T Ito, N Matsumura, S David, Y Cheng, et al. "The miR-106b-25 Polycistron, Activated by Genomic Amplification, Functions as an Oncogene by Suppressing p21 and Bim." *Gastroenterology* 136 (2009): 1689–1700. https://doi.org/10.1053/j.gastro.2009.02.002

[32] Selaru, FM, S David, SJ Meltzer, and JP Hamilton. "Epigenetic Events in Gastrointestinal Cancer." *American Journal of Gastroenterology* 104 (2009): 1910–1912. https://doi.org/10.1038/ajg.2008.145

[33] D'Urso, A, and JH Brickner. "Mechanisms of Epigenetic Memory." *Trends in Genetics* 30, no. 6 (2014): 230–236. https://doi.org/10.1016/j.tig.2014.04.004

[34] Klar, AJS. "Propagating Epigenetic States Through Meiosis: Where Mendel's Gene Is More than a DNA Moiety." *Trends in Genetics* 14 (1998): 299–301. https://doi.org/10.1016/s0168-9525(98)01535-2

[35] Jenuwein, T, and CD Allis. "Translating the Histone Code." *Science* 293 (2001): 1074–1080. https://doi.org/10.1126/science.1063127

[36] Rice, JC, and CD Allis. "Histone Methylation Versus Histone Acetylation: New Insights into Epigenetic Regulation." *Current Opinion in Cell Biology* 13 (2001): 263–273. https://doi.org/10.1016/s0955-0674(00)00208-8

[37] Suganuma, T, and JL Workman. "Signals and Combinatorial Functions of Histone Modifications." *Annual Review of Biochemistry* 80 (2011): 473–499. https://doi.org/10.1146/annurev-biochem-061809-175347

[38] Zhu, B, and D Reinberg. "Epigenetic Inheritance: Uncontested?" *Cell Research* 21 (2011): 435–441. https://doi.org/10.1038/cr.2011.26

[39] Kiefer, J. "Epigenetics in Development." *Developmental Dynamics* 2007. [Accessed via Wiley Online Library].

[40] Bantignies, F, et al. "Inheritance of Polycomb-Dependent Chromosomal Interactions in Drosophila." *Genes & Development* 17 (2003): 2406–2420. https://doi.org/10.1101/gad.269503

[41] Youngson, NA, and E Whitelaw. "Transgenerational Epigenetic Effects." *Annual Review of Genomics and Human Genetics* 9 (2008): 233–257. https://doi.org/10.1146/annurev.genom.9.081307.164445

[42] Jablonka, E, and G Raz. "Transgenerational Epigenetic Inheritance: Prevalence, Mechanisms, and Implications for the Study of Heredity and Evolution." *Quarterly Review of Biology* 84 (2009): 131–176. https://doi.org/10.1086/598822

[43] Youngson, NA, and E Whitelaw. "Transgenerational Epigenetic Effects." *Annual Review of Genomics and Human Genetics* 9 (2008): 233–257. https://doi.org/10.1146/annurev.genom.9.081307.164445

[44] Bertogliat, MJ, KC Morris-Blanco, and R Vemuganti. "Epigenetic Mechanisms of Neurodegenerative Diseases and Acute Brain Injury." *Neurochemistry International* 133 (2020): 104642. https://doi.org/10.1016/j.neuint.2019.104642

[45] Wu, YL, ZJ Lin, CC Li, et al. "Epigenetic Regulation in Metabolic Diseases: Mechanisms and Advances in Clinical Study." *Signaling Transduction and Targeted Therapy* 8 (2023): 98. https://doi.org/10.1038/s41392-023-01333-7

[46] Mazzone, R, C Zwergel, M Artico, et al. "The Emerging Role of Epigenetics in Human Autoimmune Disorders." *Clinical Epigenetics* 11 (2019): 34. https://doi.org/10.1186/s13148-019-0632-2

[47] Yu, X, H Zhao, R Wang, et al. "Cancer Epigenetics: From Laboratory Studies and Clinical Trials to Precision Medicine." *Cell Death Discovery* 10 (2024): 28. https://doi.org/10.1038/s41420-024-01803-z

[48] Tiffon, C. "The Impact of Nutrition and Environmental Epigenetics on Human Health and Disease." *International Journal of Molecular Sciences* 19, no. 11 (2018): 3425. https://doi.org/10.3390/ijms19113425

[49] Alegría-Torres, JA, A Baccarelli, and V Bollati. "Epigenetics and Lifestyle." *Epigenomics* 3, no. 3 (2011): 267–277. https://doi.org/10.2217/epi 11.22

[50] Tando, Y, and Y Matsui. "Inheritance of Environment-Induced Phenotypic Changes Through Epigenetic Mechanisms." *Environmental Epigenetics* 9, no. 1 (2023): dvad008. https://doi.org/10.1093/eep/dvad008

2 Epigenetic Disease Mechanisms

Aman Khandelwal, Suguna Kotte, and S. Ponnusankar
JSS Academy of Higher Education & Research, Ooty, Tamil Nadu, India

2.1 INTRODUCTION

Within the interesting topic of genetics, epigenetic illnesses investigate how environmental influences affect gene function and lead to a variety of health issues. Epigenetic alterations modify gene expression without changing the underlying genetic code, in contrast to genetic mutations that require changes in the DNA sequence itself (Weinhold 2006). Understanding the principles of epigenetic illnesses and their processes is critical for deciphering the complex interaction of genes and the environment in determining human health (Zoghbi & Beaudet 2016).

At its heart, epigenetics includes alterations to the structure of DNA or its associated proteins that influence gene accessibility for transcription and as a result protein creation (Moosavi & Motevalizadeh Ardekani 2016). Environmental exposures, lifestyle choices, and even experiences from previous generations all have impacts on these changes. Our understanding of how environmental influences can permanently alter our genetic makeup and perhaps contribute to the emergence of many health disorders is growing as scientists explore the processes behind epigenetic illnesses (Tiffon 2018). This investigation into the fundamentals of epigenetic illnesses will offer light on our genetic makeup's dynamic and sensitive nature, revealing insights that go beyond standard genetic explanations. Understanding these underlying principles allows researchers and healthcare practitioners to investigate innovative treatment methods and prevention techniques for epigenetic-influenced illnesses.

2.2 UNDERSTANDING EPIGENETIC MECHANISMS

2.2.1 Epigenetic Regulation Overview

The field of epigenetics studies how genes are regulated and inherited without affecting the DNA molecule (Zhao et al. 2022). Genes are like a perfectly tuned symphony: They switch on and off at precisely the appropriate periods to ensure that the organism functions as intended (Cedar & Bergman 2009). These modifications are important for the development of distinct cell types and even disease onset. In epigenetics, the primary methods of controlling genes include DNA tagging and modifying the structure of proteins (Hamilton 2011). These function similarly to signals, facilitating the accessibility and situational expression of genes. There are exciting new developments in the field of epigenetics as researchers investigate the potential effects of environmental factors and diet on gene regulation.

2.2.2 DNA Methylation in Gene Expression

DNA methylation is crucial for cell development (Cedar & Bergman 2009) and differentiation (Robertson 2005). Cell differentiation produces the DNA methylation pattern, but it also renders cells incapable of dividing (Chen et al. 2014). DNA methylation either downregulates or silences

DOI: 10.1201/9781032706511-2

promoters for genes that certain tissues are not supposed to produce, allowing cells to differentiate and preserve lineage. As these cells replicate, DNA methylases enable cells to retain their differentiated identities by transmitting their DNA methylation patterns during mitosis (Stephens et al. 2013).

DNA methylation, which is created during DNA replication, is a stable method of gene silence (Chen et al. 2014). It is an enzymatic procedure that modifies gene expression by covalently adding a methyl group (CH_3) to the fifth position of a cytosine residue (5-methylcytosine, 5mC) (Hamilton 2011; Stephens et al. 2013; Lacal & Ventura 2018). DNA methylation affects the ability of transcription factors and other DNA-binding proteins to recognize nucleotide sequences that regulate gene expression. The suppression of gene transcription is linked to this mechanism (Jones & Baylin 2007).

All living organisms undergo DNA methylation, but different species exhibit different patterns in different locations and types of DNA (Hattman 2005). Methylation is a normal DNA function, a typical alteration seen in animals, plants, and microscopic creatures. In cells with a nucleus, it occurs at a precise location known as the cytosine-phosphodiester-guanine (CpG) dinucleotide that is immediately at the 5′ end of the DNA molecule. S-adenosyl-methionine helps to add a methyl group—a small chemical group—on the cytosine portion of the DNA during this process. Similar to a carer, DNMT1 ensures that the DNA copies itself while maintaining its methylation patterns. However, DNMT3A and DNMT3B function more like architects, creating new methylation patterns from the ground up.

Methylation mostly takes place in CpG dinucleotides inside human cells (Weber et al. 2007; Ehrlich & Wang 1981), and asymmetric methylation is seldom seen (Chen & Li 2004). In humans, it takes place when the cytosine nucleotide is positioned directly next to a guanine nucleotide, namely in the 5′ direction. The linkage of nucleotides inside the DNA molecule is facilitated by phosphate groups, leading to the generally used term "CpG dinucleotide" to describe the cytosine–guanine sequence. The CpG sequence forms a base pair with the GpC sequence on the complementary strand, and methylation exhibits symmetrical patterns. Hence, methylation leads to the modification of cytosine when it is coupled with guanine on the complementary DNA strand (Stephens et al. 2013). See Figure 2.1 for an example of DNA methylation.

2.2.3 Histone Modifications and Chromatin Structure

Histone modification and DNA methylation are natural biological processes that aid in cell formation by giving stem cells the instructions they need to differentiate into a variety of specialized cell

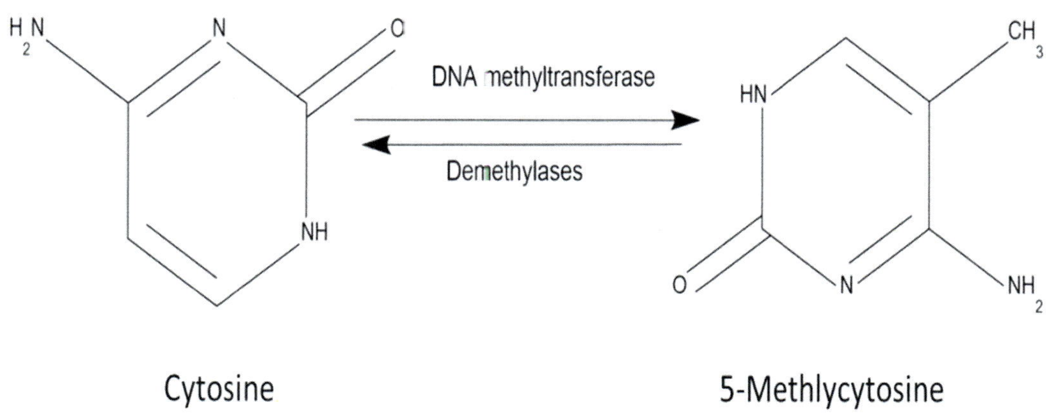

FIGURE 2.1 Conversion of cytosine to 5-methylcytosine by DNA methyltransferases.

types, including skin and brain cells (Miller & Grant 2013). Histones are a class of proteins that play a crucial role in the organization of DNA, acting as a spool around which the DNA molecule is tightly coiled. The complex formed by DNA, and proteins is often referred to as chromatin. The DNA molecule is enveloped by histone proteins, resulting in the formation of recurrent nucleosome structures like a string of beads (Völker-Albert et al. 2020).

A chromosome is created by further condensing chromatin. The nucleus of humans normally has 46 chromosomes in total. One nucleosome's core is made up of DNA encircled by an octameric proteinaceous structure, or histone octamer, which is made up of two copies of each of the main histone types (Misteli 2020). Four proteins—H2A, H2B, H3, and H4—compose the core histones. The H1 linker histones help to stabilize the octamer structure of histones, which is made up of 146 base pairs of DNAs wrapped 1.65 times around it (Turinetto & Giachino 2015).

Histone modification is a posttranslational modification occurring on the side chains of amino acids inside a histone protein. Various posttranslational modifications impact histone tails, including methylation and acetylation (Liu et al. 2023). Additional histone modifications include phosphorylation, ubiquitination, sumoylation, histone poly-ADP ribosylation, biotinylation, citrullination, and proline isomerization. These processes influence the interactions between DNA molecules and histone proteins that influence cellular processes such as gene transcription, DNA repair and replication, and chromosomal organization (Bannister & Kouzarides 2011). Histone modification entails enzymatically adding acetyl or methyl groups onto the tails of histone proteins, hence inducing alterations in gene expression (Berger 2007). For instance, histone 3K9 (H3K9) acetylation exhibits a positive correlation with transcription activation that indicates the presence of active chromatin. Conversely, histone 3K-27 trimethylation (H327Kme3) demonstrates an association with transcription repression, signifying the existence of repressed chromatin (Bannister & Kouzarides 2011; Barski et al. 2007).

The acetylation of the lysine residue is a well-documented histone modification that has been extensively investigated. Typically, there is a correlation between histone acetylation and the activation of transcription, whereas deacetylation represses transcription. Acetylation counterbalances the positive charge of the lysine residue found in the histone tail, diminishing the affinity between the histone tails and DNA. This process facilitates the accessibility of the DNA/histone complex to transcription factors (Hamilton 2011; Gibney & Nolan 2010). Histone methylation has both positive and negative regulatory effects on transcription (Bannister & Kouzarides 2011). There are synergistic relationships between DNA methylation and histone alterations. The intricacy of chromatin modifications is noteworthy due to the many changes and combinations of alterations (Hamilton 2011; Rose & Klose 2014).

The contact between the negatively charged DNA phosphate backbone and the substantially positively charged amino acids of nucleosomes is characterized by a strong ionic connection. The condensed and constricted configuration of this structure hinders the interaction of other proteins with DNA (Yusufaly et al. 2014). Not all DNA molecules are connected to histones; instead, segments of chromosomal DNA known as linker DNA are inserted between every pair of nucleosomes. The accessibility of this DNA to proteins and transcription factors is significantly enhanced (Zentner & Henikoff 2013).

2.2.4 MicroRNAs and Long Noncoding RNAs

A non-coding RNA (ncRNA) is a new-generation epigenetic mechanism; it is a transcribed functional RNA molecule that does not translate into proteins. Merely 2% of RNA gets converted to mRNA, which serves as a structural and functional element of the cell. But the remaining 98% is not "junk," as was previously thought. Numerous genes translate to ncRNA (Liu et al. 2013). Although ncRNA molecules were formerly thought to be a waste product of the genome, they play a significant role in the epigenetic regulation of genes and are probably responsible for the significant variations in phenotype seen in human groups and across animals, even if their encoded proteins are very identical (Ramasamy et al. 2014).

NcRNAs that are not translated into proteins are either housekeeping or regulatory ncRNAs (Zhang et al. 2019a). RNAs with a regulatory function are classified into two groups according to length: long ncRNAs (lncRNAs) and short-chain ncRNAs (siRNAs, miRNAs, and piRNAs). Numerous studies have shown that ncRNAs are important for epigenetic alteration and affect gene and chromosomal expression to govern cell development (Rinn & Chang 2012; Amaral et al. 2008).

2.2.4.1 microRNA

MiRNAs are small RNA molecules, about 22 nucleotides in length of 22, that can adversely regulate the expression of their target genes posttranslation (Ranganathan & Sivasankar 2014). To date, about 2300 human miRNAs have been identified (Alles et al. 2019). There is evidence linking the expression of miRNA to the development of cancer (Ranganathan & Sivasankar 2014). Before being processed into precursor miRNAs in the nucleus by RNase III Drosha and DGCR8 (DiGeorge syndrome critical region 8), miRNAs are first transcribed by RNA polymerase II into primary miRNAs. Then, exportin 5 exports them into the cytoplasm, where Dicer cleaves them into final, mature, functioning miRNAs (Ranganathan & Sivasankar 2014; Alles et al. 2019). Both complete and partial complementarity are possible when miRNAs bind to their target. While partial complementarity, which frequently occurs in the 3′ untranslated regions (UTRs), results in translational repression of their target genes, complete complementarity can cause target degradation. Multiple miRNAs may regulate a single mRNA (Ying et al. 2008).

MiRNAs and small interfering RNAs (siRNAs) are key players in the process of RNA-mediated silencing, where they exert their regulatory effects by suppressing gene expression at the post-transcription level. The potential mechanism of RNA silencing involves the binding of non-coding RNA to the 3′ UTRs of target mRNAs at the post-transcriptional level (Ranganathan & Sivasankar 2014; Ying et al. 2008). Roughly 30% of genes are subject to regulation by miRNA, accounting for a mere 1% of the whole genome. These RNAs finely regulate gene expression in both normal developmental processes and pathological conditions. Furthermore, miRNAs have a significant influence on tumor suppression, programmed cell death (apoptosis), cell proliferation, and cell motility (Shamsi et al. 2017).

2.2.4.2 Long Non-coding RNAs

NcRNAs with a length beyond 200 nucleotides are classified as long ncRNAs, and lncRNAs are also involved in a diverse range of pathological processes (Shamsi et al. 2017) such as the upregulation of the lncRNA MALAT1 in cancerous tissues. Malat1 promotes cellular growth in human cells by affecting the expression and/or pre-mRNA processing of transcription factors that regulate the cell cycle (Fu et al. 2020). Hox genes are linked with several lncRNAs that influence histone methylation and RNA polymerase accessibility throughout human development along both spatial and temporal dimensions (Rinn et al. 2007). It is well accepted in the science community that lncRNAs can interact with chromatin and subsequently attract chromatin-modifying complexes (Marchese & Huarte 2014). lncRNAs function by facilitating the activation of gene expression via trithorax-group complexes (Mattick et al. 2023). LncRNAs are often linked to the occurrence of genomic imprinting, a process that guarantees the epigenetic silencing of one of the two parental alleles of certain autosomal genes (Prabhakar et al. 2017).

LncRNAs and short RNAs play significant roles in mechanisms pertaining to nuclear architecture, chromatin regulation, transcriptional regulation, and RNA processing; lncRNAs control epigenetic mechanisms via the alteration of chromatin structure, whereas miRNAs in epigenetic processes directly or indirectly modulate the expression of DNMTs throughout embryonic development and in somatic cells (Statello et al. 2021).

2.3 EPIGENETIC CHANGES IN DISEASE PATHOGENESIS

2.3.1 DNA Methylation Alterations: Hypomethylation and Hypermethylation

2.3.1.1 Hypomethylation

DNA hypomethylation, also referred to as demethylation, is characterized by the depletion of the methyl group inside the 5-methylcytosine nucleotide (Ehrlich 2009). Methylation is an inherent process of DNA alteration that primarily impact the cytosine base (C) in mammals when it is juxtaposed with guanosine (G). Hypomethylation is the absence of methylation at the majority of CpG sites in a typically methylated region or as a widespread occurrence that impacts a significant percentage of the genome and results in a lower ratio of methylated to unmethylated cytosines (Peinado 2011).

Under typical physiological conditions, around 80% of CpG sites located outside of promoter regions in human DNA undergo methylation (Peinado 2011). Nevertheless, the process of methylation mostly takes place in regions characterized by low CpG density or repetitive DNA sequences, such as Alu elements (Lee et al. 2009). Global hypomethylation is a common occurrence in aging cells, as well as in neoplasia, where it is seen as an early event (Li & Zhang 2014). In the 1980s, low methylcytosine content was established as a significant characteristic of colorectal cancer and several other tumor types (Peinado 2011).

2.3.1.2 Hypermethylation

Hypermethylation, characterized by the introduction of methyl groups, exhibits a notable degree of specificity towards individual genes. Hypermethylation in the CpG islands found in gene promoter regions silences gene transcription, which eventually results in a lack of protein production (Herman & Baylin 2003). Additionally, more tissue-specific DNA hypermethylation is seen in gene bodies that are actively transcribed, as well as in enhancers located inside or between genes, than at promoters (Ehrlich et al. 2016; Neri et al. 2017).

Methyl groups extend into the DNA's main groove, altering its biophysical characteristics and influencing protein binding affinity (Dantas et al. 2015). For instance, the transcription of DNA into RNA is impacted if RNA polymerase 1 is unable to bind to DNA appropriately. Not only do cells replicate their DNA during mitosis, they also replicate the methylation patterns on the DNA, which function as tiny tags. As a devoted emissary, DNMT1 makes certain that these markers are faithfully transferred from the mother's DNA to the daughter's (Liyanage et al. 2014).

DNA methylation acts as a regulatory switch that controls cell processes such as growth, DNA repair, blood vessel formation, toxin response, programmed cell death, and communication between cells. In cancer development, certain genes undergo excessive methylation (Nogueira da Costa & Herceg 2012). It's important to know that increased methylation isn't inherently harmful; our bodies use this process normally for certain functions such as X chromosome inactivation in women (forming the Barr body) (Chaligné et al. 2015). Additionally, this methylation process is simply a part of aging; one critical function is silencing repetitive DNA sequences, which helps maintain chromosome integrity (Johnson et al. 2012; Ciccarone et al. 2018). Table 2.1 provides a brief explanation of the differences between DNA hypermethylation and hypomethylation.

2.3.2 Histone Modification Dysregulation in Disease States

Histone modification is like a makeover for histone proteins and essential parts of chromatin. Chromatin is like the book where our genetic information is written (Bannister & Kouzarides 2011). Dysregulation tampers with this mechanism and plays a major role in the onset and progression of various diseases (Liu et al. 2023). For instance, numerous cancer types have decreased histone acetylation, which alters the chromatin structure and suppresses the transcription of downstream tumor suppressor genes. The observed epigenetic alterations contribute to oncogenesis, and aberrant patterns of histone methylation are often seen in cancer (Audia & Campbell 2016).

TABLE 2.1
Differential Effects of DNA Hypermethylation and Hypomethylation

Methylation Status	Hypomethylation	Hypermethylation
Definition	Reduced methylation	Increased methylation
Effect on gene expression	Increased expression	Decreased expression
Associated with diseases	Gene activation	Gene silencing
Role in cancer	Oncogene activation	Tumor suppressor inactivation
Example	Cancer progression	Loss of tumor suppressor function

For example, elevated levels of histone H3 lysine 27 trimethylation (H3K27me3) correlate with gene repression in different types of cancer (Zhao & Shilatifard 2019; Audia & Campbell 2016). Cardiac hypertrophy, heart failure, and other cardiovascular illnesses are correlated with modified histone alterations that influence the transcription and translation of genes that are implicated in the regulation of heart function and remodeling (Funamoto et al. 2023). Individuals with Alzheimer's disease show altered patterns of histone acetylation and methylation that affect the transcription of genes associated with memory and cognitive processes. The etiology of Huntington's disease, a neurodegenerative condition, is associated with dysregulation of histone acetylation and methylation (Santana et al. 2023).

2.3.3 Noncoding RNAs in Disease Mechanisms

NcRNAs such as miRNAs and lncRNAs are significant regulators of gene expression and are connected to the onset and course of several illnesses (Rinn & Chang 2012). For instance, in cancer, increased miR-21 often facilitates tumor development by inhibiting tumor suppressor genes (Rhim et al. 2022). MiRNAs are also involved in cardiac remodeling, atherosclerosis, and heart failure. For instance, miR-1 and miR-133 are correlated with the development of cardiac hypertrophy, a condition characterized by enlarged heart muscle cells, and miR-155 is connected with the inflammatory processes seen in atherosclerosis, a disease characterized by the buildup of plaque in the arteries (Cao et al. 2016). Another significant molecule involved in neuronal differentiation is miR-124; dysregulation of miR-124 is associated with the development and progression of Alzheimer's disease (Zhang et al. 2023; Jiao et al. 2017).

Numerous lncRNAs have a significant role in the progression of cancer. For example, the lncRNA HOTAIR is correlated with the metastasis of breast, gastric, cervical cancer, and colorectal cancer (Hajjari & Salavaty 2015), and MALAT1 is connected with lung cancer metastasis (Sun & Ma 2019). BACE1-AS, NEAT1, and other lncRNAs have also been associated with Alzheimer's disease (Sayad et al. 2022; Dong et al. 2021), while ANRIL is associated with atherosclerosis and coronary artery disease (Ghafouri-Fard et al. 2021).

2.4 EPIGENETIC DISEASES

2.4.1 Introduction

Epigenetic diseases are distinguished by epigenetic modifications of DNA and chromatin that lead to atypical gene expression. These alterations do not include mutations to the fundamental DNA sequence, but their impacts on gene expression result in a range of health issues (Zoghbi & Beaudet 2016). Epigenetic modifications arise as a consequence of genetic mutations, environmental influences, or a synergistic interplay between the two. Epigenetic alterations mostly include DNA methylation, histone modifications, and non-coding RNA molecules and are critical in typical development and cell activities (Hamilton 2011). Table 2.2 elucidates many epigenetic disorders.

TABLE 2.2
Epigenetic Diseases

Disease(s)	Characteristics	Examples
Cancer	Aberrant DNA methylation, histone modification dysregulation	Breast cancer, colorectal cancer
Neurodevelopmental disorders	Altered DNA methylation patterns in specific gene regions	Rett syndrome, fragile X syndrome
Cardiovascular diseases	Histone alterations leading to cardiovascular complications	Atherosclerosis, hypertension
Metabolic disorders	Dysregulation of ncRNA expression influencing metabolic pathways	Obesity, type 2 diabetes
Neurodegenerative diseases	Epigenetic changes impacting neuronal function and survival	Alzheimer's disease, Parkinson's disease
Autoimmune disorders	DNA methylation alterations influencing immune cell function	Systemic lupus erythematosus, rheumatoid arthritis

2.4.2 Cancer as an Epigenetic Disease

Cancer is a prominent example of an epigenetic disorder (Lu et al. 2020). Cancer-related epigenetic alterations involve abnormal DNA methylation and histone modifications that disrupt normal functioning such as cell proliferation, DNA repair, and apoptosis (Chen et al. 2014; Peinado 2011; Zhao & Shilatifard 2019).

2.4.2.1 DNA Hypermethylation

Hypermethylation of promoter regions silences tumor suppressor genes in cancer cells (Herman & Baylin 2003). Hypermethylation is seen in several types of tumors and affects genes such as CDKN2A, as well as MLH1 specifically in colorectal cancer (Jung et al. 2020). Tumor suppressor genes that are affected by epigenetic silencing include p16INK4a, p15INK4b, and BRCA1 (Jung et al. 2020).

2.4.2.2 Histone Modifications

Reduced histone acetylation is commonly associated with tighter chromatin and decreased gene expression (Bannister & Kouzarides 2011). Certain cancers show reduced histone acetylation markers near the promoter regions of tumor suppressor genes. Elevated trimethylation of histone H3 lysine 27 (H3K27me3) can suppress tumor suppressor genes (Zhao & Shilatifard 2019; Morlando & Fatica 2018).

2.4.3 Neurodevelopmental Disorders: Epigenetic Contributions

Neurodevelopmental diseases impact the development and functioning of the nervous system (Morris-Rosendahl & Crocq 2020). The genesis of these ailments is influenced not just by genetics but also by epigenetic factors including changes in DNA methylation patterns, modifications to histone proteins, and variations in the production of ncRNA molecules (Moosavi & Motevalizadeh Ardekani 2016). This section discusses many examples of neurodevelopmental diseases and the corresponding epigenetic factors that are implicated in their development.

2.4.3.1 Autism Spectrum Disorder

Autism spectrum disorder (ASD) is a multifaceted neurodevelopmental condition that has a significant hereditary influence. It is widely believed that epigenetic processes contribute to the regulation of gene expression in persons diagnosed with ASD (Yoon et al. 2020). Numerous investigations have shown variations in the histone protein modifications and DNA methylation patterns in genes linked to ASD (Day & Sweatt 2011) These modifications to the regulation of our genes may have an impact on the function of synapses, the structure of our brains, and our social behaviors (Rylaarsdam & Guemez-Gamboa 2019).

2.4.3.2 Schizophrenia

Schizophrenia is a complex condition that arises from a combination of hereditary and environmental factors. The control of genes important to schizophrenia is influenced by epigenetic factors, such as DNA methylation and histone alterations (Brown 2011). Epigenetic modifications associated with neurotransmitter function, brain development, and synaptic plasticity have been observed in patients diagnosed with schizophrenia. These modifications are linked to altered brain connectivity and cognitive deficits that are commonly associated with the condition (Akbarian 2014).

2.4.4 Cardiovascular Diseases and Epigenetic Influences

Cardiovascular diseases (CVDs) are a spectrum of pathological illnesses that impact the cardiac muscle and the circulatory system. The development of CVDs is known to be influenced by recognized risk factors such as food, exercise, and heredity (Sharifi et al. 2020), but epigenetic factors also have a significant impact on this process (Moosavi & Motevalizadeh Ardekani 2016). This section presents a number of CVDs and their corresponding epigenetic influences (Abi Khalil 2014).

2.4.4.1 Atherosclerosis

Atherosclerosis is a pathological condition that is distinguished by the accumulation of plaque inside the arterial walls, resulting in constricted blood circulation (Rafieian et al. 2014), and epigenetic modifications in gene expression play a role in the formation and progression of atherosclerotic plaques. DNA methylation and histone modifications affect genes associated with inflammation, lipid metabolism, vascular inflammation, and atherosclerosis (Tang et al. 2021; Bergonzini et al. 2023).

2.4.4.2 Hypertension

Hypertension develops from intricate interplays among genetic, environmental, and epigenetic elements. It is connected to changes in how DNA is tagged within genes related to how blood vessels manage salt. Epigenetic adjustments can also disrupt the renin–angiotensin system, a key regulator of blood pressure (Pratamawati et al. 2023; Wise & Charchar 2016).

2.4.4.3 Heart Failure

Heart failure is a condition where the heart struggles to pump blood effectively throughout the body. Epigenetic mechanisms influence genes responsible for heart function, and in heart failure, changes in DNA tagging contribute to heart muscle stiffening, enlargement, impaired contracting, and heart failure disease progression (Ameer et al. 2020; Kim et al. 2016).

2.4.4.4 Diabetes-Related Cardiovascular Complications

The risk of developing and exacerbating cardiac issues is significantly increased by having diabetes (Martín et al. 2014). Epigenetic modifications contribute to cardiac complications in diabetes, associated with the body's processes for regulating sugar, inflammation, and oxidative stress (Villeneuve et al. 2011).

2.5 EPIGENETIC DYSREGULATION

2.5.1 Causes of Abnormal DNA Methylation Patterns

Atypical DNA methylation is a common source of epigenetic disorder. Many illnesses including cancer (Ghavifekr Fakhr et al. 2013), developmental abnormalities (Cervantes-Pérez et al. 2021), and aging-related brain issues (Johnson et al. 2012) have been connected to this instability. These odd tagging patterns are caused by a variety of factors, including genetic mutations, environmental factors, stress, viral infections, hormone fluctuations, aging, issues with the enzymes that process epigenetic modifications, and the interactions between distinct epigenetic modifications (Ho et al. 2012; Samanta et al. 2017).

2.5.2 Altered Histone Modifications in Disease Contexts

Histones, the proteins that organize and compact DNA into chromatin, play a key role in gene regulation. Alterations in histone structure such as acetylation, methylation, phosphorylation, and ubiquitination—which add or remove tiny chemical groups—contribute to dysfunction in many illnesses (Zhao & Shilatifard 2019). By varying the accessibility of the DNA, these modifications may improve or impair the way our genes are read (Liyanage et al. 2014). These changed histones disrupt the way our genes are normally regulated in a number of ways across many diseases.

2.5.2.1 Cancer

- **Histone Acetylation:** Low histone acetylation in certain malignancies leads to tighter DNA structures and slower gene activity that would otherwise prevent tumor growth. Without adequate regulation, this causes cells to proliferate and expand (Audia & Campbell 2016; Oliveto et al. 2017).
- **Histone Methylation:** Cancer is often characterized by the presence of abnormal patterns of histone methylation (Feinberg & Vogelstein 1983). Heightened trimethylation of H3K27me3 suppresses transcriptional activity in tumor suppressor genes (Chen et al. 2015).
- **Histone Mutations:** An association has been observed between juvenile brain cancers and mutations in histone H3 variations (Liu et al. 2014).

2.5.2.2 Neurological Disorders

- **Histone Acetylation and Methylation:** The dysregulation of histone acetylation and methylation affects the gene expression that plays a crucial role in the formation and functioning of neurons. This in turn contributes to the manifestation of neurological illnesses, including autism, schizophrenia, and Alzheimer's disease (Park et al. 2022).
- **Histone Deacetylase Inhibitors:** The potential therapeutic efficacy of histone deacetylase (HDAC) inhibitors has been shown in preclinical investigations for the management of neurodegenerative disorders via the facilitation of histone acetylation and subsequent gene expression (Shukla & Tekwani 2020).

2.5.2.3 CVDs

Altered histone modifications have been implicated in cardiovascular diseases like heart failure and atherosclerosis. Changes in histone acetylation and methylation can affect genes involved in cardiac function and vascular health (Shi et al. 2022).

2.5.3 Dysfunctional NcRNAs and Gene Expression

NcRNAs that are dysfunctional play a substantial role in the dysregulation of epigenetics, hence exerting severe impacts on the expression of genes. ncRNAs are a class of RNA molecules that lack the ability to encode proteins, but they possess significant regulatory functions in a multitude of cellular processes, such as the modulation of gene expression (Kaikkonen et al. 2011). Dysfunctional ncRNAs have the potential to perturb regular epigenetic regulation via the following mechanisms:

2.5.3.1 MiRNAs

- **MiRNA Dysregulation:** MiRNAs are a class of tiny ncRNAs that exert post-transcriptional control on gene expression. They do this by specifically binding to the 3′ UTRs of target mRNAs, either degrading the mRNA or inhibiting its translation (Ying et al. 2008).
- **Aberrant MiRNA Expression:** The dysregulation of miRNA expression induces aberrant gene expression. The upregulation of miRNAs limit the expression of genes associated with tumor suppression, while their downregulation can lead to the upregulation of oncogenes, therefore playing a role in the progression of cancer (Ling et al. 2013).

2.5.3.2 LncRNAs

- **LncRNA-Mediated Regulation:** LncRNAs interact with DNA, RNA, and proteins to exert regulatory control on gene expression at many hierarchical levels (Segal & Dostie 2023).
- **Altered LncRNA Expression:** The improper manifestation of lncRNAs can disrupt epigenetic control. As an example, some lncRNAs support complexes that modify chromatin, influencing histone modifications and chromatin structure (Zhang et al. 2019b).

2.6 ENVIRONMENTAL INFLUENCES ON EPIGENETICS

Environmental variables that change gene expression without changing the underlying DNA sequence are known as epigenetic impacts. These factors include stress, nutrition, exposure to chemicals, and even social interactions. Chemical changes to DNA or histone proteins caused by these environmental conditions affect the on/off states of genes (Tiffon 2018). Risk assessment and prevention saw a significant advance with the revelation that DNA mutations might result from exposure to the environment (Tiffon 2018). Consequently, the ability of chemical agents to alter the DNA sequence led to its classification (Baccarelli et al. 2009). The identification of environmental risks and the development of current regulatory efforts to limit exposure have benefited greatly from the use of this data. Environmental exposures may disrupt the growing organism's epigenome, altering the fetus's susceptibility to illness in later life (Tiffon 2018). Epigenetic pathways are also implicated throughout development in utero and at the cellular level (Thornburg et al. 2010).

Despite having the same genotype, monozygotic twins in research often show notable phenotypic discordance, such as differences in anthropomorphic traits and susceptibility to certain diseases. This discrepancy led researchers to look into possible causes, with an emphasis on epigenetic variations. The investigations revealed patterns of histone acetylation and DNA methylation in a large group of monozygotic twins (Jirtle & Skinner 2007). The results showed that whereas older monozygotic twins have significant changes in 5-methylcytosine DNA and histone acetylation patterns that affect their gene expression profiles, they are epigenetically identical in their early years. These findings highlight the importance of epigenetics in explaining how different phenotypes emerge from the same genome (Fraga et al. 2005; Castillo et al. 2014).

Epidemiological studies demonstrate how unique exposures have affected the health of specific groups through time, notably between mother and child. The Dutch Hunger Winter is a well-known example. A famine devastated the western Netherlands in 1944, forcing residents, especially pregnant women, to live on 400 to 800 calories per day. Scientists discovered heightened rates of obesity, changed lipid profiles, and CVD in adulthood among the children conceived, carried, or delivered during this time period (Bleker et al. 2021). Table 2.3 lists these effects.

2.6.1 Epigenetic Alterations Due to Diet

The epigenetic enzymes DNMT, HDAC, and HAT can be directly inhibited by nutrients, or they can influence the availability of substrate needed for certain enzymatic activity. This ultimately alters the way vital genes are expressed, which has an impact on our general health. Researchers on the impacts of nutrition on changing epigenetic systems have found, for example, that a diet high in

TABLE 2.3
Effects of Environmental Factors on Epigenetic Changes and Gene Expression

Environmental Factor	Impact on Epigenetic Changes	Impact on Gene Expression and Epigenetic Regulation
Diet	Alters DNA methylation patterns, histone modifications, and miRNA levels	Regulates the expression of genes involved in metabolism, inflammation, and other physiological processes
Toxins	Induces DNA methylation and histone modifications, leading to gene silencing	Downregulates key genes associated with detoxification, immune response, and cell growth
Stress	Induces DNA methylation and histone modifications, leading to gene silencing	Modulates the expression of stress-responsive genes, impacting neural and endocrine functions
Environmental pollutants	Disrupt DNA methylation and histone acetylation, leading to gene dysregulation	Alters the expression of genes involved in immune response, respiratory function, and development
Lifestyle factors	Affects DNA methylation, histone modifications, and chromatin remodeling	Modulates the expression of genes associated with aging, disease susceptibility, and cellular homeostasis

polyunsaturated fatty acids (PUFAs) produces carcinogenic free radicals and oxidative stress, both of which have been associated with epigenetic changes. In human endothelial cells treated with arachidonic acid, altered gene methylation activated a pro-angiogenic pathway. PUFAs, in contrast, suppress tumorigenic processes by dampening inflammation and the NF-κB pathway (Tiffon 2018). The next subsections delve into how these dietary agents can lead to epigenetic alterations and briefly explain their mechanisms.

2.6.1.1 Folate (Vitamin B9)

Folate donates methyl in the one-carbon metabolism pathway; it supports proper DNA methylation by providing methyl groups for DNA methyltransferases. Adequate folate intake ensures correct DNA methylation patterns, essential for gene regulation. Folate deficiency can lead to hypomethylation, affecting gene expression and increasing disease susceptibility. Folate deficiency-induced hypomethylation can lead to the inappropriate activation of genes that should be silenced, including genes associated with cell growth, inflammation, and immune responses. This dysregulation of gene expression can increase an individual's susceptibility to various diseases, including cancer, CVDs, neurological disorders, and developmental abnormalities (Crider et al. 2012; Mahmoud & Ali 2019).

2.6.1.2 Vitamin B12

Vitamin B12 collaborates with folate in one-carbon metabolism, helping regenerate methionine from homocysteine and ensuring a continuous supply of methyl groups for DNA methylation, and hypomethylation alters gene expression (Lyon et al. 2020; Froese et al. 2019).

2.6.1.3 Choline

Choline facilitates epigenetic processes, participating in DNA methylation as a methyl donor after changing into betaine. Choline is particularly essential during fetal development because it promotes appropriate DNA methylation. Low choline interferes with DNA methylation, which affects gene expression and development (Zeisel 2017; Blusztajn & Mellott 2012).

2.6.1.4 Polyphenols

Polyphenols like resveratrol and quercetin are found in fruits, vegetables, tea, and red wine (Manach et al. 2004) and can promote histone acetylation, a modification associated with gene activation. They can inhibit HDACs, and this increase in histone acetylation can alter gene activity and contribute to the health benefits associated with polyphenol-rich diets (Waldecker et al. 2008). Polyphenols regulate genes linked to cancer, heart issues, and neurological disorders, supporting overall health and reducing disease risk (Bag & Bag 2018).

2.6.1.5 Omega-3 Fatty Acids

Omega-3s are found in fatty fish, flaxseeds, and walnuts (Covington 2004). They influence gene expression by changing how MiRNAs regulate gene expression after the genetic code is transcribed. Specifically, they shift the expression of these miRNAs to control specific genes related to inflammation, lipid metabolism, and cell growth (Albracht et al. 2018). Omega-3s regulate miRNA expression, reducing pro-inflammatory gene activity and enhancing anti-inflammatory (Albracht et al. 2018). Omega-3s also influence histone modifications in the brain, affecting genes linked to brain protection, flexibility, and cognitive function. This suggests a potential role in maintaining brain health and reducing disease risk (Dighriri et al. 2022).

2.6.1.6 Sulforaphane

Sulforaphane, found in cruciferous vegetables like broccoli, can reverse excessive DNA methylation, reactivating silenced tumor suppressor genes in cancer (Nandini et al. 2020; Su et al. 2018). In human breast cancer, sulforaphane inhibits HDAC activity, demonstrating anti-cancer properties. This suppression leads to cell cycle arrest and death while influencing the expression of proteins in breast cancer development (Kim et al. 2010).

2.6.1.7 Curcumin

This component of turmeric influences DNA methylation and histone modifications, showing potential in epigenetic cancer therapies. Curcumin can regulate genes linked to cancer, heart problems, brain diseases, inflammation, and inflammation, supporting disease prevention and treatment efficacy (Fabianowska et al. 2021; Hassan et al. 2019).

2.6.1.8 Selenium

The essential trace element selenium contributes to DNA methylation and histone modifications by participating in the synthesis and function of selenoproteins, proteins that contain selenium in the form of the amino acid selenocysteine (Hariharan & Dharmaraj 2020). Selenoproteins such as selenoprotein P and glutathione peroxidases support redox balance and protect DNA from oxidative damage. Their roles in epigenetic regulation influence gene expression and cellular health. Selenium deficiency disrupts these epigenetic processes, increasing disease risk (Genchi et al. 2023).

Selenium deficiency and altered epigenetic marks contribute to cancer risk by disrupting DNA methylation and histone modifications, leading to tumor suppressor gene silencing and oncogene activation (Genchi et al. 2023). Disrupted epigenetic regulation due to selenium deficiency can contribute to the development of CVDs by affecting genes involved in inflammation, oxidative stress, and lipid metabolism. Adequate selenium intake is essential for maintaining proper epigenetic marks related to disease prevention (Zhu et al. 2022).

2.6.2 Epigenetic Alterations Due to Toxins

2.6.2.1 Tobacco Smoke

Tobacco smoke contains carcinogens like benzene and PAHs that lead to histone modifications and alterations in DNA methylation. These epigenetic alterations disrupt gene regulation, affecting critical pathways related to inflammation, oxidative stress, and cancer susceptibility and increasing the risk of lung cancer, CVDs, and respiratory disorders (Caliri et al. 2021).

2.6.2.2 Air Pollution

Air pollution consists of fine particulate matter such as $PM_{2.5}$, volatile organic compounds, and heavy metals. These induce epigenetic alterations by initiating oxidative stress and inflammation. $PM_{2.5}$ and toxic metals generate reactive oxygen species (ROS), leading to DNA damage, histone modifications, and altered gene expression. Epigenetic changes induced by air pollutants are associated with an increased risk of respiratory diseases (e.g., asthma, lung cancer), CVDs, and developmental issues in children (Thangavel et al. 2022; Li et al. 2018).

2.6.2.3 Pesticides

Pesticides such as organophosphates alter many physiological processes including gene activity, for instance by tampering with DNA methylation patterns, an essential regulatory mechanism for gene expression (Zhang et al. 2012; Paul et al. 2018). Pesticides disrupt gene expression by inhibiting DNMT enzymes and particularly affect histone acetylation and methylation. Epigenetic changes influence gene accessibility and chromatin structure, and some of these modifications contribute to neurological conditions, cancer, and other health issues connected to pesticide exposure (Song et al. 2010).

2.6.2.4 Heavy Metals

Heavy metals such as arsenic, cadmium, mercury, and lead induce epigenetic alterations through two mechanisms. First, they directly bind to and disrupt the activity of enzymes involved in DNA methylation and histone modification, leading to changes in epigenetic marks. Second, they generate oxidative stress, causing DNA damage and triggering repair processes that alter DNA methylation patterns and histone modifications (Balali-Mood et al. 2021). These epigenetic changes can affect gene expression, contributing to diseases including cancer, neurodevelopmental disorders, and CVDs (Ho et al. 2012).

2.6.2.5 Radiation

Radiation exposure, particularly ionizing radiation from sources like X-rays and gamma rays, induces epigenetic alterations through DNA damage: Ionizing radiation directly causes single- and double-strand breaks, base modifications, and DNA cross-linking. These DNA lesions trigger cellular repair mechanisms that can inadvertently lead to epigenetic changes (Belli et al. 2020). Radiation-induced DNA damage alters DNA methylation patterns and histone modifications (Shumei et al. 2010), and these epigenetic alterations influence gene expression and increase the risk of cancer and genetic disorders associated with radiation exposure (Bourguignon et al. 2005).

2.6.2.6 Dietary Toxins

These toxins are produced by molds and are present in tainted food. For instance, aflatoxin B1-8,9-exo-epoxide, one of its byproducts, covalently binds to DNA, particularly at guanine bases, forming DNA adducts. These chemical alterations disrupt the normal DNA structure and function and cause errors during DNA repair. This can disrupt DNA methylation, leading to abnormal gene regulation. Furthermore, aflatoxin exposure alters histones, proteins that aid in packing DNA (Dai et al. 2017). These genetic modifications increase the risk of liver cancer and other health problems (Okeke et al. 2020).

2.6.2.7 Alcohol and Drugs

Alcohol and illicit drugs induce epigenetic alterations through various mechanisms. Alcohol metabolism generates acetaldehyde, a carcinogenic compound that can lead to DNA damage and affect DNA methylation patterns. Additionally, drugs like opioids and amphetamines can influence DNA methylation and histone modifications through changes in neurotransmitter signaling. Chronic substance abuse can disrupt normal epigenetic regulation, contributing to addiction-related behaviors and increasing the risk of mental health disorders and other health complications (Alegría et al. 2011).

2.6.2.8 Infections

Human papillomavirus (HPV) and the hepatitis B and C viruses induce epigenetic alterations primarily through interactions with host cell machinery. HPV oncoproteins E6 and E7 interact with and disrupt the function of key epigenetic regulators, leading to DNA methylation changes and histone modifications in host cells (Ghiani & Chiocca 2022). In the case of hepatitis B and C, viral proteins and chronic inflammation induce oxidative stress and activate signaling pathways that influence epigenetic marks (Gao et al. 2019). These virus-induced epigenetic alterations contribute to the development of cervical and liver cancers, respectively, by dysregulating the expression of critical genes involved in tumor suppression and immune response (Ghiani & Chiocca 2022; Gao et al. 2019).

2.6.2.9 Medications

Medications can influence epigenetics through various mechanisms. Some drugs, particularly chemotherapy agents, can directly damage DNA, leading to epigenetic changes during the repair process (Cao & Yan 2020). Others affect epigenetic marks by interacting with enzymes involved in DNA methylation or histone modification. Additionally, HDAC inhibitors and other medications can indirectly influence epigenetics by altering cellular signaling (Lakshmaiah et al. 2014). These changes can have therapeutic effects but also carry risks, highlighting the need for careful monitoring (Xu et al. 2019b).

2.6.3 Epigenetic Alterations Due to Stress

Psychological stressors such as workplace or relationship stress can elicit changes in the epigenome (Gottschalk et al. 2020) including triggering the secretion of cortisol and adrenaline. Elevated cortisol alters histone acetylation and methylation via interacting with glucocorticoid receptors (Lemche et al. 2016). Stress can influence the expression of miRNAs, small RNAs involved in post-transcriptional gene regulation that affect pathways related to mood, immunity, and cell stress responses (Penner-Goeke & Binder 2019).

2.7 THERAPEUTIC STRATEGIES TARGETING EPIGENETIC MODIFICATIONS

2.7.1 Epigenetic Therapy

Epigenetic treatment is a novel medical strategy that seeks to alter or manage epigenetic modifications on genes to address a range of illnesses, including cancer, neurological disorders, and autoimmune conditions (Moosavi & Motevalizadeh Ardekani 2016). Epigenetic modifications include DNA methylation, changes in histones, and non-coding RNA molecules, and all serve pivotal functions in gene expression (Handy et al. 2011). Various epigenetic treatment modalities and their uses in certain medical conditions are presented in Table 2.4.

2.7.2 DNA Methylation Inhibitors

DNA methylation inhibitors are used in the epigenetic treatment of a range of pathological conditions (Joseph et al. 2018; Uddin & Fandy 2021). The inhibitors can selectively interact with enzymes referred to as DNA methyltransferases (DNMTs), inducing modifications in DNA methylation patterns, and thereby resulting in alterations in the expression of genes (Joseph et al. 2018; Heerboth et al. 2014).

TABLE 2.4
Epigenetic Therapies and Their Mechanisms

Therapy	Mechanism of Action	Targeted Epigenetic Modifications	Potential Applications
DNA methylation inhibitors	Inhibit DNA methyltransferases, leading to demethylation of specific gene regions	Reversal of aberrant DNA methylation	Cancer, neurodevelopmental disorders
Histone deacetylase inhibitors	Block histone deacetylase enzymes, resulting in increased histone acetylation and gene expression	Modulation of histone acetylation	Neurodegenerative diseases, cancer
ncRNA modulators	Regulate the expression of specific ncRNAs involved in epigenetic regulation	Control of ncRNA-mediated gene silencing	CVDs, metabolic disorders
Chromatin remodeling agents	Modify chromatin structure and accessibility through ATP-dependent chromatin remodeling complexes	Alteration of chromatin conformation and gene accessibility	Autoimmune disorders, CVDs

2.7.3 Histone Deacetylase Inhibitors

HDAC inhibitors regulate gene expression by targeting HDACs, enzymes that remove acetyl groups from histones, in the process compacting chromatin and influencing gene regulation (Seto & Yoshida 2014). HDAC inhibitors enhance histone acetylation, relax chromatin structure, and modify gene expression profiles (Chueh et al. 2015).

2.8 CHALLENGES AND FUTURE DIRECTIONS

2.8.1 Current Limitations in Understanding Epigenetic Mechanisms

Comprehending the functioning of epigenetic mechanisms is like navigating an intricate and swiftly changing scientific terrain. Even while researchers have come a long way, there are still difficulties in fully comprehending epigenetics, including the complexity of epigenetic regulation, tissue and cell type specificity, temporal dynamics, functional annotation, technological limitations, data integration, and ethical and legal issues (Santaló & Berdasco 2022). See Table 2.5 for more discussion of the challenges and potential solutions.

2.8.2 Emerging Technologies for Studying Epigenetic Regulation

The study of epigenetic regulation has seen significant changes as a result of advancements in cutting-edge technology. Researchers now have better equipment to examine the complicated topic of epigenetics. For instance, the following tools hold great promise for potentially transforming our understanding of how epigenetics influences gene expression, development, diseases, and evolution: single-cell epigenomics (Godfrey & Rodriguez-Meira 2022); nanopore sequencing, long-read sequencing with pacbio and hifi sequencing, multi-omics integration, 3D epigenomics (Li 2021); CUT&RUN and CUT&TAG (Fallet et al. 2023); high-resolution

TABLE 2.5
Challenges and Solutions in Understanding Epigenetic Mechanisms

Challenge	Future Direction
Lack of comprehensive epigenetic biomarkers	Identify novel epigenetic markers for early disease detection
Limited understanding of epigenetic regulatory networks	Elucidate complex epigenetic regulatory networks governing specific cellular processes
Incomplete knowledge of environmental epigenetic influences	Investigate the long-term effects of environmental exposures on epigenetic landscapes
Technical limitations in epigenetic profiling techniques	Develop advanced high-throughput technologies for precise epigenetic profiling
Challenges in translating epigenetic findings into therapies	Integrate epigenetic knowledge into targeted therapeutic interventions for personalized medicine

microscopy (Xu et al. 2019a); CRISPR-based epigenome editing (Nakamura et al. 2021); liquid biopsies (Fox-Fisher et al. 2023); and AI and machine learning (De Riso & Cocozza 2021). These emerging technologies hold great promise in overcoming current limitations in epigenetic research.

2.8.3 Potential Role of Epigenetics in Precision Medicine

Precision medicine is focused on customizing medical treatments for individual patients based on their genetic, epigenetic, and environmental characteristics (Delpierre et al. 2023). Epigenetics can significantly advance precision medicine in the fields of disease risk prediction, diagnostics, treatment response prediction, drug development, personalized therapies, aging and longevity, environmental exposures, and rare diseases are all linked to each other (Tammen et al. 2013). Through the incorporation of epigenetic data into precision medicine, we may realize the possibility of more individualized, focused, and effective treatment plans that take into account the distinct features of every person's biology and surroundings (Gray et al. 2019; Strianese et al. 2020). This strategy is a big step toward a more efficient and individualized medical age.

REFERENCES

Abi Khalil C. The emerging role of epigenetics in cardiovascular disease. *Ther Adv Chronic Dis*. 2014;5(4):178–187.

Akbarian S. Epigenetic mechanisms in schizophrenia. *Dialogues Clin Neurosci*. 2014;16(3):405–417.

Albracht-Schulte K, Kalupahana NS, Ramalingam L, et al. Omega-3 fatty acids in obesity and metabolic syndrome: A mechanistic update. *J Nutr Biochem*. 2018;58:1–16.

Alegría-Torres JA, Baccarelli A, Bollati V. Epigenetics and lifestyle. *Epigenomics*. 2011;3(3):267–277.

Alles J, Fehlmann T, Fischer U, et al. An estimate of the total number of true human miRNAs. *Nucleic Acids Res*. 2019;47(7):3353–3364.

Amaral PP, Dinger ME, Mercer TR, Mattick JS. The eukaryotic genome as an RNA machine. *Science*. 2008;319(5871):1787–1789.

Ameer SS, Hossain MB, Knöll R. Epigenetics and heart failure. *Int J Mol Sci*. 2020;21(23):9010.

Audia JE, Campbell RM. Histone modifications and cancer. *Cold Spring Harb Perspect Biol*. 2016; 8(4):a019521.

Baccarelli A, Bollati V. Epigenetics and environmental chemicals. *Curr Opin Pediatr*. 2009;21(2):243–251.

Bag A, Bag N. Tea polyphenols and prevention of epigenetic aberrations in cancer. *J Nat Sci Biol Med*. 2018;9(1):2–5.

Balali-Mood M, Naseri K, Tahergorabi Z, Khazdair MR, Sadeghi M. Toxic mechanisms of five heavy metals: Mercury, lead, chromium, cadmium, and arsenic. *Front Pharmacol*. 2021;12:643972.

Bannister AJ, Kouzarides T. Regulation of chromatin by histone modifications. *Cell Res.* 2011;21(3):381–395.

Barski A, Cuddapah S, Cui K, et al. High-resolution profiling of histone methylations in the human genome. *Cell.* 2007;129(4):823–837.

Belli M, Tabocchini MA. Ionizing radiation-induced epigenetic modifications and their relevance to radiation protection. *Int J Mol Sci.* 2020;21(17):5993.

Berger SL. The complex language of chromatin regulation during transcription. *Nature.* 2007;447(7143): 407–412.

Bergonzini M, Loreni F, Lio A, et al. Panoramic on epigenetics in coronary artery disease and the approach of personalized medicine. *Biomedicines.* 2023;11(10):2864.

Bleker LS, de Rooij SR, Painter RC, Ravelli AC, Roseboom TJ. Cohort profile: The Dutch famine birth cohort (DFBC)- a prospective birth cohort study in the Netherlands. *BMJ Open.* 2021;11(3):e042078.

Blusztajn JK, Mellott TJ. Choline nutrition programs brain development via DNA and histone methylation. *Cent Nerv Syst Agents Med Chem.* 2012;12(2):82–94.

Bourguignon MH, Gisone PA, Perez MR, et al. Genetic and epigenetic features in radiation sensitivity. Part II: Implications for clinical practice and radiation protection. *Eur J Nucl Med Mol Imaging.* 2005;32(3): 351–368.

Brown AS. The environment and susceptibility to schizophrenia. *Prog Neurobiol.* 2011;93(1):23–58.

Caliri AW, Tommasi S, Besaratinia A. Relationships among smoking, oxidative stress, inflammation, macromolecular damage, and cancer. *Mutat Res Rev Mutat Res.* 2021;787:108365.

Cao J, Yan Q. Cancer epigenetics, tumor immunity, and immunotherapy. *Trends Cancer.* 2020;6(7):580–592.

Cao RY, Li Q, Miao Y, et al. The emerging role of MicroRNA-155 in cardiovascular diseases. *Biomed Res Int.* 2016;2016:9869208.

Castillo-Fernandez JE, Spector TD, Bell JT. Epigenetics of discordant monozygotic twins: Implications for disease. *Genome Med.* 2014;6(7):60.

Cedar H, Bergman Y. Linking DNA methylation and histone modification: Patterns and paradigms. *Nat Rev Genet.* 2009;10(5):295–304.

Cervantes-Pérez SA, Yong-Villalobos L, Florez-Zapata NMV, et al. Atypical DNA methylation, sRNA-size distribution, and female gametogenesis in Utricularia gibba. *Sci Rep.* 2021;11(1):15725.

Chaligné R, Popova T, Mendoza-Parra MA, et al. The inactive X chromosome is epigenetically unstable and transcriptionally labile in breast cancer. *Genome Res.* 2015;25(4):488–503.

Chen HP, Zhao YT, Zhao TC. Histone deacetylases and mechanisms of regulation of gene expression. *Crit Rev Oncog.* 2015;20(1–2):35–47.

Chen QW, Zhu XY, Li YY, Meng ZQ. Epigenetic regulation and cancer (review). *Oncol Rep.* 2014;31(2): 523–532.

Chen T, Li E. Structure and function of eukaryotic DNA methyltransferases. *Curr Top Dev Biol.* 2004;60:55–89.

Chueh AC, Tse JW, Tögel L, Mariadason JM. Mechanisms of histone deacetylase inhibitor-regulated gene expression in cancer cells. *Antioxid Redox Signal.* 2015;23(1):66–84.

Ciccarone F, Tagliatesta S, Caiafa P, Zampieri M. DNA methylation dynamics in aging: How far are we from understanding the mechanisms? *Mech Ageing Dev.* 2018;174:3–17.

Covington MB. Omega-3 fatty acids. *Am Fam Physician.* 2004;70(1):133–140.

Crider KS, Yang TP, Berry RJ, Bailey LB. Folate and DNA methylation: A review of molecular mechanisms and the evidence for folate's role. *Adv Nutr.* 2012;3(1):21–38.

Dai Y, Huang K, Zhang B, Zhu L, Xu W. Aflatoxin B1-induced epigenetic alterations: An overview. *Food Chem Toxicol.* 2017;109(Pt 1):683–689.

Dantas Machado AC, Zhou T, Rao S, et al. Evolving insights on how cytosine methylation affects protein-DNA binding. *Brief Funct Genomics.* 2015;14(1):61–73.

Day JJ, Sweatt JD. Epigenetic mechanisms in cognition. *Neuron.* 2011;70(5):813–829.

Delpierre C, Lefèvre T. Precision and personalized medicine: What their current definition says and silences about the model of health they promote. Implication for the development of personalized health. *Front Sociol.* 2023;8:1112159.

De Riso G, Cocozza S. Artificial intelligence for epigenetics: Towards personalized medicine. *Curr Med Chem.* 2021;28(32):6654–6674.

Dighriri IM, Alsubaie AM, Hakami FM, et al. Effects of omega-3 polyunsaturated fatty acids on brain functions: A systematic review. *Cureus.* 2022;14(10):e30091.

Dong LX, Zhang YY, Bao HL, Liu Y, Zhang GW, An FM. LncRNA NEAT1 promotes Alzheimer's disease by down regulating micro-27a-3p. *Am J Transl Res.* 2021;13(8):8885–8896.

Ehrlich KC, Paterson HL, Lacey M, Ehrlich M. DNA hypomethylation in intragenic and intergenic enhancer chromatin of muscle-specific genes usually correlates with their expression. *Yale J Biol Med.* 2016;89(4): 441–455.

Ehrlich M, Wang RY. 5-methylcytosine in eukaryotic DNA. *Science.* 1981;212(4501):1350–1357.

Ehrlich M. DNA hypomethylation in cancer cells. *Epigenomics.* 2009;1(2):239–259.

Fabianowska-Majewska K, Kaufman-Szymczyk A, Szymanska-Kolba A, Jakubik J, Majewski G, Lubecka K. Curcumin from turmeric rhizome: A potential modulator of DNA methylation machinery in breast cancer inhibition. *Nutrients.* 2021;13(2):332.

Fallet M, Blanc M, Di Criscio M, Antczak P, Engwall M, Guerrero Bosagna C, Rüegg J, Keiter SH. Present and future challenges for the investigation of transgenerational epigenetic inheritance. *Environ Int.* 2023;172:107776.

Feinberg AP, Vogelstein B. Hypomethylation distinguishes genes of some human cancers from their normal counterparts. *Nature.* 1983;301(5895):89–92.

Fox-Fisher I, Shemer R, Dor Y. Epigenetic liquid biopsies: A novel putative biomarker in immunology and inflammation. *Trends Immunol.* 2023;44(5):356–364.

Fraga MF, Ballestar E, Paz MF, et al. Epigenetic differences arise during the lifetime of monozygotic twins. *Proc Natl Acad Sci U S A.* 2005;102(30):10604–10609.

Froese DS, Fowler B, Baumgartner MR. Vitamin B12, folate, and the methionine remethylation cycle-biochemistry, pathways, and regulation. *J Inherit Metab Dis.* 2019;42(4):673–685.

Fu S, Wang Y, Li H, Chen L, Liu Q. Regulatory networks of LncRNA MALAT-1 in cancer. *Cancer Manag Res.* 2020;12:10181–10198.

Funamoto M, Imanishi M, Tsuchiya K, Ikeda Y. Roles of histone acetylation sites in cardiac hypertrophy and heart failure. *Front Cardiovasc Med.* 2023;10:1133611.

Gao Q, Zhu H, Dong L, et al. Integrated proteogenomic characterization of HBV-related hepatocellular carcinoma. *Cell.* 2019;179(2):561–577.e22.

Genchi G, Lauria G, Catalano A, Sinicropi MS, Carocci A. Biological activity of selenium and its impact on human health. *Int J Mol Sci.* 2023;24(3):2633.

Ghafouri-Fard S, Gholipour M, Taheri M. The emerging role of long non-coding RNAs and circular RNAs in coronary artery disease. *Front Cardiovasc Med.* 2021;8:632393.

Ghavifekr Fakhr M, Farshdousti Hagh M, Shanehbandi D, Baradaran B. DNA methylation pattern as important epigenetic criterion in cancer. *Genet Res Int.* 2013;2013:317569.

Ghiani L, Chiocca S. High risk-human papillomavirus in HNSCC: Present and future challenges for epigenetic therapies. *Int J Mol Sci.* 2022;23(7):3483.

Gibney ER, Nolan CM. Epigenetics and gene expression. *Heredity (Edinb).* 2010;105(1):4–13.

Godfrey LC, Rodriguez-Meira A. Viewing AML through a New Lens: Technological advances in the study of epigenetic regulation. *Cancers (Basel).* 2022;14(23):5989.

Gottschalk MG, Domschke K, Schiele MA. Epigenetics underlying susceptibility and resilience relating to daily life stress, work stress, and socioeconomic status. *Front Psychiatry.* 2020;11:163.

Gray ID, Kross AR, Renfrew ME, Wood P. Precision medicine in lifestyle medicine: The way of the future? *Am J Lifestyle Med.* 2019;14(2):169–186.

Hajjari M, Salavaty A. HOTAIR: An oncogenic long non-coding RNA in different cancers. *Cancer Biol Med.* 2015;12(1):1–9.

Hamilton JP. Epigenetics: Principles and practice. *Dig Dis.* 2011;29(2):130–135.

Handy DE, Castro R, Loscalzo J. Epigenetic modifications: Basic mechanisms and role in cardiovascular disease. *Circulation.* 2011;123(19):2145–2156.

Hariharan S, Dharmaraj S. Selenium and selenoproteins: It's role in regulation of inflammation. *Inflammo pharmacology.* 2020;28(3):667–695.

Hassan FU, Rehman MS, Khan MS, et al. Curcumin as an alternative epigenetic modulator: Mechanism of action and potential effects. *Front Genet.* 2019;10:514.

Hattman S. DNA-[adenine] methylation in lower eukaryotes. *Biochemistry (Mosc).* 2005;70(5):550–558.

Heerboth S, Lapinska K, Snyder N, Leary M, Rollinson S, Sarkar S. Use of epigenetic drugs in disease: An overview. *Genet Epigenet.* 2014;6:9–19.

Herman JG, Baylin SB. Gene silencing in cancer in association with promoter hypermethylation. *N Engl J Med.* 2003;349(21):2042–2054.

Ho SM, Johnson A, Tarapore P, Janakiram V, Zhang X, Leung YK. Environmental epigenetics and its implication on disease risk and health outcomes. *ILAR J.* 2012;53(3–4):289–305.

Jiao S, Liu Y, Yao Y, Teng J. miR-124 promotes proliferation and differentiation of neuronal stem cells through inactivating Notch pathway. *Cell Biosci.* 2017;7:68.

Jirtle RL, Skinner MK. Environmental epigenomics and disease susceptibility. *Nat Rev Genet.* 2007;8(4): 253–262.

Johnson AA, Akman K, Calimport SR, Wuttke D, Stolzing A, de Magalhães JP. The role of DNA methylation in aging, rejuvenation, and age-related disease. *Rejuvenation Res.* 2012;15(5):483–494.

Jones PA, Baylin SB. The epigenomics of cancer. *Cell.* 2007;128(4):683–692.

Joseph DB, Strand DW, Vezina CM. DNA methylation in development and disease: An overview for prostate researchers. *Am J Clin Exp Urol.* 2018;6(6):197–218.

Jung G, Hernández-Illán E, Moreira L, Balaguer F, Goel A. Epigenetics of colorectal cancer: Biomarker and therapeutic potential. *Nat Rev Gastroenterol Hepatol.* 2020;17(2):111–130.

Kaikkonen MU, Lam MT, Glass CK. Non-coding RNAs as regulators of gene expression and epigenetics. *Cardiovasc Res.* 2011;90(3):430–440.

Kim GW, Gocevski G, Wu CJ, Yang XJ. Dietary, metabolic, and potentially environmental modulation of the lysine acetylation machinery. *Int J Cell Biol.* 2010;2010:632739.

Kim SY, Morales CR, Gillette TG, Hill JA. Epigenetic regulation in heart failure. *Curr Opin Cardiol.* 2016; 31(3):255–265.

Lacal I, Ventura R. Epigenetic inheritance: Concepts, mechanisms and perspectives. *Front Mol Neurosci.* 2018; 11:292.

Lakshmaiah KC, Jacob LA, Aparna S, Lokanatha D, Saldanha SC. Epigenetic therapy of cancer with histone deacetylase inhibitors. *J Cancer Res Ther.* 2014;10(3):469–478.

Lee HS, Kim BH, Cho NY, et al. Prognostic implications of and relationship between CpG island hypermethylation and repetitive DNA hypomethylation in hepatocellular carcinoma. *Clin Cancer Res.* 2009;15(3): 812–820.

Lemche E, Chaban OS, Lemche AV. Neuroendocrinological and epigenetic mechanisms subserving autonomic imbalance and HPA dysfunction in the metabolic syndrome. *Front Neurosci.* 2016;10:142.

Li E, Zhang Y. DNA methylation in mammals. *Cold Spring Harb Perspect Biol.* 2014;6(5):a019133.

Li T, Hu R, Chen Z, et al. Fine particulate matter (PM2.5): The culprit for chronic lung diseases in China. *Chronic Dis Transl Med.* 2018;4(3):176–186.

Li Y. Modern epigenetics methods in biological research. *Methods.* 2021;187:104–113.

Ling H, Fabbri M, Calin GA. MicroRNAs and other non-coding RNAs as targets for anticancer drug development. *Nat Rev Drug Discov.* 2013;12(11):847–865.

Liu G, Mattick JS, Taft RJ. A meta-analysis of the genomic and transcriptomic composition of complex life. *Cell Cycle.* 2013;12(13):2061–2072.

Liu R, Wu J, Guo H, et al. Post-translational modifications of histones: Mechanisms, biological functions, and therapeutic targets. *MedComm (2020).* 2023;4(3):e292.

Liu X, McEachron TA, Schwartzentruber J, Wu G. Histone H3 mutations in pediatric brain tumors. *Cold Spring Harb Perspect Biol.* 2014;6(4):a018689.

Liyanage VR, Jarmasz JS, Murugeshan N, Del Bigio MR, Rastegar M, Davie JR. DNA modifications: Function and applications in normal and disease States. *Biology (Basel).* 2014;3(4):670–723.

Lu Y, Chan YT, Tan HY, Li S, Wang N, Feng Y. Epigenetic regulation in human cancer: The potential role of epi-drug in cancer therapy. *Mol Cancer.* 2020;19(1):79.

Lyon P, Strippoli V, Fang B, Cimmino L. B vitamins and one-carbon metabolism: Implications in human health and disease. *Nutrients.* 2020;12(9):2867.

Mahmoud AM, Ali MM. Methyl donor micronutrients that modify DNA methylation and cancer outcome. *Nutrients.* 2019;11(3):608.

Manach C, Scalbert A, Morand C, Rémésy C, Jiménez L. Polyphenols: Food sources and bioavailability. *Am J Clin Nutr.* 2004;79(5):727–747.

Marchese FP, Huarte M. Long non-coding RNAs and chromatin modifiers: Their place in the epigenetic code. *Epigenetics.* 2014;9(1):21–26.

Martín-Timón I, Sevillano-Collantes C, Segura-Galindo A, Del Cañizo-Gómez FJ. Type 2 diabetes and cardiovascular disease: Have all risk factors the same strength? *World J Diabetes.* 2014;5(4):444–470.

Mattick JS, Amaral PP, Carninci P, et al. Long non-coding RNAs: Definitions, functions, challenges and recommendations. *Nat Rev Mol Cell Biol.* 2023;24:430–447.

Miller JL, Grant PA. The role of DNA methylation and histone modifications in transcriptional regulation in humans. *Subcell Biochem.* 2013;61:289–317.

Misteli T. The self-organizing genome: Principles of genome architecture and function. *Cell.* 2020;183 (1):28–45.

Moosavi A, Motevalizadeh Ardekani A. Role of epigenetics in biology and human diseases. *Iran Biomed J.* 2016;20(5):246–258.

Morlando M, Fatica A. Alteration of epigenetic regulation by long noncoding RNAs in cancer. *Int J Mol Sci.* 2018;19(2):570.

Morris-Rosendahl DJ, Crocq MA. Neurodevelopmental disorders-the history and future of a diagnostic concept. *Dialogues Clin Neurosci.* 2020;22(1):65–72.

Nakamura M, Gao Y, Dominguez AA, Qi LS. CRISPR technologies for precise epigenome editing. *Nat Cell Biol.* 2021;23(1):11–22.

Nandini DB, Rao RS, Deepak BS, Reddy PB. Sulforaphane in broccoli: The green chemoprevention!! Role in cancer prevention and therapy. *J Oral Maxillofac Pathol.* 2020;24(2):405.

Neri F, Rapelli S, Krepelova A, et al. Intragenic DNA methylation prevents spurious transcription initiation. *Nature.* 2017;543(7643):72–77. https://doi.org/10.1038/nature21373

Nogueira da Costa A, Herceg Z. Detection of cancer-specific epigenomic changes in biofluids: Powerful tools in biomarker discovery and application. *Mol Oncol.* 2012;6(6):704–715.

Okeke E, Davwar PM, Roberts L, et al. Epidemiology of liver cancer in Africa: Current and future trends. *Semin Liver Dis.* 2020;40(2):111–123.

Oliveto S, Mancino M, Manfrini N, Biffo S. Role of microRNAs in translation regulation and cancer. *World J Biol Chem.* 2017;8(1):45–56.

Park J, Lee K, Kim K, Yi SJ. The role of histone modifications: From neurodevelopment to neurodiseases. *Signal Transduct Target Ther.* 2022;7(1):217.

Paul KC, Chuang YH, Cockburn M, Bronstein JM, Horvath S, Ritz B. Organophosphate pesticide exposure and differential genome-wide DNA methylation. *Sci Total Environ.* 2018;645:1135–1143.

Peinado MA. Hypomethylation of DNA. In: Schwab M. (eds) *Encyclopedia of Cancer.* Springer, Berlin, Heidelberg, 2011.

Penner-Goeke S, Binder EB. Epigenetics and depression. *Dialogues Clin Neurosci.* 2019;21(4):397–405.

Prabhakar B, Zhong XB, Rasmussen TP. Exploiting long noncoding RNAs as pharmacological targets to modulate epigenetic diseases. *Yale J Biol Med.* 2017;90(1):73–86.

Pratamawati TM, Alwi I, Asmarinah. Summary of known genetic and epigenetic modification contributed to hypertension. *Int J Hypertens.* 2023;2023:5872362.

Rafieian-Kopaei M, Setorki M, Doudi M, Baradaran A, Nasri H. Atherosclerosis: Process, indicators, risk factors and new hopes. *Int J Prev Med.* 2014;5(8):927–946.

Ramasamy A, Trabzuni D, Guelfi S, et al. Genetic variability in the regulation of gene expression in ten regions of the human brain. *Nat Neurosci.* 2014;17(10):1418–1428.

Ranganathan K, Sivasankar V. MicroRNAs—Biology and clinical applications. *J Oral Maxillofac Pathol.* 2014; 18(2):229–234.

Rhim J, Baek W, Seo Y, Kim JH. From molecular mechanisms to therapeutics: Understanding MicroRNA-21 in cancer. *Cells.* 2022;11(18):2791.

Rinn JL, Chang HY. Genome regulation by long noncoding RNAs. *Annu Rev Biochem.* 2012;81:145–166.

Rinn JL, Kertesz M, Wang JK, et al. Functional demarcation of active and silent chromatin domains in human HOX loci by noncoding RNAs. *Cell.* 2007;129(7):1311–1323.

Robertson KD. DNA methylation and human disease. *Nat Rev Genet.* 2005;6(8):597–610.

Rose NR, Klose RJ. Understanding the relationship between DNA methylation and histone lysine methylation. *Biochim Biophys Acta.* 2014;1839(12):1362–1372.

Rylaarsdam L, Guemez-Gamboa A. Genetic causes and modifiers of autism spectrum disorder. *Front Cell Neurosci.* 2019;13:385.

Samanta S, Rajasingh S, Cao T, Dawn B, Rajasingh J. Epigenetic dysfunctional diseases and therapy for infection and inflammation. *Biochim Biophys Acta Mol Basis Dis.* 2017;1863(2):518–528.

Santaló J, Berdasco M. Ethical implications of epigenetics in the era of personalized medicine. *Clin Epigenetics.* 2022;14(1):44.

Santana DA, Smith MAC, Chen ES. Histone modifications in Alzheimer's disease. *Genes (Basel).* 2023;14(2):347.

Sayad A, Najafi S, Hussen BM, et al. The emerging roles of the β-secretase BACE1 and the long non-coding RNA BACE1-AS in human diseases: A focus on neurodegenerative diseases and cancer. *Front Aging Neurosci.* 2022;14:853180.

Segal D, Dostie J. The talented LncRNAs Meshing into transcriptional regulatory networks in cancer. *Cancers (Basel).* 2023;15(13):3433.

Seto E, Yoshida M. Erasers of histone acetylation: The histone deacetylase enzymes. *Cold Spring Harb Perspect Biol.* 2014;6(4):a018713.

Shamsi MB, Firoz AS, Imam SN, Alzaman N, Samman MA. Epigenetics of human diseases and scope in future therapeutics. *J Taibah Univ Med Sci*. 2017;12(3):205–211.

Sharifi-Rad J, Rodrigues CF, Sharopov F, et al. Diet, lifestyle and cardiovascular diseases: Linking pathophysiology to cardioprotective effects of natural bioactive compounds. *Int J Environ Res Public Health*. 2020;17(7):2326.

Shi Y, Zhang H, Huang S, et al. Epigenetic regulation in cardiovascular disease: Mechanisms and advances in clinical trials. *Signal Transduct Target Ther*. 2022;7(1):200.

Shukla S, Tekwani BL. Histone deacetylases inhibitors in neurodegenerative diseases, neuroprotection and neuronal differentiation. *Front Pharmacol*. 2020;11:537.

Shumei M, Xin L, Benzheng J, Yu Y, Xiaodong L. Low-dose radiation-induced responses: Focusing on epigenetic regulation. *Int J Radiation Biol*. 2010;86(7):517–528.

Song C, Kanthasamy A, Anantharam V, Sun F, Kanthasamy AG. Environmental neurotoxic pesticide increases histone acetylation to promote apoptosis in dopaminergic neuronal cells: Relevance to epigenetic mechanisms of neurodegeneration. *Mol Pharmacol*. 2010;77(4):621–632.

Statello L, Guo CJ, Chen LL, Huarte M. Gene regulation by long non-coding RNAs and its biological functions. *Nat Rev Mol Cell Biol*. 2021;22(2):96–118.

Stephens KE, Miaskowski CA, Levine JD, Pullinger CR, Aouizerat BE. Epigenetic regulation and measurement of epigenetic changes. *Biol Res Nurs*. 2013;15(4):373–381.

Strianese O, Rizzo F, Ciccarelli M, et al. Precision and personalized medicine: How genomic approach improves the management of cardiovascular and neurodegenerative disease. *Genes (Basel)*. 2020; 11(7):747.

Su X, Jiang X, Meng L, Dong X, Shen Y, Xin Y. Anticancer activity of sulforaphane: The epigenetic mechanisms and the Nrf2 signaling pathway. *Oxid Med Cell Longev*. 2018;2018:5438179.

Sun Y, Ma L. New insights into long non-coding RNA MALAT1 in cancer and metastasis. *Cancers (Basel)*. 2019;11(2):216.

Tammen SA, Friso S, Choi SW. Epigenetics: The link between nature and nurture. *Mol Aspects Med*. 2013; 34(4):753–764.

Tang H, Zeng Z, Shang C, Li Q, Liu J. Epigenetic regulation in pathology of atherosclerosis: A novel perspective. *Front Genet*. 2021;12:810689.

Thangavel P, Park D, Lee YC. Recent insights into particulate matter (PM2.5)-mediated toxicity in humans: An overview. *Int J Environ Res Public Health*. 2022;19(12):7511.

Thornburg KL, Shannon J, Thuillier P, Turker MS. In utero life and epigenetic predisposition for disease. *Adv Genet*. 2010;71:57–78.

Tiffon C. The impact of nutrition and environmental epigenetics on human health and disease. *Int J Mol Sci*. 2018;19(11):3425.

Turinetto V, Giachino C. Histone variants as emerging regulators of embryonic stem cell identity. *Epigenetics*. 2015;10(7):563–573.

Uddin MG, Fandy TE. DNA methylation inhibitors: Retrospective and perspective view. *Adv Cancer Res*. 2021;152:205–223.

Villeneuve LM, Reddy MA, Natarajan R. Epigenetics: Deciphering its role in diabetes and its chronic complications. *Clin Exp Pharmacol Physiol*. 2011;38(7):451–459.

Völker-Albert M, Bronkhorst A, Holdenrieder S, Imhof A. Histone modifications in stem cell development and their clinical implications. *Stem Cell Reports*. 2020;15(6):1196–1205.

Waldecker M, Kautenburger T, Daumann H, Busch C, Schrenk D. Inhibition of histone-deacetylase activity by short-chain fatty acids and some polyphenol metabolites formed in the colon. *J Nutr Biochem*. 2008;19(9):587–593.

Weber M, Hellmann I, Stadler MB, et al. Distribution, silencing potential and evolutionary impact of promoter DNA methylation in the human genome. *Nat Genet*. 2007;39(4):457–466.

Weinhold B. Epigenetics: The science of change. *Environ Health Perspect*. 2006;114(3):A160–A167.

Wise IA, Charchar FJ. Epigenetic modifications in essential hypertension. *Int J Mol Sci*. 2016;17(4):451.

Xu J, Liu Y. A guide to visualizing the spatial epigenome with super-resolution microscopy. *FEBS J*. 2019a;286(16):3095–3109.

Xu W, Chen W, Chen G, et al. Drug-induced epigenetic alterations and long-term health outcomes. *Pharmacol Ther*. 2019b;200:22–33.

Ying SY, Chang DC, Lin SL. The microRNA (miRNA): Overview of the RNA genes that modulate gene function. *Mol Biotechnol*. 2008;38(3):257–268.

Yoon SH, Choi J, Lee WJ, Do JT. Genetic and epigenetic etiology underlying Autism spectrum disorder. *J Clin Med*. 2020;9(4):966.

Yusufaly TI, Li Y, Singh G, Olson WK. Arginine-phosphate salt bridges between histones and DNA: Intermolecular actuators that control nucleosome architecture. *J Chem Phys*. 2014;141(16):165102.
Zeisel S. Choline, other methyl-donors and epigenetics. *Nutrients*. 2017;9(5):445.
Zentner GE, Henikoff S. Regulation of nucleosome dynamics by histone modifications. *Nat Struct Mol Biol*. 2013;20(3):259–266.
Zhang P, Wu W, Chen Q, Chen M. Non-coding RNAs and their integrated networks. *J Integr Bioinform*. 2019a;16(3):20190027.
Zhang WH, Jiang L, Li M, Liu J. MicroRNA-124: An emerging therapeutic target in central nervous system disorders. *Exp Brain Res*. 2023;241(5):1215–1226.
Zhang X, Wallace AD, Du P, et al. DNA methylation alterations in response to pesticide exposure in vitro. *Environ Mol Mutagen*. 2012;53(7):542–549.
Zhang X, Wang W, Zhu W, et al. Mechanisms and functions of long non-coding RNAs at multiple regulatory levels. *Int J Mol Sci*. 2019b;20(22):5573.
Zhao K, Mao Y, Li Y, Yang C, Wang K, Zhang J. The roles and mechanisms of epigenetic regulation in pathological myocardial remodeling. *Front Cardiovasc Med*. 2022;9:952949.
Zhao Z, Shilatifard A. Epigenetic modifications of histones in cancer. *Genome Biol*. 2019;20(1):245.
Zhu H, Wang X, Meng X, et al. Selenium supplementation improved cardiac functions by suppressing DNMT2-mediated GPX1 promoter DNA methylation in AGE-induced heart failure. *Oxid Med Cell Longev*. 2022;2022:5402997.
Zoghbi HY, Beaudet AL. Epigenetics and human disease. *Cold Spring Harb Perspect Biol*. 2016;8(2):a019497.

3 Epigenetic Drug Discovery and Its Importance in Precision Medicine

Mohammad Aleemuddin[1] *and Ayesha Sultana*[2]
[1]Yenepoya Dental College, Mangalore, Karnataka, India
[2]Yenepoya Pharmacy College & Research Centre, Mangalore, Karnataka, India

3.1 INTRODUCTION

The word "epigenetic" was coined by the renowned biologist Conrad Waddington to describe how an organism's phenotype adapts to its environment. Over time, the concept of epigenetics evolved and now refers to phenotypic variations that occur without changing an organism's genotype. As a result, multicellular eukaryotic organisms frequently exhibit the clear epigenetic modification of their genomes [1]. Every cell in our body has the same DNA, and these cells differentiate to become different tissues and organs based on differences in gene expression at the cellular level [2].

Epigenetics is the study of DNA alterations that affect gene expression without altering a person's genetic coding. These modifications influence whether specific genes are turned on or off. An epigenetic modification occurs when a chemical binds to a gene and affects its activity. All of these genome-wide epigenetic modifications are collectively referred to as the epigenome. Although these alterations do not alter the DNA sequence, they can persist in the cells during division and even be passed on to offspring. Factors such as diet and toxin exposure can influence the epigenome. Because of its role in gene activity, protein synthesis, and, ultimately, human health and disease, epigenetics become a popular research area in biology and medicine.

Epigenome dysregulation contributes to diseases such as cancer, autoimmune disorders, CVDs, and neurological conditions. The epigenetic effectors—writers, erasers, and readers—modify chromatin to regulate gene expression. In simplest terms, writers add new epigenetic markers, erasers remove epigenetic marks, and readers interpret to alter the structure of DNA or histones and, if necessary, enlist more machinery.

At the molecular level, epigenetic changes involve reversible modifications to nucleic acids and histone proteins that together form nucleosomes. These chemical changes are mediated by enzymes called writers, which not only attract histone proteins to DNA but also trigger the activity of macromolecules like chromatin remodelers and non-coding RNAs (ncRNAs). Reader domains that recognize distinct activities in chemically modified proteins and nucleic acids coordinate these binding interactions. Finally, to ensure reversibility, eraser enzymes catalyze the removal of written information there by leading to the development of dynamic character [3].

Epigenetics plays a significant role in organisms' development and adaptive responses to the environment and can unexpectedly contribute to disease progression. Many diseases have been linked to epigenetic abnormalities, which can also be environmentally induced. These epigenetic changes have the potential for use as disease biomarkers [4].

DOI: 10.1201/9781032706511-3

3.2 EARLY DEVELOPMENT OF EPIGENETIC DRUGS

While the development of epigenetic drugs using specific targets might seem like a virtual concept, it is an emerging reality. Their ultimate success depends more on clinical efficacy than on detailed molecular targeting, with discovery driven by phenotypic observations.

3.2.1 DNA Methyltransferase Inhibitors

DNA methyltransferases (DNMTs), a class of cytosine methylases, regulates epigenetics by controlling DNA methylation. This process epigenetically influences cancer-related gene expression and plays a significant role in cancer genesis. In 1980, researchers found that nucleotide analogs could inhibit DNA methylation, leading to the discovery of cytidine modifications such as 5-azacitidine and decitabine [4].

Nucleosides serve as essential building blocks for DNA and RNA, and structural analogs can disrupt metabolic pathways and replication. Drugs like 5-azacytidine and 5-aza-2′-deoxycytidine have been created through modifications of the pyrimidine nucleoside base in cytidine [4], and because of their structural resemblance to cytidine, polymerases can recognize both of them and incorporate them into the expanding chain of nucleic acids. While the ribose sugar in 5-Aza-C is merged into DNA and RNA following the removal of 3′ OH (hydroxy groups), the DNA sugar in 5-Aza-dC contains deoxyribose and is merged into DNA only. In contrast to the building blocks of nucleic acids, only nitrogen replaces carbon, but this has detrimental biological effects [5].

3.2.2 Histone Deacetylase Inhibitors

Enzymes known as histone deacetylases (HDACs) work by removing acetyl markers from lysine residues on histones. This tightens the chromatin structure and decreases transcription accessibility [6].

Vorinostat and romidepsin were among the first drugs to demonstrate epigenetic post-translational modification of histone proteins. Vorinostat's potential began in 1971 with observations that dimethysulfoxide (DMSO) enabled murine erythroleukemia to proliferate. Ronald Breslow proposed a mechanism of action for DMSO that involved hydrogen bonding or metal coordination to target proteins. This discovery revealed multiple interacting analogs including SAHA (suberoylanilide hydroxamic acid) whose exposure followed the same mechanism. Phenotypic screening and in vivo experiments validated the DMSO ↔ SAHA pathway [4].

HDACs identify small compounds that inhibit enzyme activity such as trichostatin A and trapoxin A (found in natural products). SAHA and trichostatin A have comparable structures, and Breslow stated that the latter was also an inhibitor. As X-ray crystallographic studies have accurately reported, hydroxyamic acid in SAHA is a metal chelator that reversibly attaches to the zinc cation in the HDAC active site [7].

In summary, all epigenetic medications were identified and studied via phenotypic studies, and it takes a long time for a drug to be approved for use in a clinical setting. Although vorinostat and romidepsin are examples of HDAC inhibitors, their use in treating malignancies is limited to cutaneous T cell lymphomas. SAHA is utilized clinically not just in cancer treatment but also in the treatment of many neurological conditions, including Huntington's disease [4].

Hexamethylamine bisacetamide proved to be a more promising candidate after utilizing a traditional medicinal chemistry tactic known as "multivalency." It proceeded to Phase I/II clinical trials, where patients with acute myeloid leukemia and myelodysplastic syndrome showed partial responses. After tri- or tetra-amides failed to boost the valency, the approach shifted to considering hydroxamic acids as an amide isostere. It is true that suberic bishydroxamic acid was more powerful than compound Hexamethylamine bisacetamide. The new candidate SAHA (Vorinostat, Zolinza) was chosen as a result of this approach because it had the ideal ratio of toxicity to potency [4].

3.3 SECOND-GENERATION EPIGENETIC DRUGS

3.3.1 DNMTs

There are certain disadvantages to the 5-azac (5-azacytidine) and 5-aza-dc (5-aza-2′-deoxycytidine) that were eventually approved. Zebularine and guadecitabine, two nucleoside analogue DNMT inhibitors, have reached the clinical development stage. Second-generation DNMTs that use the bi-substrate approach—which combines a methyl group donor with cytosine—produce some of the most potent DNMTi molecules. These compounds reactivate genes in cancerous cells via promoter demethylation [6].

Zebularine is stable in acidic and neutral pH and inhibits the action of both DNMTs (Table 3.1) and cytidine deaminase [8]. Furthermore, a number of small-molecule non-nucleosides have been identified as DNMT inhibitors and are employed as preclinical tool compounds against the target. These derivatives demonstrate significant effects in a number of cancer models and human illnesses including parasite infections [4].

3.3.2 HDAC Inhibitors

HDAC inhibitors require a zinc-binding surface for binding to the active site with high affinity [4] [6]. Hydroxyamic acid has been demonstrated to be the most effective option, and many synthetic HDAC inhibitors containing this have been found. The focus has been on optimizing vorinostat's pharmacokinetic characteristics, which include its quick metabolism and low bioavailability. The company Prolifix concentrated on trichostatin A and vorinostat analogues with rigid sulfonamide linkers, and in 2014, saw the FDA approve the medication belinostat for the treatment of peripheral T-cell lymphoma [4].

Dacinostat was discontinued because of its high toxicity. In 2015, the FDA approved panobinostat for use in treating patients with refractory multiple myeloma who had not responded to previous treatments [4]. Panobinostat is unique among HDAC inhibitors in that it is approved for use in cases other than cutaneous T cell lymphoma and peripheral T cell lymphoma and can help identify multiple myeloma more accurately.

Even though a large number of Hydroxyamic acid HDAC inhibitors have been brought close to clinical development, questions remain about their pharmacokinetics. Hydroxamic acids can be eliminated by phase I conjugation and metabolism to toxic products. They can also have effects from nonselective metal binding. This prompted further research into and testing of metal-binding functional groups, but none of the results were satisfactory. The only recognized and approved HDAC inhibitor with a zinc-binding group that contains thiol is the previously discussed romidepsin. Despite the discovery of additional similar natural products and synthetic thiols that contain effective HDAC inhibitors, none of them have advanced to the point of clinical trials [4].

TABLE 3.1
DNMTs

S. no	Year	Main events
1	1964	Synthesis of azacitidine and decitabine.
2	1967	Azacitidine enters the phases of clinical trials.
3	1980	Azacitidine and decitabine are labeled DNA methyl transferases inhibitors.
4	1980	Decitabine enters the phases of clinical trials and its progression.
5	2004	Azacitidine earns FDA clearance
6	2006	Decitabine earns FDA clearance.

In addition to thiols, carboxylic acid is another zinc binding site with a high affinity for HDAC inhibition. The first substance that was shown to impede histone deacetylation in early 1978 was butyric acid as the sodium salt. Butyric acid excretes in minutes, and its inhibitory action sparked interest in short-chain carboxylic acids, particularly phenyl butyric acid and valproic acid, which were previously approved medications for the treatment of epilepsy and urea cycle disorders, respectively. After receiving attention, Pivanex moved on to phase II clinical trials.

Expanding the use of HDAC inhibitors (Table 3.2) through the screening of second-generation compounds in non-hematological cancers was a stronger and more advanced development. They were not very effective, though, as they are typically single agents. In patients with solid tumors, namely advanced stages of breast cancer and stages of metastatic lung cancer, the combination of HDAC and DNMT was observed to be beneficial.

- Belinostat, dacinostat, panobinostat, quisinostat, and tefinostat are a few examples of hydroxamic acid.
- Benzamide HDAC inhibitors, such as tacedinaline, entinostat, mocetinostat, and chidamide have advanced to the clinical stage.
- HDAC inhibitors for carboxylic acid, including butyric acid (Pivanex), valproic acid, and phenyl butyric acid [4].

3.4 THIRD-GENERATION EPIGENETIC DRUGS

The third wave of epigenetic drug discovery identified three more targets: bromodomain inhibitors, lysine demethylases (KDMs), and lysine histone methyltransferases (KMTs). Epigenetic therapy has become popular due to the approval of seven small molecule inhibitors of HDACs and DNMTs [9].

3.4.1 Histone Methyltransferase Inhibitors

Protein lysine residues can have one to three methyl groups added to the post-translational by KMTs. The capacity to create strong, selective SAM-mimetic (S-adenosyl methionine) by utilizing variations in the cofactor binding site has been crucial to the advancement of KMT drug research [6]. The first KMT inhibitor to reach the phase I stage of clinical trials for the treatment of leukemia was pinometostat, a DOT1L inhibitor (disruptor of telomeric silencing 1-like). The clinical candidates sinefungin, pinometostat, tazemetostat are a few examples of KMT inhibitors [4].

TABLE 3.2
HDAC Inhibitors

S. no	Year	Main Events
1	1971	DMSO (dimethyl sulfoxide) was identified as a differentiating agent.
2	1976	Vorinostat, the bisamide precursor to SAHA, was found to be a differentiation agent.
3	1990	Trichostatin A is discovered to be a HDAC inhibitor.
4	1990	Romidepsin is identified as a natural product.
5	1996	Synthesis of vorinostat.
6	1997	Romidepsin enters the clinical trial phase.
7	1998	HDAC inhibitors include romidepsin and vorinostat.
8	2000	Clinical trials are conducted on vorinostat.
9	2006	Vorinostat receives FDA approval.
10	2009	Romidepsin gets an FDA authorization.

3.4.2 KDM Inhibitors

KDMs reverse the lysine methylation of DNA and histone proteins. This alters the chromatin structure or the transcription of genes at their promoter. The KDMs are classified into two categories according to their catalytic mechanism and ability to reverse lysine methylation. The most successful example is tranylcypromine, which was initially prescribed as an antidepressant and is an approved monoamine oxidase inhibitor [10].

3.4.3 Bromodomains

The previously mentioned targets for drug discovery are either writers (DNMTs and KMTs, methylate DNA and histones, respectively) or erasers (HDACs and KDMs, deacetylate and demethylate histones, respectively). Through chromatin remodeling, histone acetylation, and the recruitment of additional transcriptional machinery, bromodomain proteins are essential for the regulation of transcription. The most studied epigenetic reader in drug discovery is the BET (bromo and extra terminal) family [4,6,11].

The bromodomains are the first epigenetic readers to have a significant impact on drug discovery. The BET, which consists of BRD2, BRD3, BRD4, and BRDT, has been a primary target for drug discovery within the bromodomains due to its possible relevance to cancer. BET inhibitors are also being researched for a number of diseases, such as diabetes or the reactivation of latent HIV-1 in human cells.

Clinical prospects based on benzodiazepine scaffolds are being studied in cancer, Resverlogix's RVX-208 was found in an atherosclerosis program by phenotypic screening. The discovery of three new targets—bromodomains, KMTs, and KDMs—has sped up the third wave of epigenetic drug discovery. The bromodomain efforts were accelerated by reporter-based phenotypic assays, which also identified high-affinity ligands, the mechanism of which was only later discovered.

3.5 FOURTH-GENERATION EPIGENETIC DRUGS

Although the newly developed CRISPR/Cas technology has attracted a lot of attention as a tool for correcting genetic mutations, it also possesses the potent ability to render any given gene inactive. Furthermore, CRISPR/Cas can be designed to direct effectors to that genomic locus in order to re-express the silent gene if loss of function is caused by epigenetic silencing of expression. The only clinical conditions for which epigenetic drugs are commonly used are hematological malignancies. In certain cancer forms, the response to chemotherapy can be predicted by both genetic and epigenetic modifications of epigenetic enzymes (i.e., loss or gain of function), such as promoter hypermethylation of tumor suppressor genes [4,12].

3.6 EPIGENETICS IN PRECISION MEDICINE

By identifying new mechanisms controlling epigenetic expression and creating cutting-edge epigenetic biomarkers, therapeutic targets, and potentially useful epigenetic medications for clinical use, the field of epigenetic research aims to advance precision medicine. Patients, healthcare systems, and clinical practice will all be impacted by precision medicine. Early diagnosis, monitoring the course of a disease, forecasting its course, risk-based patient selection and stratification, anticipating future comorbidities, and even assessing the benefits and drawbacks of therapeutic interventions in particular patient subgroups can all be aided by epigenetic biomarkers.

In 2008, the National Research Council published *Towards Precision Medicine*, which defined precision medicine as "the tailoring of medical treatment to the individual characteristics of each patient" to divide people into smaller groups according to how they react to a given treatment or how susceptible they are to a certain disease. Ginsburg and Phillips (2018) stated that in order to save costs and reduce side effects, therapeutic or preventative interventions can then be targeted toward people who will benefit and spare those who won't. Numerous applications of precision medicine include improving tailored treatments, facilitating precise disease diagnosis, and identifying specific disease subtypes or stages [14].

Advantages of epigenetic biomarkers over biomarkers based on genes or proteins [14]:

- Their dynamic character.
- Their ability to reveal gene function.
- Their ability to reveal the particular genetic programmes that change throughout illness.
- Most techniques (like RT-qPCR) for examining epigenetic biomarkers have already been applied in clinical settings.

Any one of the following traits would be present in the ideal epigenetic biomarker [14]:

- Predicts the likelihood that a disease will develop in the future (risk).
- Specifies an illness (detection).
- Gives details on the natural history of the illness.
- Predicts the trajectory of the illness (prognostic).
- Reacts predictably to therapy.
- Monitors how patients react to treatment or medication (therapy monitoring).
- Facilitates the execution of targeted therapy and diagnosis concurrently.

Due to their greater stability compared to RNA and proteins, microRNAs (miRNA) and DNA methylation are more practical and useful as biomarkers in therapeutic contexts. Most commercial in vitro diagnostic tests that are currently on the market are based on measuring the relative expression of microRNAs or analyzing the DNA methylation of specific genes. RT-qPCR-based methods like methylight, methyl-specific PCR, and methylation-sensitive high-resolution melting, as well as pyrosequencing technologies, make these methods easily quantifiable [15].

3.7 IN VITRO DIAGNOSTIC TESTS BASED ON EPIGENETICS

3.7.1 Colorectal Cancer

New epigenetic biomarkers for colorectal cancer (CRC) diagnosis, prognosis, and treatment response prediction have been found worldwide; some of these have been approved and are presently available in market. This is because the development and course of CRC are significantly influenced by epigenetic changes in DNA methylation, miRNA expression, and posttranslational modifications in histones.

The first DNA methylation assay authorized by the FDA for general colorectal cancer screening in persons over 50 at average risk was called Cologuard® (Exact Sciences Corp., Madison, WI, USA). A mutation in the KRAS gene (exon 2, codons 12, 13) was discovered to be a viral oncogene that causes Kirsten rat sarcoma). The Cologuard® in vitro diagnostic (IVD) test uses three key components to detect colon neoplasia: 1) methylation levels of N-Myc downstream-regulated gene 4 (NDRG4) and bone morphogenetic protein 3 (BMP3), with ß-actin as the reference gene; (2) a non-DNA immunochemical assay for human hemoglobin, with ACTB (ß-actin) as the reference gene; and (3) an analytical technique for obtaining fecal hemoglobin biomarker values. Cologuard's® sensitivity and specificity for colorectal cancer diagnosis were 92.3% and 86.6%, respectively, in a trial with 9,989 individuals [13,16].

The FDA approved Epi proColon®2.0 (Epigenomics Inc., Berlin, Germany) in 2016 as the first blood test designed for early CRC diagnosis. The test uses peripheral blood samples to determine the SEPT9 gene's methylation status. SEPT9 hypermethylation was identified as a crucial component of CRC, as septins—proteins—are required for cell division [13,16].

Recently, the EarlyTect® CRC test (Genomictree Inc., Daejeon, South Korea) received CE-IVD certification. The EarlyTect GI Syndecan2 methylation assay is an IVD test that assesses the methylation status of SDC2 (Syndecan-2) using cfDNA (circulating free DNA) isolated from 0.5 ml of serum for early detection of CRC [13].

A CE-IVD registered theranostic test called the miRpredX-31-3p kit (Integra Gen S.A., Evry, France) was created to find people with metastatic CRC who might benefit from anti-EGFR (epidermal growth factor receptor) therapy [17].

A new test based on these biomarkers is being developed, but currently, the NuQ® tests (Volition SA, Namur, Belgium) diagnose CRC by analyzing characteristics including DNA methylation of DNA bound to nucleosomes, post-translational histone modifications and variants, and the presence of cell-free nucleosomes [18]. Only a drop of blood (10 µl) is needed for the NuQ® tests, which use ELISA. The most sophisticated test currently available is the Nu.QTM Colorectal Cancer Screening Triage Test, which combines several NuQ® tests that were CE-IVD marked in the past. The NuQ®X test determines the concentration of 5-methylcytosine in DNA attached to free-floating nucleosomes, the Nu.Q®T test identifies cell-free nucleosomes, also known as total free circulating nucleosomes, and Nu.Q®V uses histone variant analysis to identify CRC [13].

3.7.2 Breast Cancer

Anomalies related to epigenetic modifications are often noted at different phases of breast cancer. Breast cancer starts and spreads because of specific gene expressions that are influenced by chromatin remodeling, non-coding RNA-mediated gene regulation, and DNA methylation. Potential biomarkers for early cancer detection, prognosis, treatment response monitoring, and risk prediction of breast cancer recurrence have been examined in relation to epigenetic modifications. Precision medicine and the use of biomarkers have given patients with breast cancer more treatment options. Epigenetic biomarkers in particular can be used to further improve the in vitro tests used to manage breast cancer [13,20].

The Therascreen PITX2 RGQ PCR kit (Qiagen, Dusseldorf, Germany) uses a methylation-based CE-IVD marked assay to predict the response of high-risk breast cancer patients who are ER-positive (estrogen receptor), HER2-negative (human epidermal growth factor receptor), and lymph node positive. The test, which was given the CE-IVD mark in 2018 [13,19], identifies patients who are likely to respond to anthracycline chemotherapy.

3.7.3 Cervical Cancer

Several genes, including SOX1, ZNF582 (zinc finger protein 582), and PAX1, have been found to be hypermethylated in cervical cancer compared with normal cervical tissue. Shen-Gunter et al. evaluated DNA methylation at the ADCY8, CDH8, and ZNF582 promoters, along with the HPV (human papillomavirus) genotype, in relation to the cytological grade. Their findings indicate the potential of these markers as significant for the molecular classification of Pap smears [22]. Commercialization of a single epigenetic test based on ZNF582 and PAX1 gene methylation analysis is underway [13]. The CE-IVD has approved the use of the Cervi-M® and Oral-M® DNA assays (developed by Epigene, iStat Biomedical Co., Taiwan) in the diagnosis of cervical and oral cancers. Oral and cervical malignancies have significant levels of methylation in these genes [13,21].

3.7.4 Glioblastoma

The molecular and clinical significance of EGFR amplification, MGMT promoter methylation(O6-methylguanine-DNA methyltransferase), and IDH1/2 (isocitrate dehydrogenases 1 and 2) mutations has led to their proposal as predictive biomarkers. One of the most significant prognostic factors for gliablastoma multiforme (GBM) is MGMT promoter methylation; it can forecast the therapeutic outcome of a regimen that includes alkylating drugs like temozolomide (Temodar®) and carmustine (BCNU, Gliadel®). As a result, this gene's DNA methylation was proposed as a biomarker for glioma classification and therapy guidance. Two kits are used: PyroMark Q96 cytosine-phosphodiester-guanine (CpG) MGMT and PyroMark Therascreen MGMT. DNA taken from FFPE (formalin-fixed paraffin embedded) tissues and blood can be tested for GBM using the Therascreen MGMT Pyro kit [13,23].

3.7.5 Lung Cancer

Patients with lung cancer are currently experiencing better clinical outcomes as a result of novel immunotherapy approaches based on anti-PD1/PD-L1 mAb and the use of particular tyrosine kinase inhibitors in EGFR-mutated, ALK/ROS1-rearranged NLSC. It has been discovered that short stature homeobox gene two methylation is a biomarker that can reliably distinguish between lung tumor and normal tissue using the Epi proLung BL Reflex Assay® [13,24].

3.7.6 Unknown Cancers

EPICUP™ (Ferrer, Spain) is a CE-IVD test that determines the biological origin of tissue in cases of CUP (Cancer of unknown primary site). The first epigenetic test designed to detect tumors with an unknown origin was EPICUP™. It can identify up to 87% of cancer patients with an unknown origin. EPICUP™ compares the methylation patterns of malignancies with known and unknown primary tumors, based on a CpG site analysis with the Infinium Human Methylation 450 Beadchip microarray (Illumina), a human methylation matrix [13,25].

3.7.7 Prostate Cancer

Over the last ten years, we have discovered that interactions with various epigenetic enzymes and epigenetic changes strongly affect the androgen receptor and its signaling cascade. Consequently, we now know that epigenetic modifications play an important role in the onset and spread of prostate cancer.

3.7.8 Epithelioid Sarcoma

The FDA recently approved the use of an EZH2 inhibitor (enhancer of zeste 2 polycomb repressive complex 2 subunit) for the treatment of epithelioid sarcoma. Monitoring the methylation of the RASSF genes can also lead to an epigenetic prognosis in Ewing sarcomas that distinguishes benign from aggressive tumors.

3.7.9 Gynecological Cancer

Research on the DNA methylome of various diseases may help find disease-associated genes and biomarkers for diagnosis and prognosis.

3.7.10 Pancreatic Cancer

The disease has been linked to four significant gene mutations: KRAS, TP53 (tumor protein p53), SMAD4, and CDKN2A (cyclin dependent kinase inhibitor 2A). However, new research indicates that metastasis progression may not be caused by specific gene mutations but rather by modifications to epigenetic marks.

The importance of epigenetic modifications in the development of pancreatic cancer is becoming increasingly evident. A plethora of new options for the discovery of biomarkers and the development of cutting-edge treatments have been made possible by our growing understanding of these epigenetic changes and their roles in the development of pancreatic cancer.

3.7.11 Brain Cancer

Our comprehension of the biology of brain tumors has improved due to our understanding of epigenetic pathways, which has allowed us to find new targets for diagnostics and treatment. Comprehending the multiple epigenetic mechanisms, which include controlling miRNA, histone acetylation, and DNA sequence methylation, is extremely difficult because G-quadruplex structures are involved, acting as a regulatory network for the biology of tumors.

This understanding has resulted in the creation and application of medicinal compounds such as silencing RNAs, antagomirs, and histone deacetylase inhibitors. The objective is to deliver cancer therapies that are oriented around the patient, but this has proven to be a challenging endeavor and will probably stay so for some time to come.

3.7.12 Skin Diseases

The relevance of epigenetic modifications in human illnesses has been thoroughly investigated in relation to cancer, given that global hypomethylation of DNA and promoter hypermethylation are indicators of carcinogenesis. Cutting-edge drugs that modify the epigenetic map have demonstrated potential in the treatment of hematologic cancers and cutaneous T-cell lymphoma. Numerous studies are now being conducted to identify the epigenetic profiles of different skin conditions as a result of these findings.

An increasing body of research suggests that epigenetic modifications play a significant role in the pathogenic events of connective tissue disorders and inflammatory skin diseases like psoriasis and atopic dermatitis. This gives skin disorders new options for therapy. Since epigenetic changes can be altered and even reversed, they are excellent candidates for therapeutic intervention in contrast to genetics. Determining epigenetic profiles may also help with diagnosis, prognosis, and therapy response evaluation, all of which may enhance the way skin diseases are managed.

3.8 CONCLUSION

Current advancements center on how crucial it is to combine epidrugs in order to find better treatment strategies for both cancer and non-cancerous illnesses. Authorization of epigenetic biomarkers will help define the medication response, identify responsive individuals, and hasten the process of arriving at an appropriate diagnosis. To optimize drug engineering, read-out assays, and clinical trial design, researchers from many domains, including medicine, clinical practice, and research, must collaborate.

To incorporate these novel epigenetic tests into clinical practices, the majority have employed methodology that utilizes easily accessible, affordable analytical instruments such as microarrays and RT-qPCR (Real-time quantitative reverse transcription PCR) for DNA methylation and miRNA analysis. It will take some time before these epigenetic tests are fully incorporated into standard clinical practice. The companies that are creating epigenetic tests ought to concentrate on streamlining the technology by employing affordable, easily navigable technology that is used to analyze epigenetic biomarkers in a clinical laboratory setting. Additionally, to improve patient adherence to screening programs and predict cancer diagnosis, companies should concentrate on finding biomarkers in non-invasive biospecimens.

REFERENCES

[1] Waddington CH. The epigenotype. *Endeavour.* 1942;1:18–20.

[2] Deans C, Margaret KA. What do you mean, "Epigenetic?" *Genetics*. 2015;199:887–896.

[3] Biswas S, Rao CM. Epigenetic tools (The Writers, The Readers and The Erasers) and their implications in cancer therapy. *Eur J Pharmacol*. 2018;837:8–24.

[4] Ganesan A, Arimondo PB, Rots MG. The timeline of epigenetic drug discovery: From reality to dreams. *Clin Epigenetic*. 2019;11:174.

[5] Pískala AŠF. Nucleic acids components and their analogues. LI. Synthesis of1-glycosyl derivatives of 5-azauracil and 5-azacytosine. *Coll Czech Chem Commun*. 1964;29:2060–2076.

[6] Feehley T, O'Donnell CW. Drugging the epigeneome in the age of precision medicine. *Clin Epigenetic*. 2023;15:6.

[7] Kijima M, Yoshida M, Sugita K, Horinouchi S, Beppu T. Trapoxin, an antitumor cyclic tetrapeptide, is an irreversible inhibitor of mammalian histone deacetylase. *J Biol Chem*. 1993;268:22429–22435.

[8] Yoo CB, Cheng JC, Jones PA. Zebularine: A new drug for epigenetic therapy. *Biochem Soc Trans*. 2004;32:910–912.

[9] Kaniskan HÜ, Konze KD, Jin J. Selective inhibitors of protein methyltransferases. *J Med Chem*. 2015; 58:1596–1629.

[10] Hauser A-T, Robaa D, Jung M. Epigenetic small molecule modulators of histone and DNA methylation. *Curr Opin Chem Biol*. 2018;45:73–85.

[11] Cochran AG, Conery AR, Sims RJ. Bromodomains: A new target class for drug development. *Nat Rev DrugDiscov*. 2019;18:609–628.

[12] Esteller M. Epigenetic drugs: More than meets the eye. *Epigenetic*. 2017;12:307.

[13] Beltrán-García J, Osca-Verdegal R. Epigenetic IVD tests for personalized precision medicine in cancer. *Front Genet*. 2019:10.

[14] Ginsburg GS, Phillips KA. Precision medicine: From science to value. *Health Aff. (Millwood)*. 2018;37: 694–701.

[15] García-Giménez JL, Mena-Mollá S, Beltrán-García J, Sanchis-Gomar F. Challenges in the analysis of epigenetic biomarkers in clinical samples. *Clin. Chem. Lab. Med*. 2017a;55(10):1–4.

[16] Imperiale TF, Ransohoff DF, Itzkowitz SH, Levin TR, Lavin P, Lidgard GP, et al. Multitarget stool DNA testing for colorectal-cancer screening. *N Engl J Med*. 2014;370:1287–1297.

[17] Ramon L, David C, Fontaine K, Lallet E, Marcaillou C, Martin-Lannerée S, et al. Technical validation of a reverse-transcription quantitative polymerase chain reaction in vitro diagnostic test for the determination of MiR-31-3p expression levels in formalin-fixed paraffin-embedded metastatic colorectal cancer tumor specimens. *Biomark Insights*. 2018;13:1177271918763357.

[18] Holdenrieder S, Dharuman Y, Standop J, Trimpop N, Herzog M, Hettwer K, et al. Novel serum nucleosomics biomarkers for the detection of colorectal cancer. *Anticancer Res*. 2014;34(5):2357–2362.

[19] Aubele M, Schmitt M, Napieralski R, Paepke S, Ettl J, Absmaier M, et al. The predictive value of PITX2 DNA methylation for high-risk breast cancer therapy: Current guidelines, medical needs, and challenges. *Dis Markers* 2017;2017;1–14.

[20] Meisel JL, Venur VA, Gnant M, Carey L. Evolution of targeted therapy in breast cancer: Where precision medicine began. *Am Soc Clin Oncol Educ Book*. 2018;38:78–86.

[21] Bakkum-Gamez JN, Wentzensen N, Maurer MJ, Hawthorne KM, Voss JS, Kroneman TN, et al. Detection of endometrial cancer via molecular analysis of DNA collected with vaginal tampons. *Gynecol Oncol*. 2015;137:14–22.

[22] Shen-Gunther J, Wang C-M, Poage GM, Lin C-L, Perez L, Banks NA, et al. Molecular Pap smear: HPV genotype and DNA methylation of ADCY8, CDH8, and ZNF582 as an integrated biomarker for high-grade cervical cytology. *Clin Epigenetics*. 2016;8:96.

[23] Quillien V, Lavenu A, Ducray F, Meyronet D, Chinot O, Fina F, et al. Clinical validation of the CE-IVD marked Therascreen MGMT kit in a cohort of glioblastoma patients. *Cancer Biomarkers*. 2017;20:435–441.

[24] Lewin J, Plum A, Hildmann T, Rujan T, Eckhardt F, Liebenberg V, et al. Comparative DNA methylation analysis in normal and tumour tissues and in cancer cell lines using differential methylation hybridisation. *Int J Biochem Cell Biol*. 2007;39:1539–1550.

[25] Moran S, Martínez-Cardús A, Sayols S, Musulén E, Balañá C, Estival-Gonzalez A, et al. Epigenetic profiling to classify cancer of unknown primary: A multicentre, retrospective analysis. *Lancet Oncol*. 2016;17:1386–1395.

4 Traditional Medicine and Its Importance in Epigenetic Healing

Venkateswara Rao Amara,[1] Ajay Sharma,[2] Abhishek Tiwari,[3] Varsha Tiwari,[3] Vineet Mittal,[4] Mohan Gandhi Bonthu,[5] Bhanu Sirisha Kagitapalli,[6] and Arya Lakshmi Marisetti[2]

[1]National Institute of Pharmaceutical Education and Research, Hajipur, India

[2]Delhi Institute of Pharmaceutical Sciences and Research, Pushp Vihar, New Delhi, India

[3]IFTM University, Moradabad, UP, India

[4]Maharshi Dayanand University, Rohtak, India

[5]V.V. Institute of Pharmaceutical Sciences, Gudlavalleru, Andhra Pradesh

[6]Vanderbilt University Medical Center, Nashville, TN, USA

4.1 INTRODUCTION

Various native healthcare systems are popular throughout the world for improving the well-being of mankind. The usage of traditional medicine to manage diseases predates the current conventional (also known as allopathic or Western) medicine. Traditional medicine applies knowledge, skill, and practices based on the theories, beliefs, and experiences indigenous to different cultures, whether explicable or not, incorporating plant, animal and mineral based medicines, spiritual therapies, manual techniques, and exercises, used singularly or in combination to maintain health and manage illnesses (Fokunang et al. 2011). Traditional medicine is a broad term that denotes treatments that are not generally applied in conventional Western culture medicines; it is also referred to as complementary, alternative, or integrative medicine and employed interchangeably with conventional medicine in some nations. The US National Center for Complementary and Integrative Health (NCCIH) has clarified the terminology and classification of these health approaches. When a non-traditional method is combined with standard medical treatment, it is referred to as "complementary." Conversely, when a non-traditional method is used as a substitute for conventional medical treatment, it is classified as "alternative." Collectively, these two approaches are known as complementary and alternative medicine (CAM). The NCCIH currently categorizes CAM into three distinct groups (NCCIH 2021a), starting with nutritional approaches include natural products such as herbs (botanicals), vitamins, minerals, and probiotics.

The broadest category, psychological and physical approaches, encompasses therapies and practices such as yoga, massage, tai chi, qigong, acupuncture, chiropractic care, osteopathic manipulation, art therapy, music therapy, dance therapy, mindfulness-based stress reduction, meditation, relaxation techniques like breathing exercises and guided imagery, hypnotherapy, the Feldenkrais method, the Alexander technique, Pilates, Rolfing structural integration, Trager psychophysical

DOI: 10.1201/9781032706511-4

integration, and progressive muscle relaxation. Sometimes referred to as meditative movement or mind–body practices, yoga, tai chi, and qigong incorporate both meditative and physical elements in their routines. In the third category, some complementary practices do not neatly fit into either of the previous groups. Examples include the practices of traditional healers, Ayurveda, Unani, Siddha, homeopathy, naturopathy, Chinese or Oriental medicine, and functional medicine. For the scope of this chapter, CAM is referred to as traditional medicine. Traditional medicine has gained attention for its low toxicity and effectiveness in treating chronic illnesses.

4.2 EPIGENETICS AND EPIGENETIC HEALING

Epigenetics is the study of how environment and behaviors affect gene expression (at the transcriptional and post-transcriptional levels or at the translation levels and post-translational modifications) by changing how the body reads a DNA sequence without changing the DNA sequence. This process changes the phenotype without changing the genotype. Unlike genetic changes, epigenetic changes are mostly reversible and some of them are even heritable. Epigenetic modifications encompass a range of common types, including chromatin folding and attaching to the nuclear matrix; DNA packaging; DNA methylation; and various histone modifications such as acetylation, methylation, and phosphorylation, non-coding RNA regulation (miRNA, pina, lncRNA, etc.). To modulate DNA accessibility, proteins known as chromatin remodelers or modifiers alter chromatin structure by changing nucleosome positioning and histone–DNA contacts. These modifications help properly prepare genomic loci for replication, repair, recombination, or transcription; some common epigenetic modifications and their effect of transcription are depicted in Figure 4.1 and Table 4.1 (Han et al. 2011; Webster et al. 2013)

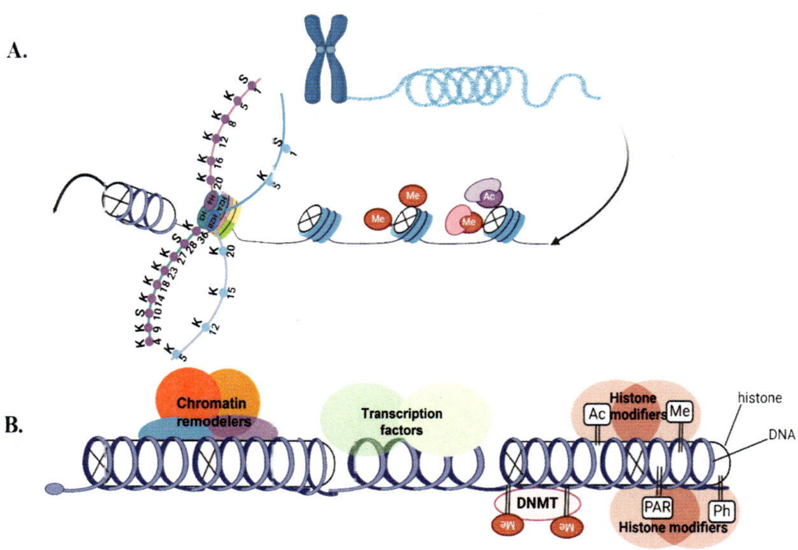

FIGURE 4.1 Epigenetic modifications: A) nucleosome structure, B) interactions between chromatin regulators and transcription factors for controlling gene expression. Chromatin remodelers include ISWI, CHD, IN0808, SWI/SNF, etc. Transcription factors include CREB, NF-κB, GATAs, etc. Histone modifiers include the enzymes that carry out acetylation, methylation, phosphorylation, etc.

Notes: K, lysine; S, serine amino acids of modified histones; number corresponds to the position of the amino acid residue in the histone protein. DNMT, DNA methyltransferase; Ac, acetylation; Me, methylation; PAR, poly-ADP ribosylation; Ph, phosphorylation.

TABLE 4.1
Epigenetic Modifications and Their Associated Transcriptional Regulatory Effects

Mechanism	Position	Effect on Transcription
H3 acetylation	K9, 14	↑
H3 methylation	K4, 36	↑
	K9, 27	↓
H3 phosphorylation	S10	↑/↓
H4 acetylation	K5, K8, K12, K16	↑
H4 methylation	K20	↓
RNA-based mechanisms	Activating noncoding RNAs–HOTTIP	↑
	Repressive noncoding RNAs–Xist, ANRIL	↓
DNA base modifications	Methylation (CpG dinucleotides)	↓

Note: ANRIL, antisense noncoding RNA in the INK4 locus; HOTTIP, HOXA distal transcript antisense RNA; K, lysine; S, serine.

Epigenetics plays a vital role in cellular differentiation, regulating the selective expression of proteins in different cell types. It also influences organisms' development from single-celled eukaryotes to multicellular beings including complex mammals like humans. Epigenetics are also vital in physiology and disease (Reik 2007). Its manipulation was effective in treating many types of cancers and metabolic and developmental disorders (Egger et al. 2004).

MicroRNAs are small, single-stranded, non-coding RNA molecules, typically 21–23 nucleotides in length. They function by pairing with corresponding base sequences in the 3′ untranslated regions of specific target mRNAs, resulting in either degradation or translational repression of the targeted mRNAs. Under normal physiological conditions, miRNAs function in feedback mechanism by safeguarding key biological processes such as cell proliferation, differentiation, and apoptosis (Esquela-Kerscher et al. 2006).

Deregulation of these miRNAs, whether singular or multiple, were reported to influence the expression pattern of several hundreds of mRNAs and actuates the cells toward transformation, carcinogenesis, and other diseases. Researchers have unraveled the roles of more than 2000 human miRNAs discovered so far, their role in various diseases like cancer, metabolic, inflammatory, neurological and cardiovascular diseases including cardiac hypertrophy, arrhythmia, myocardial infarction, and heart failure (Tufekci et al. 2014; Romaine et al. 2015).

Extracellular (circulating) miRs are a novel biomarker, well established clinically and preclinically for cancers, heart and liver disease, and more. They also satisfy the criteria for a reliable diagnostic and prognostic biomarker: non-invasive accessibility, disease specificity, sensitivity to changes in pathology (disease progression and therapeutic response), easy assessment via RT-qPCR or microarray, predictive and robust stability (protected from ribonucleases by vesicles), and seamless translation from preclinical settings to clinical platforms. Circulating miRNAs serve as mediators for inter- and extracellular communication (Etheridge et al. 2011).

Conventional healthcare providers have long been unsure of the underlying mechanisms of traditional medicines due to limited solid scientific literature. In this review, our objective is to highlight that a major part of these unraveled mechanisms involves epigenetics, the ability of external or endogenous influences to control gene expression reversibly without directly changing the underlying genetic code. Over the past two decades, research has shown that environmental factors shape the epigenome and directly influence health (Skinner 2011). The first human disease to be linked to epigenetics was cancer, in 1983. Over the last four decades, physiology and pathophysiology including diabetes, obesity, cardiovascular diseases, PCOS, autoimmune/inflammatory diseases (such as arthritis, SLE), psychiatric, and many more disorders were linked to epigenetics, making it an excellent therapeutic target. Scientists liken the genome and epigenome to the hardware and operating system of a computer. Like the software in an operating system, the epigenome determines which functions the genetic "hardware" does and does not perform (Child 2010). It is assessed that 90% of life is governed by epigenetics. Epigenetic healing refers to the treatment of a disease by targeting its epigenetic components, and we aim to provide insights into the epigenetic effects of traditional medicine in this review.

New scientific research indicates that environmental factors, through epigenetic mechanisms, can strongly affect children's brain development and their likelihood of growing up to be healthy and productive members of society. For example, adverse childhood experiences such as abuse, neglect, household dysfunction; unhealthy habits; and low socio-economic status in early life increase the risk for diseases linked to neuroendocrine dysregulation and inflammation in later life (Child 2010).

4.3 MINDFULNESS AND MEDITATION

Mindfulness is a mental state of equanimity that can be achieved through the practice of meditation and other mind–body practices that effectively manage stress (Househam et al. 2017). By producing a state of physical and mental relaxation, which can improve self-awareness and emotional regulation, meditation and mindfulness-based practices support the cooperation between brain and body to bring physical and emotional healing and well-being (Verdone et al. 2023). One day of intensive meditation reduced expression of chromatin–modulatory histone deacetylase genes (HDAC2, 3 and 9), alterations in global modification of histones (H4ac; H3K4me3), and decreased expression of pro-inflammatory genes (RIPK2 and COX2) (Kaliman et al. 2014).

Chaix et al. (2020) found differentially methylated DNA sites (DMS), with a 3% change across 61 genes (57 showing increased methylation and 4 showing decreased methylation). These affected genes are associated with immune cell metabolism, aging, and binding sites for transcription factors involved in immune response and inflammation. In contrast with controls, mindfulness meditators showed 31 differentially methylated DNA regions (DMRs) in peripheral blood leukocytes. These DMRs were associated with psychiatric and neurodegenerative disorders, cardiovascular diseases, and cancer (García-Campayo et al. 2018). A month-long retreat attended by experienced meditators ($n = 28$) was associated with downregulated TNF and NF-kB pathways in PBMCs, indicating reduced inflammation. These effects were associated with increased expression of the methyltransferases KMT2G, DNMT3A, KMT5B, and PRMT5 and decreased expression of the deacetyltransferase HDAC9. DNMT3A interacts with HDACs, PRMT5 and the repressive histone-methyltransferase KMT2G (also known as SETD1B) to silence gene transcription (Fuks et al. 2001; Li et al. 2006; Tsutsui et al. 2013). KMT5B (also known as SUV420H1) is involved in the methylation of histone 4, and the age-related loss of this methylation is associated with human cancers (Fraga et al. 2005; Fraga et al. 2007).

4.4 YOGA

Yoga is an ancient Indian practice that combines physical and mental disciplines. It includes various elements such as physical postures (asanas), breathing exercises (pranayama), and meditative techniques. The practice of yoga encompasses a wide range of styles, varying from soft and delicate approaches to highly strenuous and physically challenging methods. The benefits of yoga are bountiful, ranging from improving general wellness and musculo-skeletal and lifestyle diseases to deterring smoking and many chronic diseases and improving quality of life (NCCIH 2023b).

Infertile individuals showed improved fertility after 21 days of yoga, with 147 promoters of fertility-related genes undergoing hypermethylation and 229 undergoing hypomethylation. These changes were linked to genomic integrity in spermatozoa (Bisht et al. 2020). Women who attended eight weeks of twice-weekly, hour-long yoga classes showed less DNA methylation of tumour necrosis factor (TNF) in their whole blood cells (Harkess et al. 2016). Indeed, methylation levels within TNF show strong gender-specific correlations with certain long-chain fatty acids. In women, omega-6 essential fatty acid, PUFA γ-linolenic acid was negatively correlated with TNF methylation; adrenic and eicosadienoic acids were positively correlated (Hussey et al. 2021).

In another study, eight weeks of yoga practice significantly reduced rheumatoid arthritis disease based on DAS28-ESR score and inflammation (downregulation of pro-inflammatory RORγt, IL-17, CXCR2, IL-6, and CXCL2 and upregulation of anti-inflammatory FoxP3 and TGF-β) and sustained Th17/Treg cell homeostasis. The yoga group also showed higher global 5-mC (5-methyl cytosine) and lower global 5-hmC (5-hydroxymethyl cytosine) and HDAC1, which regulated the aforementioned gene expression patterns, than in the non-yoga group (Gautam et al. 2023).

A nine-week yoga practice significantly upregulated expressions of PPARG, INSR, IRS2, and CAPN10 in women with polycystic ovarian syndrome (PCOS). These augmented gene expressions improved BMI and insulin resistance and reduced anti-Müllerian hormone, luteinizing hormone, testosterone, and triglyceride levels, thereby improving PCOS. The yoga group showed decreased miR-21 and miR-128 and increased miR-320. These changes contribute to a higher apoptosis rate among dysfunctional granulosa cells in PCOS patients and enhance insulin receptor expression. Elevated miR-320 levels could play a role in inhibiting insulin resistance in individuals with PCOS and in regulating the ERK1/2 signaling pathway (Kumari et al. 2023).

4.5 TAI CHI

Tai chi is a traditional Chinese mind–body exercise that involves a series of slow gentle movements and physical postures, a meditative state of mind, and controlled breathing (NCCIH 2023a). Tai chi slowed the decline in age-related trends for DNA methylation losses or gains of six CpG sites from three different chromosomes in the tai chi cohort, in contrast with the control cohort (Ren et al. 2012). Tai chi practitioners also have less DNA damage, stronger DNA repair, advanced levels of lymphocyte renewal, and active endogenous antioxidant enzymes (Goon et al. 2008; Goon et al. 2009).

4.6 AYURVEDA

Ayurvedic medicine, or Ayurveda, is one of the oldest codified medical systems in the world. Grounded in historical texts, its principles promote a natural and holistic strategy for maintaining physical and mental well-being. Ayurvedic treatments encompass a range of modalities, including products primarily derived from botanical sources but also potentially incorporating animal, metal, and mineral components, in addition to dietary recommendations, exercise regimens, and lifestyle modifications (NCCIH 2019).

Nutraceuticals modify gene expression by altering epigenetic mechanisms (Sharma & Keith Wallace 2020). Curcumin, a polyphenolic molecule extracted from the *Curcuma longa* plant, is a renowned traditional herbal medicine that has demonstrated antioxidant, anti-inflammatory, anti-fibrosis, and anti-diabetes characteristics by virtue of its effects on DNA methylation, miRNAs, and histone-modifying enzymes in addition to its nonepigenetic mechanisms. A large number of studies indicated that curcumin could be utilized to treat diabetes and associated complications (Tang et al. 2022). The hallmarks of almost all cancer types, including breast cancer, include promoters of the hypomethylation for oncogenes and prometastatic genes as well as global DNA demethylation, and hypermethylation of tumour suppressor related genes. Curcumin has been shown to modulate epigenetic events that are dysregulated in cancer cells and possess the potential to prevent cancer or enhance the effects of conventional anti-cancer therapy (Wu et al. 2020; Fabianowska-Majewska et al. 2021).

Thymoquinone is a phytochemical mainly found in black cumin or black seed (*Nigella sativa*; family Ranunculaceae) that has been used in Ayurvedic and Unani systems for its anti-inflammatory, antidiabetic, antihypertensive, antimicrobial, analgesic, immunomodulatory, spasmolytic, hepatoprotective, renal-protective, gastroprotective, bronchodilatory, antioxidant, and antineoplastic effects in vitro and in vivo (Khan et al. 2019). Lu et al. reported that thymoquinone activates SIRT1, a histone deacetylase that suppresses p53 acetylation in myocardial ischemia/reperfusion injury, leading to reduced mitochondrial oxidative stress damage and apoptosis in cardiomyocytes (Lu et al. 2018).

Quercetin (3, 3′, 4′, 5, 7-pentahydroxyflvanone) is a flavonoid found in citrus fruits, red grapes, apples, onions, and tea (Oliveira et al. 2015; Quintanilha et al. 2017). This phenolic molecule is known for its chemopreventive, anti-inflammatory, antioxidant, and neuroprotective properties (Fürst & Zündorf 2014). Quercetin and its metabolite, isorhamnetin, downregulate the mRNA and protein levels of the proinflammatory mediators, including TNF-α, IL-6, and IL-1β by downregulating miR-155 and upregulating miR-146a, which may be how this polyphenol inhibits the activation of NF-κB, thereby containing the inflammatory process (Boesch-Saadatmandi et al. 2011; Lam et al. 2012).

4.7 DIET

Lifestyle dietary changes affect epigenetics, and some clinical studies have been reported for the use of epigenetics-related changes as analytical tools for glycaemia, blood pressure, and inflammation control after dietary tweak. Results from the POUNDS LOST trial show effects on insulin resistance of an important interplay between dietary protein consumption and DNA methylation. Mean protein intake (15% energy) versus high protein intake (25%) was associated with a greater decrease in insulin and HOMA-IR in overweight or obese individuals with the highest DNA methylation at the TXNIP gene (Li et al. 2022). The same POUNDS LOST trial witnessed plummeted systolic and diastolic blood pressures in association with higher DNA methylation of LINC00319 gene (Kou et al. 2023) in the mean protein intake group. Administering eight weeks of supplementation of 300 g of vegetables and legumes containing on average 191 μg/day of folate and one hazelnut oil capsule to overweight/obese women resulted in decreased oxidative stress corresponding to a decrease in ADRB3 DNAm levels (Lima et al. 2017).

Cardiac dysfunction and injury correlated with increased levels of miR-451, miR-25, and miR-155 and decreased miR-99 in the hearts of uninephrectomized rats fed a high salt diet. Each of these miRs inversely affected the expression of proteins involved in the physiology and pathophysiology of heart (Amara et al. 2017). In HFD-fed rats, endothelial dysfunction and increased blood pressure were associated with increased DNMT1 expression and decreased expression of HSP72, SIRT1, and eNOS (Karpe & Tikoo 2014).

Curcumin and other bioactive dietary components have been shown to influence epigenetic targets involved in breast cancer. These bioactive ingredients include dietary polyphenols, epigallocatechin gallate from green tea, genistein from soybean, isothiocyanates from plant foods, resveratrol from grapes, and sulforaphane from cruciferous vegetables (Khan et al. 2012; Shukla et al. 2014). In light of these findings, an “epigenetics diet” has been proposed that utilizes these bioactive dietary compounds to neutralize epigenetic aberrations as a form of cancer treatment and be utilized as cancer prevention (Lewis & Tollefsbol 2017).

Glioma is one of the most common and aggressive brain tumours in children and adults (Valvona et al. 2016). For these cancers, there are very few effective treatment methods; even after active surgery, chemotherapy, and radiotherapy, the patient survival rate is still very low (Gutmann & Kettenmann 2019). Since 2014, Skala and Sitarek et al. have investigated the effects of Chinese herbal medicine on glioma, and initially they found that *Leonurus sibiricus* transgenic roots and *Rhaponticum carthamoides* transformed root promote glioma cell apoptosis and inhibit their viability via DNA damage, PARP cleavage/increased H2A.X histone levels, and downregulated UHRF-1/DNMT1 (Sitarek et al. 2018).

4.8 HOMEOPATHY

More than 200 years ago, the medical system known as homeopathy, or homeopathic medicine, was created in Germany. It is predicated upon two nontraditional theories: that "like cures like," a drug that causes comparable symptoms in healthy individuals may also treat a sickness and that a drug is more effective at lower dosages (the law of minimum dose). The original component is so diluted in many homeopathic remedies that it is no longer present in any molecules. Homeopathic remedies utilize a wide range of materials, including plant-based substances like red onions, arnica, belladonna, and stinging nettle; minerals such as white arsenic; and animal-derived products, for example, crushed whole bees (NCCIH 2021b).

Homeopathic medications might be in the shape of pills, ointments, gels, liquids, creams, or sugar pellets that are supposed to be inserted under the tongue. Different persons with the same ailment often receive different therapies because they are customized for each person. Homeopathy uses a different diagnostic system from that of conventional medicine for assigning treatments to individuals and recognizes different clinical patterns of signs and symptoms (NCCIH 2021b).

HeLa cells, a cervical carcinoma cell line used extensively in cancer research, when treated with the homeopathic drugs condurango and *Hydrastis canadensis*, displayed a pattern of over 100 differentially expressed genes, suggesting the power of ultra-diluted homeopathic medicines in reversing aberrant epigenetic cancer marks (Khuda-Bukhsh et al. 2013). Condurango-treated lung cancer cells showed a significant decrease in hypermethylation associated with tumour suppressor genes p15 and p53 in vitro, as well as inhibition of hypermethylation of p15 in post-cancer treated rat models (Sikdar et al. 2014).

4.9 NATUROPATHY

Naturopathy, or naturopathic medicine, is a clinical framework that has developed from a mix of customary practices and medical care. It is directed by an exceptional arrangement of rules that perceive the body's intrinsic recuperating limit, underline infection counteraction, and urge individual changes and efforts for ideal well-being. Naturopathic treatment modalities include dietary and lifestyle changes, stress reduction, herbs, dietary supplements, homeopathy, manipulative therapies, exercise therapy, practitioner-guided detoxification, psychotherapy, counselling, hydrotherapy, and pharmaceuticals (NCCIH 2024).

Propolis, a honeybee determined naturopathic plan, includes caffeic corrosive phenethyl ester (CAPE) as a functioning fixing that has displayed prompt cell separation, cell cycle capture, and apoptosis within the malignant breast cell lines MCF-7 and MDA-231. CAPE acts as an HDAC inhibitor, leading to an accumulation of acetylated histone proteins, and in turn resulting in the downregulation of both progesterone- and oestrogen-related receptors. This leads to cell differentiation and inhibition of the mdr-1 gene, which would otherwise render chemotherapeutic drugs nonfunctional in the cell lines (Omene et al. 2013).

4.10 UNANI

Unani, also known as Greek–Arab medicine, originated in Greece and was further developed by Arab scholars into an elaborate form of medical science based on the teachings of Hippocrates (Buqrat) and Galen (Jalinoos) (Ministry of Ayush 2023). Withaferin A (WA) is a bioactive compound isolated from *Withania somnifera* leaves and roots (alternatively called ashwagandha or Indian ginseng in traditional Ayurvedic and Unani medicine) is used to treat inflammatory ailments and has been a potential anti-cancer drug (Hassannia et al. 2020). Small amounts of WA (700 nM) changes how genes are expressed in drug-resistant triple-negative breast cancer cells, making them behave more like treatment-sensitive luminal breast cancer cells. This change is due to a specific type of gene silencing, where DNA becomes more tightly packed, turning off certain genes like PLAU, TNSF12, ADAM8, and ME3. This process involves the control of enzymes like DNMT1, KDM demethylases, and HDACs (Szarc vel Szic et al. 2014; Szarc vel Szic et al. 2017; Royston et al. 2018). In addition, a specific enzyme called HDAC6 is also a direct target of WA in multiple myeloma (Dom et al. 2018).

Onosma bracteatum, an important herbal drug of Unani system, has been used to treat Alzheimer's disease, arrhythmias, and hypertension (Fareed et al. 2012). The benzoquinone-type molecule ehretiquinone, isolated from *O. bracteatum* wall, exerts an anti-aging effect in yeast, mammalian cells, and mice by inducing antioxidative stress and autophagy through histone deacetylase Sir2 (silent information regulator 2) regulation. Sir2 is known to slow down aging (Pan et al. 2021).

4.11 COMPLEMENTARY AND ALTERNATIVE MEDICINES

Sowa Rigpa, also called Tibetan, Bhutanese, or Amchi medicine, is the traditional medicine of many regions around the Himalayas. The treatment regimens of Bhutanese Sowa Rigpa medicine include changes in food and lifestyle (as many illnesses arise from improper diet and lifestyle); complex polyingredient medicinal formulations; and five external therapies: bloodletting (therapeutic phlebotomy), gold/silver needle, heated oil cauterization and moxibustion, medicinal bath and steam, and nasal irrigation (Yeshi et al. 2019). Moxibustion, a component of Chinese and Tibetan medicine, is a type of therapy that involves burning mugwort leaves close to the skin's surface area using a stick to apply heat; mugwort is a small, spongy herb that is believed to enhance healing with acupuncture. Herb-partitioned moxibustion has proven effective in treating Crohn's disease (Joos et al. 2004; Bao et al. 2014). A preclinical study showed that herb-partitioned moxibustion attenuated inflammatory responses and colonic damages in Crohn's disease in rats via up-regulating the expression of miR-147 and miR-205, which down-regulated the expression of inflammation-related mRNA cytokines (Wei et al. 2015).

REFERENCES

Amara, V. R., S. K. Surapaneni, and K. Tikoo. 2017. "Dysregulation of MicroRNAs and Renin-Angiotensin System in High Salt Diet-Induced Cardiac Dysfunction in Uninephrectomized Rats." *PLoS ONE* 12 (7):e0180490. http://doi.org/10.1371/journal.pone.0180490

Bao, C. H., J. M. Zhao, H. R. Liu, Y. Lu, Y. F. Zhu, Y. Shi, Z. J. Weng, H. Feng, X. Guan, J. Li, W. F. Chen, L. Y. Wu, X. M. Jin, C. Z. Dou, and H. G. Wu. 2014. "Randomized Controlled Trial: Moxibustion and Acupuncture for the Treatment of Crohn's Disease." *World J Gastroenterol* 20 (31):11000–11011. http://doi.org/10.3748/wjg.v20.i31.11000

Bisht, S., S. Banu, S. Srivastava, R. U. Pathak, R. Kumar, R. Dada, and R. K. Mishra. 2020. "Sperm Methylome Alterations Following Yoga-Based Lifestyle Intervention in Patients of Primary Male Infertility: A Pilot Study." *Andrologia* 52 (4):e13551. http://doi.org/10.1111/and.13551

Boesch-Saadatmandi, C., A. Loboda, A. E. Wagner, A. Stachurska, A. Jozkowicz, J. Dulak, F. Döring, S. Wolffram, and G. Rimbach. 2011. "Effect of Quercetin and Its Metabolites Isorhamnetin and Quercetin-3-Glucuronide on Inflammatory Gene Expression: Role of miR-155." *J Nutr Biochem* 22 (3):293–299. http://doi.org/10.1016/j.jnutbio.2010.02.008

Chaix, R., M. Fagny, M. Cosin-Tomás, M. Alvarez-López, L. Lemee, B. Regnault, R. J. Davidson, A. Lutz, and P. Kaliman. 2020. "Differential DNA Methylation in Experienced Meditators after an Intensive Day of Mindfulness-Based Practice: Implications for Immune-Related Pathways." *Brain Behav Immun* 84:36–44. http://doi.org/10.1016/j.bbi.2019.11.003

Child, National Scientific Council on the Developing. 2010. *Early Experiences Can Alter Gene Expression and Affect Long-Term Development*. Cambridge, MA: National Scientific Council on the Developing Child, Center on the Developing Child at Harvard University.

Dom, M., F. Offner, W. Vanden Berghe, and X. Van Ostade. 2018. "Proteomic Characterization of Withaferin A-Targeted Protein Networks for the Treatment of Monoclonal Myeloma Gammopathies." *J Proteomics* 179:17–29. http://doi.org/10.1016/j.jprot.2018.02.013

Egger, Gerda, Gangning Liang, Ana Aparicio, and Peter A. Jones. 2004. "Epigenetics in Human Disease and Prospects for Epigenetic Therapy." *Nature* 429 (6990):457–463.

Esquela-Kerscher, Aurora, and Frank J. Slack. 2006. "Oncomirs—MicroRNAs with a Role in Cancer." *Nat Rev Cancer* 6 (4):259–269.

Etheridge, A., I. Lee, L. Hood, D. Galas, and K. Wang. 2011. "Extracellular MicroRNA: A New Source of Biomarkers." *Mutat Res* 717 (1–2):85–90. http://doi.org/10.1016/j.mrfmmm.2011.03.004

Fabianowska-Majewska, K., A. Kaufman-Szymczyk, A. Szymanska-Kolba, J. Jakubik, G. Majewski, and K. Lubecka. 2021. "Curcumin from Turmeric Rhizome: A Potential Modulator of DNA Methylation Machinery in Breast Cancer Inhibition." *Nutrients* 13 (2). http://doi.org/10.3390/nu13020332

Fareed, Sheeba, Hefazat H. Siddiqui, Syed E. Haque, Mohammad Khalid, and Juber Akhtar. 2012. "Psycho-immunomodulatory Effects of Onosma bracteatum Wall. (Gaozaban) on Stress Model in Sprague Dawley Rats." *J Clin Diagn Res* 6 (7).

Fokunang, C. N., V. Ndikum, O. Y. Tabi, R. B. Jiofack, B. Ngameni, N. M. Guedje, E. A. Tembe-Fokunang, P. Tomkins, S. Barkwan, F. Kechia, E. Asongalem, J. Ngoupayou, N. J. Torimiro, K. H. Gonsu, V. Sielinou, B. T. Ngadjui, F. Angwafor, A. Nkongmeneck, O. M. Abena, J. Ngogang, T. Asonganyi, V. Colizzi, J. Lohoue, and Kom Kamsu. 2011. "Traditional Medicine: Past, Present and Future Research and Development Prospects and Integration in the National Health System of Cameroon." *Afr J Tradit Complement Altern Med* 8 (3):284–295. http://doi.org/10.4314/ajtcam.v8i3.65276

Fraga, Mario F., Ruben Agrelo, and Manel Esteller. 2007. "Cross-Talk between Aging and Cancer." *Ann N Y Acad Sci* 1100 (1):60–74. https://doi.org/10.1196/annals.1395.005

Fraga, Mario F., Esteban Ballestar, Ana Villar-Garea, Manuel Boix-Chornet, Jesus Espada, Gunnar Schotta, Tiziana Bonaldi, Claire Haydon, Santiago Ropero, Kevin Petrie, N. Gopalakrishna Iyer, Alberto Pérez-Rosado, Enrique Calvo, Juan A. Lopez, Amparo Cano, Maria J. Calasanz, Dolors Colomer, Miguel Ángel Piris, Natalie Ahn, Axel Imhof, Carlos Caldas, Thomas Jenuwein, and Manel Esteller. 2005. "Loss of Acetylation at Lys16 and Trimethylation at Lys20 of Histone H4 is a Common Hallmark of Human Cancer." *Nat Genet* 37 (4):391–400. http://doi.org/10.1038/ng1531

Fuks, François, Wendy A. Burgers, Nadia Godin, Masataka Kasai, and Tony Kouzarides. 2001. "Dnmt3a Binds Deacetylases and Is Recruited by a Sequence-Specific Repressor to Silence Transcription." *EMBO J* 20 (10):2536–2544. http://doi.org/10.1093/emboj/20.10.2536

Fürst, R., and I. Zündorf. 2014. "Plant-Derived Anti-Inflammatory Compounds: Hopes and Disappointments Regarding the Translation of Preclinical Knowledge into Clinical Progress." *Mediators Inflamm* 2014:146832. http://doi.org/10.1155/2014/146832

García-Campayo, Javier, Marta Puebla-Guedea, Alberto Labarga, Amaya Urdánoz, Miren Roldán, Laura Pulido, Xabier Martínez de Morentin, Álvaro Perdones-Montero, Jesús Montero-Marín, and Maite Mendioroz. 2018. "Epigenetic Response to Mindfulness in Peripheral Blood Leukocytes Involves Genes Linked to Common Human Diseases." *Mindfulness* 9 (4):1146–1159. http://doi.org/10.1007/s12671-017-0851-6

Gautam, Surabhi, Romsha Kumar, Uma Kumar, Sanjeev Kumar, Kalpana Luthra, and Rima Dada. 2023. "Yoga Maintains Th17/Treg Cell Homeostasis and Reduces the Rate of T Cell Aging in Rheumatoid Arthritis: A Randomized Controlled Trial." *Sci Rep* 13 (1):14924. http://doi.org/10.1038/s41598-023-42231-w

Goon, J. A., A. H. Noor Aini, M. Musalmah, M. Y. Yasmin Anum, W. M. Wan Nazaimoon, and W. Z. Wan Ngah. 2009. "Effect of Tai Chi Exercise on DNA Damage, Antioxidant Enzymes, and Oxidative Stress in Middle-Age Adults." *J Phys Act Health* 6 (1):43–54.

Goon, J. A., A. H. Noor Aini, M. Musalmah, M. Y. Yasmin Anum, and W. Z. Wan Ngah. 2008. "Long Term Tai Chi Exercise Reduced DNA Damage and Increased Lymphocyte Apoptosis and Proliferation in Older Adults." *Med J Malaysia* 63 (4):319–324.

Gutmann, D. H., and H. Kettenmann. 2019. "Microglia/Brain Macrophages as Central Drivers of Brain Tumor Pathobiology." *Neuron* 104 (3):442–449. http://doi.org/10.1016/j.neuron.2019.08.028

Han, Pei, Calvin T. Hang, Jin Yang, and Ching-Pin Chang. 2011. "Chromatin Remodeling in Cardiovascular Development and Physiology." *Circ Res* 108 (3):378–396.

Harkess, K. N., J. Ryan, P. H. Delfabbro, and S. Cohen-Woods. 2016. "Preliminary Indications of the Effect of a Brief Yoga Intervention on Markers of Inflammation and DNA Methylation in Chronically Stressed Women." *Transl Psychiatry* 6 (11):e965. http://doi.org/10.1038/tp.2016.234

Hassannia, B., E. Logie, P. Vandenabeele, T. Vanden Berghe, and W. Vanden Berghe. 2020. "Withaferin A: From Ayurvedic Folk Medicine to Preclinical Anti-Cancer Drug." *Biochem Pharmacol* 173:113602. http://doi.org/10.1016/j.bcp.2019.08.004

Househam, A. M., C. T. Peterson, P. J. Mills, and D. Chopra. 2017. "The Effects of Stress and Meditation on the Immune System, Human Microbiota, and Epigenetics." *Adv Mind Body Med* 31 (4):10–25.

Hussey, B., R. P. Steel, B. Gyimah, J. C. Reynolds, I. M. Taylor, M. R. Lindley, and S. Mastana. 2021. "DNA Methylation of Tumour Necrosis Factor (TNF) Alpha Gene Is Associated with Specific Blood Fatty Acid Levels in a Gender-Specific Manner." *Mol Genet Genomic Med* 9 (12):e1679. http://doi.org/10.1002/mgg3.1679

Joos, S., B. Brinkhaus, C. Maluche, N. Maupai, R. Kohnen, N. Kraehmer, E. G. Hahn, and D. Schuppan. 2004. "Acupuncture and Moxibustion in the Treatment of Active 'Crohn's Disease: A Randomized Controlled Study." *Digestion* 69 (3):131–139. http://doi.org/10.1159/000078151

Kaliman, P., M. J. Alvarez-López, M. Cosín-Tomás, M. A. Rosenkranz, A. Lutz, and R. J. Davidson. 2014. "Rapid Changes in Histone Deacetylases and Inflammatory Gene Expression in Expert Meditators." *Psychoneuroendocrinology* 40:96–107. http://doi.org/10.1016/j.psyneuen.2013.11.004

Karpe, Pinakin Arun, and Kulbhushan Tikoo. 2014. "Heat Shock Prevents Insulin Resistance-Induced Vascular Complications By Augmenting Angiotensin-(1–7) Signalling." *Diabetes* 63:1124–1139. http://doi.org/10.2337/db13-1267

Khan, Md Asaduzzaman, Mousumi Tania, and Junjiang Fu. 2019. "Epigenetic Role of Thymoquinone: Impact on Cellular Mechanism and Cancer Therapeutics." *Drug Discov Today* 24 (12):2315–2322. https://doi.org/10.1016/j.drudis.2019.09.007

Khan, S. I., P. Aumsuwan, I. A. Khan, L. A. Walker, and A. K. Dasmahapatra. 2012. "Epigenetic Events Associated with Breast Cancer and Their Prevention by Dietary Components Targeting the Epigenome." *Chem Res Toxicol* 25 (1):61–73. http://doi.org/10.1021/tx200378c

Khuda-Bukhsh, Anisur Rahman, Santu Kumar Saha, and Sourav Roy. 2013. "Evidence in Support of Gene Regulatory Hypothesis: Gene Expression Profiling Manifests Homeopathy Effect as More Than Placebo." *Int J High Dilution Res ISSN 1982–6206* 12 (45):162–167.

Kou, Minghao, Xiang Li, Xiaojian Shao, Elin Grundberg, Xuan Wang, Hao Ma, Yoriko Heianza, J. Alfredo Martinez, George A. Bray, Frank M. Sacks, and Lu Qi. 2023. "DNA Methylation of Birthweight–Blood Pressure Genes and Changes of Blood Pressure in Response to Weight-Loss Diets in the POUNDS Lost Trial." *Hypertension* 80 (6):1223–1230. http://doi.org/10.1161/HYPERTENSIONAHA.123.20864

Kumari, Deepika, Manoj Kumar, Prabhakar Tiwari, Reeta Mahey, Neena Malhotra, Richa Mishra, and Rima Dada. 2023. "Impact of Yoga in Polycystic Ovary Syndrome." *Int J Ayurveda Res* 4 (3):132–136. http://doi.org/10.4103/ijar.ijar_133_23

Lam, Tram K., Stephanie Shao, Yingdong Zhao, Francesco Marincola, Angela Pesatori, Pier Alberto Bertazzi, Neil E. Caporaso, Ena Wang, and Maria Teresa Landi. 2012. "Influence of Quercetin-Rich Food Intake on MicroRNA Expression in Lung Cancer Tissues." *Cancer Epidemiol Biomarkers Prev* 21 (12):2176–2184. http://doi.org/10.1158/1055-9965.Epi-12-0745

Lewis, K. A., and T. O. Tollefsbol. 2017. "The Influence of an Epigenetics Diet on the Cancer Epigenome." *Epigenomics* 9 (9):1153–1155. http://doi.org/10.2217/epi-2017-0077

Li, Hongwei, Tibor Rauch, Zhao-Xia Chen, Piroska E. Szabó, Arthur D. Riggs, and Gerd P. Pfeifer. 2006. "The Histone Methyltransferase SETDB1 and the DNA Methyltransferase DNMT3A Interact Directly and Localize to Promoters Silenced in Cancer Cells*." *J Biol Chem* 281 (28):19489–19500. https://doi.org/10.1074/jbc.M513249200

Li, Xiang, Xiaojian Shao, Lydia A. Bazzano, Qiaochu Xue, Boryana S. Koseva, Elin Grundberg, Iris Shai, George A. Bray, Frank M. Sacks, and Lu Qi. 2022. "Blood DNA Methylation at TXNIP and Glycemic Changes in Response to Weight-Loss Diet Interventions: The POUNDS Lost Trial." *Int J Obes* 46 (6):1122–1127. http://doi.org/10.1038/s41366-022-01084-5

Lima, R. P. A., R. A. F. do Nascimento, R. C. P. Luna, D. C. Persuhn, A. S. da Silva, M. da Conceição Rodrigues Gonçalves, A. T. C. de Almeida, R. M. de Moraes, E. V. Junior, E. Fouilloux-Meugnier, H. Vidal, L. Pirola, M. Magnani, N. F. P. de Oliveira, P. O. Prada, and M. J. de Carvalho Costa. 2017. "Effect of a Diet Containing Folate and Hazelnut Oil Capsule on the Methylation Level of the ADRB3 Gene, Lipid Profile and Oxidative Stress in Overweight or Obese Women." *Clin Epigenetics* 9:110. http://doi.org/10.1186/s13148-017-0407-6

Lu, Yunyang, Yingda Feng, Dan Liu, Zhiran Zhang, Kai Gao, Wei Zhang, and Haifeng Tang. 2018. "Thymoquinone Attenuates Myocardial Ischemia/Reperfusion Injury Through Activation of SIRT1 Signaling." *Cell Physiol Biochem* 47 (3):1193–1206. http://doi.org/10.1159/000490216

Ministry of Ayush, Govt. of India. 2023. "Introduction to Ayush Systems of Medicine." accessed 15-01-2024. https://ayush.gov.in/#!/health

NCCIH. 2019. "Ayurvedic Medicine: In Depth." The National Center for Complementary and Integrative Health (NCCIH) Clearing House, Last Modified Jan 2019, accessed 11/11/2023. https://www.nccih.nih.gov/health/ayurvedic-medicine-in-depth

NCCIH. 2021a. "Complementary, Alternative, or Integrative Health: What's in a Name?" The National Center for Complementary and Integrative Health (NCCIH) Clearing house, Last Modified April 2021, accessed 11/11/2023. https://www.nccih.nih.gov/health/complementary-alternative-or-integrative-health-whats-in-a-name

NCCIH. 2021b. "Homeopathy: What You Need to Know." The National Center for Complementary and Integrative Health (NCCIH) Clearing House, Last Modified April 2021, accessed 01/15/2024. https://www.nccih.nih.gov/health/homeopathy

NCCIH. 2023a. "Tai Chi: What You Need to Know." The National Center for Complementary and Integrative Health (NCCIH) Clearing House, Last Modified Mar 2022. https://www.nccih.nih.gov/health/tai-chi-what-you-need-to-know

NCCIH. 2023b. "Yoga: What You Need to Know." The National Center for Complementary and Integrative Health (NCCIH) Clearing House, Last Modified Aug 2023, accessed 11/11/2023. https://www.nccih.nih.gov/health/yoga-what-you-need-to-know

NCCIH. 2024. "Naturopathy." The National Center for Complementary and Integrative Health (NCCIH) Clearing House, Last Modified Sep 2017, accessed 01/15/2024. https://www.nccih.nih.gov/health/naturopathy

Oliveira, T. T., K. M. Campos, A. T. Cerqueira-Lima, T. Cana Brasil Carneiro, E. da Silva Velozo, I. C. Ribeiro Melo, E. A. Figueiredo, E. de Jesus Oliveira, D. F. de Vasconcelos, L. C. Pontes-de-Carvalho, N. M. Alcântara-Neves, and C. A. Figueiredo. 2015. "Potential Therapeutic Effect of Allium Cepa L. and Quercetin in a Murine Model of Blomia Tropicalis Induced Asthma." *Daru* 23 (1):18. http://doi.org/10.1186/s40199-015-0098-5

Omene, C., M. Kalac, J. Wu, E. Marchi, K. Frenkel, and O. A. 'O'Connor. 2013. "Propolis and Its Active Component, Caffeic Acid Phenethyl Ester (CAPE), Modulate Breast Cancer Therapeutic Targets via an Epigenetically Mediated Mechanism of Action." *J Cancer Sci Ther* 5 (10):334–342.

Pan, Y., Y. Liu, R. Fujii, U. Farooq, L. Cheng, A. Matsuura, J. Qi, and L. Xiang. 2021. "Ehretiquinone from Onosma Bracteatum Wall Exhibits Antiaging Effect on Yeasts and Mammals Through Antioxidative Stress and Autophagy Induction." *Oxid Med Cell Longev* 2021:5469849. http://doi.org/10.1155/2021/5469849

Quintanilha, B. J., B. Z. Reis, G. B. S. Duarte, S. M. F. Cozzolino, and M. M. Rogero. 2017. "Nutrimiromics: Role of MicroRNAs and Nutrition in Modulating Inflammation and Chronic Diseases." *Nutrients* 9 (11). http://doi.org/10.3390/nu9111168

Reik, Wolf. 2007. "Stability and Flexibility of Epigenetic Gene Regulation in Mammalian Development." *Nature* 447 (7143):425–432.

Ren, H., V. Collins, S. J. Clarke, J. S. Han, P. Lam, F. Clay, L. M. Williamson, and K. H. Andy Choo. 2012. "Epigenetic Changes in Response to Tai Chi Practice: A Pilot Investigation of DNA Methylation Marks." *Evid Based Complement Alternat Med* 2012:841810. http://doi.org/10.1155/2012/841810

Romaine, S. P., M. Tomaszewski, G. Condorelli, and N. J. Samani. 2015. "MicroRNAs in Cardiovascular Disease: An Introduction for Clinicians." *Heart* 101 (12):921–928. http://doi.org/10.1136/heartjnl-2013-305402

Royston, K. J., B. Paul, S. Nozell, R. Rajbhandari, and T. O. Tollefsbol. 2018. "Withaferin A and Sulforaphane Regulate Breast Cancer Cell Cycle Progression Through Epigenetic Mechanisms." *Exp Cell Res* 368 (1):67–74. http://doi.org/10.1016/j.yexcr.2018.04.015

Sharma, H., and R. Keith Wallace. 2020. "Ayurveda and Epigenetics." *Medicina (Kaunas)* 56 (12). http://doi.org/10.3390/medicina56120687

Shukla, S., S. M. Meeran, and S. K. Katiyar. 2014. "Epigenetic Regulation by Selected Dietary Phytochemicals in Cancer Chemoprevention." *Cancer Lett* 355 (1):9–17. http://doi.org/10.1016/j.canlet.2014.09.017

Sikdar, Sourav, and Anisur Rahman Khuda Bukhsh. 2014. "Hypermethylation Involved in DNA Profiles of Lung Cancer Specific Tumour Suppressor Genes and Epigenetic Modification Caused by an Ultra-Highly Diluted Homeopathic Drug, Condurango 30C, in Vitro and in Vivo." *Int J High Dilution Res* 13 (47).

Sitarek, P., T. Kowalczyk, S. Santangelo, A. J. Białas, M. Toma, J. Wieczfinska, T. Śliwiński, and E. Skała. 2018. "The Extract of Leonurus Sibiricus Transgenic Roots with AtPAP1 Transcriptional Factor Induces Apoptosis via DNA Damage and Down Regulation of Selected Epigenetic Factors in Human Cancer Cells." *Neurochem Res* 43 (7):1363–1370. http://doi.org/10.1007/s11064-018-2551-6

Skinner, M. K. 2011. "Environmental Epigenomics and Disease Susceptibility." *EMBO Rep* 12 (7):620–622. http://doi.org/10.1038/embor.2011.125

Szarc vel Szic, K., K. Declerck, R. A. J. Crans, J. Diddens, D. B. Scherf, C. Gerhäuser, and W. Vanden Berghe. 2017. "Epigenetic Silencing of Triple Negative Breast Cancer Hallmarks by Withaferin A." *Oncotarget* 8 (25):40434–40453. http://doi.org/10.18632/oncotarget.17107

Szarc vel Szic, K., K. Op de Beeck, D. Ratman, A. Wouters, I. M. Beck, K. Declerck, K. Heyninck, E. Fransen, M. Bracke, K. De Bosscher, F. Lardon, G. Van Camp, and W. Vanden Berghe. 2014. "Pharmacological Levels of Withaferin A (Withania Somnifera) Trigger Clinically Relevant Anticancer Effects Specific to Triple Negative Breast Cancer Cells." *PLoS ONE* 9 (2):e87850. http://doi.org/10.1371/journal.pone.0087850

Tang, C., Y. Liu, S. Liu, C. Yang, L. Chen, F. Tang, F. Wang, L. Zhan, H. Deng, W. Zhou, Y. Lin, and X. Yuan. 2022. "Curcumin and Its Analogs as Potential Epigenetic Modulators: Prevention of Diabetes and Its Complications." *Pharmacology* 107 (1–2):1–13. http://doi.org/10.1159/000520311

Tsutsui, Taiki, Rikiya Fukasawa, Kaori Shinmyouzu, Reiko Nakagawa, Kazuyuki Tobe, Aki Tanaka, and Yoshiaki Ohkuma. 2013. "Mediator Complex Recruits Epigenetic Regulators via Its Two Cyclin-dependent Kinase Subunits to Repress Transcription of Immune Response Genes*." *J Biol Chem* 288 (29):20955–20965. https://doi.org/10.1074/jbc.M113.486746

Tufekci, K. U., M. G. Oner, R. L. Meuwissen, and S. Genc. 2014. "The Role of MicroRNAs in Human Diseases." *Methods Mol Biol* 1107:33–50. http://doi.org/10.1007/978-1-62703-748-8_3

Valvona, C. J., H. L. Fillmore, P. B. Nunn, and G. J. Pilkington. 2016. "The Regulation and Function of Lactate Dehydrogenase A: Therapeutic Potential in Brain Tumor." *Brain Pathol* 26 (1):3–17. http://doi.org/10.1111/bpa.12299

Verdone, L., M. Caserta, T. D. Ben-Soussan, and S. Venditti. 2023. "On the Road to Resilience: Epigenetic Effects of Meditation." *Vitam Horm* 122:339–376. http://doi.org/10.1016/bs.vh.2022.12.009

WebMD. "What Is Moxibustion?", Last Modified July 2023, accessed 11/12/2023. https://www.webmd.com/balance/what-is-moxibustion

Webster, Andrew L. H., Matthew Shu-Ching Yan, and Philip A. Marsden. 2013. "Epigenetics and Cardiovascular Disease." *Can J Cardiol* 29 (1):46–57.

Wei, K., D. Zhang, J. Hong, C. Zhang, X. Feng, Y. Huang, J. Liu, L. Wu, H. Wu, and X. Ma. 2015. "Herb-Partitioned Moxibustion and the miRNAs Related to 'Crohn's Disease: A Study Based on Rat Models." *Evid Based Complement Alternat Med* 2015:265238. http://doi.org/10.1155/2015/265238

Wu, R., L. Wang, R. Yin, R. Hudlikar, S. Li, H. D. Kuo, R. Peter, D. Sargsyan, Y. Guo, X. Liu, and A. N. Kong. 2020. "Epigenetics/Epigenomics and Prevention by Curcumin of Early Stages of Inflammatory-Driven Colon Cancer." *Mol Carcinog* 59 (2):227–236. http://doi.org/10.1002/mc.23146

Yeshi, Karma, Yangbum Gyal, Katharina Sabernig, Jigme Phuntsho, Tawni Tidwell, Tenzin Jamtsho, Rinchen Dhondup, Eliot Tokar, and Phurpa Wangchuk. 2019. "An Integrated Medicine of Bhutan: Sowa Rigpa Concepts, Botanical Identification, and the Recorded Phytochemical and Pharmacological Properties of the Eastern Himalayan Medicinal Plants." *Eur J Integr Med* 29:100927. https://doi.org/10.1016/j.eujim.2019.100927

5 Epigenetic Drug Discovery in Multiresistant Cancers and Autoimmune Diseases

Stuti Biswal, Rajeev Kumar Sahoo, Sambit Kumar Patra, Shikshya Swarupa Panda, and Bijesh K Biswal
Cancer Drug Resistance Laboratory, Department of Life Sciences, National Institute of Technology Rourkela, Odisha, India

5.1 INTRODUCTION

Cancer is a terrible wound which can never be healed. Cancer is characterized by abnormal cells, unrestricted growth, and suppresses the host's immune response[1]; because of this characteristic feature, cancer is closely related with autoimmune diseases: The dysregulation of immune reaction in turn induces these diseases[2]. Autoimmune diseases that are shown to be associated with cancer include inflammatory bowel diseases (IBDs) like Crohn's disease and ulcerative colitis and have an elevated possibility for colorectal cancer, and celiac disease is an influencing factor for cancer in the small intestine. Interestingly rheumatoid arthritis (RA) is reported as a key risk factor for cancer growth mainly in lymphoma [2]. Systemic sclerosis, dermatomyositis, and polymyositis are risk factors for hematopoietic, nasopharyngeal, and lung cancers, and B- and T-cell lymphomas are also related with autoimmune diseases [3].

Cancer treatment includes surgery, radiotherapy, chemotherapy, immunotherapy, and targeted therapy alone or in combination [4], whereas treatment modalities for autoimmune diseases include surgery, immunosuppressant drugs, anti-inflammatory agents, and corticosteroid treatments[5]. The two disease types share the use of immunosuppressants and immunomodulatory drugs. Unfortunately, despite the vast array of treatment modalities, the success rates for treating both disease types are low, in part because of resistance that develops in many patients toward the treatments[2]. Multiple causes lead to therapy resistance, and epigenetic alteration plays a crucial role. Epigenetic changes are changes to DNA without affecting the nucleotide sequences; in multiresistant cancer, these changes include DNA and histone methylation and histone acetylation [6], while in autoimmune diseases, they involve disturbances in gene expression related to immune function. These changes enhance autoimmune response, drive inflammation, and affect the function and differentiation of immune cells[7].

Therefore, to obtain better outcomes for both autoimmune diseases and cancers, a novel therapeutic option is needed that can target these alterations in epigenetics. Today, plant-based phytocompounds are attracting the attention of every researcher for developing potent anti-cancer drugs as they exhibit potent pharmacological activity with multiway targets and less or no cytotoxicity. This chapter gives a brief overview about the association between cancer and autoimmune diseases, therapy resistance because of epigenetic changes, and using phytocompounds solely or in combination with other epigenetic drugs to overcome it.

DOI: 10.1201/9781032706511-5

5.2 EPIGENETIC CHANGES IN CANCER AND THERAPY RESISTANCE

Epigenetic changes have a critical role in cancer development and progression, including the development of resistance to therapy. These changes, which affect gene expression without changing the nucleotide sequence of DNA, offer cancer cells a dynamic way to adapt, survive, and evade treatment. The key mechanisms of epigenetic changes leading to therapy resistance include DNA methylation in which the addition of methyl groups to DNA activates oncogenes and silences tumor suppressor genes[8].

In breast cancer, methylation of BRCA1, which helps repair DNA damage, can lead to increased resistance to chemotherapy and radiation therapy;[9] histone modifications including acetylation and de-acetylation in histone proteins that change the chromatin structure influence how genes respond to transcriptional machinery, thereby influencing gene expression and therapy response[10]. In colorectal cancer, specific histone modifications can promote the expression of genes involved in drug efflux pumps, which export drugs out of cancer cells, making them less effective[11].

Additionally, noncoding RNAs (ncRNAs), particularly microRNAs (miRNAs) potentially contribute to therapy resistance by targeting and silencing specific genes[12]. In melanoma, miRNAs target and silence genes involved in apoptosis (programmed cell death), leading to resistance to targeted therapies[13]. The consequences of epigenetic therapy resistance are reduced effectiveness of existing therapies, increased risk of cancer recurrence and metastasis, and difficulty in developing new effective treatment strategies.

5.3 EPIGENETIC CHANGES IN AUTOIMMUNE DISEASES AND THERAPY RESISTANCE

The complex interplay between autoimmune diseases and epigenetic changes plays a significant role in therapy resistance, posing a major challenge for treatment success. The mechanism of action in this intricate relationship is epigenetic modifications silence genes involved in drug metabolism and transport, leading to higher drug concentrations and increased toxicity [14], while aberrant histone modifications create inaccessible chromatin structures that prevent drugs from reaching their target genes. Additionally ncRNAs interfere with drug action by modulating the protein expression involved in drug response pathways[15]. In RA, DNA methylation leads to increased expression of genes associated with inflammation and resistance to anti-TNF drugs [16], and histone modifications promote the expression of genes involved in B cell activation, contributing to resistance to B cell-depleting therapies [17]. Furthermore, miRNAs downregulate the expression of genes involved in immune suppression, leading to resistance to anti-inflammatory medications [18]. These changes can be targeted to discover new drugs to overcome therapy resistance and achieve better results.

5.4 EPIGENETIC DRUGS USED IN CANCER TREATMENT

Epigenetic modifications are crucial for the development and spread of cancer [19]. Epigenetic alterations, as opposed to genetic mutations, affect the structure of DNA and histones; because mechanisms such as DNA methylation and histone modification are appealing targets for cancer treatment, the development of epigenetic medications to rectify abnormal epigenetic patterns in cancer cells has advanced significantly in recent years [20]. Often categorized as histone deacetylase (HDAC) or DNA methyltransferase (DNMT) inhibitors, these medications have great potential for reversing the epigenetic dysregulation linked to several cancers [21]. This section explores the classification, mechanism of action, and therapeutic importance of epigenetic medicines used in cancer treatment.

5.4.1 DNMT Inhibitors

DNMT inhibitors are the compounds that prevent the activity of DNMTs, the enzymes that add methyl groups to DNA molecules in mammals. DNA methylation mainly occurs at the cytosine-phosphodiester-guanine (CpG) dinucleotide sequences of the gene promoter region, where the DNMTs add a methyl group to the fifth carbon of cytosine [22]. Mammals have five main DNA methyltransferase enzymes: DNMT1, DNMT2, DNMT3A, DNMT3B, and DNMT3L. DNMT1, 3A, and 3B are involved in the de novo DNA methylation process, whereas DNMT 3L regulates the function of DNMT3A and 3B. The role of DNMT2 in DNA methylation needs to be clarified [22].

Dysregulation of DNMT activity leads to abnormal DNA methylation patterns, which include hypermethylation of tumor suppressor genes and global hypomethylation, and these are standard features in various cancers. For example, dysregulation of DNMT1 leads to hypermethylation of the CpG island of the promoter region of tumor suppressor genes (TSGs), which inactivates TSG and causes cancer. Similarly, DNMT 3A and 3B regulate the methylation process in various cancers [23, 24]. As described, DNMT inhibitors cause hypermethylation, which reactivates aberrantly repressed genes by interfering with the regular process of DNA methylation. They are listed in Table 5.1.

5.4.2 HDAC Inhibitors

Histone deacetylation is one of the main epigenetic control mechanisms and is mediated by HDACs. Histone acetylation is mediated by the interaction of histone acetyltransferases (HATs) and HDACs. HDACs are categorized into four classes based on sequence homology and cellular location. Class I, HDAC1, HDAC2, HDAC3, and HDAC8, are primarily found in the nucleus and are associated with histone deacetylation [25]. Histones and non-histone proteins are affected by class II, HDAC4, HDAC5, HDAC6, HDAC7, HDAC9, and HDAC10, which move back and forth between the cytoplasm and the nucleus. Class III HDACs, sometimes called sirtuins (SIRT1–7), are involved in several cellular functions, such as metabolism and aging; they need NAD+ as a cofactor. HDAC11, the sole member of class IV, shares a structure with classes I and II HDACs [26].

HDACIs remove acetyl groups from the lysine residues located at the NH_2 terminal tails of the core histones. This action leads to a more closed chromatin structure, resulting in the repression of gene expression [27]. Vorinostat, romidepsin, and belinostat are among many HDACIs that show anticancer effects through epigenetic regulation. Vorinostat leads to hyperacetylation of histones and non-histone proteins by inhibiting class I and II HDACs, although not class III; the alteration in the chromatin structure facilitates an open and transcriptionally active state, which increases the expression of genes linked to apoptosis and cell cycle arrest [28]. Similarly, romidepsin, a class I HDACI, influences gene expression for cell survival and proliferation by hyperacetylating the histones and is effectively used in T-cell lymphoma [29,30]. Belinostat, as a pan-HDAC inhibitor, enhances histone acetylation globally, causing the accumulation of acetyl groups of histones and promoting apoptosis and cell death [31]. Hence, HDACIs suppress tumor cell growth and survival. The approach of enhancing histone hyperacetylation highlights the potential of HDACIs as precise treatments for diverse cancers. Current research persists in revealing the precise genomic and epigenomic alterations guided by these inhibitors to utilize in the current cancer treatment approach.

5.4.3 Histone Methyltransferase Inhibitors

Histone methyltransferases (HMTs) are the enzymes that catalyze the transfer of methyl groups to particular arginine or lysine residues in histone protein, using S-adenosyl methionine (SAM) as a methyl donor [32]. This covalent modification of the histone occurs at the specific regions and results in mono-, di-, or trimethylation. These methylation patterns influence gene expression in various ways [33].

For example, the well-known HMT enhancer of zest homolog 2 (EZH2) catalyzes the trimethylation of histone H3 lysine 27 (H3K27me3), a modification linked to gene silencing [34]. Histone H3 lysine 79 (H3K79) is the target of other HMTs including DOT1L, which helps to activate genes [35]. In cancer, dysregulation of histone methylation is a common theme. The delicately balanced epigenetic landscape is upset by aberrant histone methylation patterns, which can activate or repress genes essential for cell cycle regulation, DNA repair, cellular metabolism, cellular migration, metastasis, and apoptosis. Such dysregulation plays a role in unchecked growth, resistance to cell death, and genomic instability, which are cancer cell hallmarks [36].

HMT inhibitors (HMTIs) have been developed since histone methylation was identified as a critical factor in the advancement of cancer. These tiny compounds provide a therapeutic approach to restoring normal epigenetic regulation in cancer cells by targeting HMTs. A list of HMTIs is given in Table 5.1.

5.4.3.1 Epigenetic Drugs in Autoimmune Diseases

The discovery of epigenetic drugs as potential therapeutic agents against autoimmune disorders has significantly changed the treatment options available. Immune cell dysregulation in autoimmune conditions are a result of epigenetic modification by multiple processes such as histone acetylation, methylation, DNA methylation, microRNA modifications, etc [37]. These modifications are targeted by HDACIs and DNMTIs. These compounds target the alterations in chromatin and DNA caused by deregulated DNMT and HDAC, affecting gene expression [38]. Preclinical and clinical data and molecular mechanism analyses provide significant new insights into the potential of these compounds as innovative therapeutic agents for treating autoimmune disorders. As more studies are conducted on the intricate relationships between immune system function and epigenetic regulation, immunotherapy has a promising future because these drugs can target specific autoimmune diseases.

5.5 HDACIS IN THE TREATMENT OF AUTOIMMUNE DISEASE

HDACIs have shown promise in treating autoimmune diseases as a novel approach to modulate aberrant immune responses [39]. Inhibitors such as vorinostat, romidepsin, and belinostat, as discussed, prevent histones from losing their acetyl groups, encouraging an open chromatin structure [40]. By affecting the function and differentiation of immune cells, HDACIs have immunomodulatory effects in autoimmune diseases like RA and systemic lupus erythematosus (SLE). For example, vorinostat targets T-cells, affecting the ratio of effector to regulatory T-cells (Tregs) [41]. Romidepsin impacts T-cell activity and cytokine generation, whereas belinostat might impact T-cell-mediated immune responses. HDAC inhibitors can change histone acetylation patterns to control gene expression in immune cells, restoring immunological homeostasis and suppressing pathogenic reactions [42]. They seek to minimize the widespread immunosuppression linked to conventional treatments while addressing the underlying epigenetic dysregulation in autoimmune diseases.

5.5.1 DNMT Inhibitors in Autoimmune Diseases

The investigation of therapeutic approaches for autoimmune disorders has shown promise in the area of DNMT inhibitors [43]. DNMT, in particular, is one epigenetic modification that significantly influences immune responses and self-tolerance. The dysregulation of essential immune-related genes has been linked to aberrant DNA methylation patterns, contributing to the pathophysiology of autoimmune diseases [44].

Understanding the molecular mechanisms underlying autoimmune disorders requires understanding the location-specific nature of DNA methylation alterations. Hypomethylation frequently happens at particular CpG sites in the promoters of immune regulation-related genes such as CD70 involving T-cell regulation and in IFN-γ involving cytokine signaling [45]. Hypermethylation frequently affects CpG sites within regulatory regions of genes essential to immune regulation, such as promoter regions of genes like TNF-α in RA [46]. DNA methylation alterations at particular CpG sites within genes is also linked to immune cell function (HLA genes), myelin production (PLP1), and neuroinflammation (CD40) are observed in multiple sclerosis (MS). For example, genes linked to immune regulation and B-cell function have their promoters altered by methylation patterns at CpG sites in type 1 diabetes [47]. Changes in methylation occur at CpG sites in genes related to intestinal inflammation (IL23R) and immune response (IL6) in IBDs, which includes Crohn's disease and ulcerative colitis [48]. The accurate identification of methylation sites highlights the complex regulatory processes and offers information about possible targets for treating these complicated autoimmune diseases. Researchers have looked to DNA methylation inhibitors as possible agents to modulate these epigenetic modifications to address this.

5.5.2 MicroRNA-Based Therapies

With its unique method of regulating immune responses and gene expression, miRNA-based epigenetic therapy has become a viable treatment option for autoimmune diseases. Small noncoding RNA molecules called miRNAs bind to target messenger RNAs and affect their translation and stability, which is how they play a critical role in post-transcriptional gene regulation [49]. Dysregulated miRNA expression is frequently linked to the aberrant immune responses specific to autoimmune disorders [50]. By either blocking overexpressed miRNAs linked to pro-inflammatory pathways or enhancing lacking miRNAs involved in immune regulation, miRNA-based epigenetic therapy aims to bring equilibrium back [51].

Preclinical research on various autoimmune disorders has shown promising results for using synthetic miRNA mimics or inhibitors as therapeutic agents [52]. For example, to reduce the production of pro-inflammatory cytokines, miR-146a suppresses the NF-κB signaling pathway and modifies Toll-like receptor (TLR) responses in RA and SLE. MiR-155 affects T-cell differentiation, survival pathways like PI3K-Akt, and regulatory elements like SOCS1 and SHIP1, and it is linked to RA, MS, and psoriasis [53]. MiR-21 impacts NF-κB signaling in the tissue repair processes in IBD and psoriasis [54]. MiR-326 in MS regulates Th17 cell differentiation and the Ets-1 pathway, while miR-192 in IBD influences Wnt and TGF-β signaling pathways [55]. As critical nodes in the immune system, these miRNAs are targeted by therapeutic interventions to modulate immune balance, suppress pro-inflammatory signals, and support regulatory processes disturbed in autoimmune diseases. Finding new miRNA-pathway associations as research develops could lead to more accurate and successful treatment plans customized to the complexity of autoimmune diseases.

5.5.3 Immune Cell-Specific Targeting

To precisely modulate immune responses, epigenetic drugs that target immune cell-specific pathways have emerged as promising therapeutic strategies for treating autoimmune disorders. Recognizing the heterogeneity and diverse functions of different cell populations involved in autoimmune pathogenesis has led to the introduction of immune cell-specific targeting [56]. Immune cell-specific targeting drugs work by explicitly altering the epigenetic makeup of immune cells to rectify immunological imbalances and prevent abnormal reactions [57].

DNMT and HDAC inhibitors specifically target immune cell populations in autoimmune diseases. The DNMTIs azacitidine and decitabine demonstrated promise in regulating immune dysregulation in SLE by altering DNA methylation patterns in T-cells, B-cells, and dendritic cells [58]. Similarly, in RA, the HDACIs romidepsin and vorinostat affect T-cell and macrophage function to reduce inflammatory responses [59]. Other compounds with immune cell-targeting effects are listed in Table 5.1. Immune cell-specific epigenetic targeting offers a sophisticated strategy for more precise and efficacious treatments for autoimmune diseases.

5.6 TARGETING EPIGENETIC MODIFICATIONS FOR DRUG DISCOVERY IN MULTIRESISTANT CANCERS

Multidrug resistance (MDR) has been a major barrier in the efficacy of various treatment modalities in many cancers. Epigenetic modifications like DNA methylation, modification of histone, ncRNAs, and chromatin remodulation are the major contributing reasons behind the development of MDR [78]. Along with these, genomic imprinting and nucleosome localization have been associated with the epigenetic modifications that lead to MDR [79]. Wajapaeyee et al. stated that epigenetic factors can reinforce each other, contributing to resistance. For instance, histone modifications and DNA methylation are linked to chromatin modulation and ncRNA [80]. Moreover, abnormal epigenetic changes in genomic DNA and histone proteins can disrupt signalling pathways, activating oncogenes and suppressing TSGs, causing MDR. Thus, targeting the epigenetic modifications can provide a solid opportunity for overcoming the MDR in many cancers.

5.6.1 DNA Methylation for Overcoming Drug Resistance

5.6.1.1 DNMTs

As previously discussed, DNA methylation involves the addition of a methyl group to the 5-carbon of CpG dinucleotide through DNMTs. Studies have reported that DNA methylation contributes to TSG suppression, like retinoblastoma tumour suppressor (RB1) and BRCA1 in multiple cancers due to hypermethylation [81].

Three DNMT enzymes possess the ability to catalyse methyltransferases [82]. DNMT1, a maintenance methyltransferase, transfers the methylation patterns of parental DNA to newly synthesized DNA by preferentially methylating hemi-methylated DNA, whereas DNMT3a and DNMT3b carry out de novo methylation during embryogenesis [83]. Although DNMT3L belongs to DNMT3 family, but it lacks methyl transferase activity and can increase binding capacity of DNMT3A and DNMT3B to the methyl donor SAM, enhancing their functionality [84].

Research findings indicate a strong correlation between tumour resistance and DNMT. Cancer stem cells (CSCs) play a pivotal role in cancer resistance, and under the control of DNMT1 brain-expressed X-linked protein 1 can stimulate Wnt/β-catenin signalling to preserve self-renewal capability of liver stem cells [84,85]. The slow cycling and tumour formation of glioma stem cells (GSCs) are inhibited by nuclear localization of DNMT1 in turn promoting chemoresistance of GSCs; the inhibition of the CD133 (CSC marker)–DNMT1 interaction can cause DNA damage that can suppress GSC self-renewal and tumorigenesis [86]. Targeting DNMTs can be an effective approach for radio- and chemo-sensitization because DNMT3A and DNMT3B are overexpressed in tamoxifen-resistant breast cancers and rhabdomyosarcoma [87]. In a comparable way, decitabine sensitivity and mRNA-level expression of DNMT3B are inversely connected in pancreatic cancers [88].

TABLE 5.1
Epigenetic Drugs

S.N.	Generic Name	Trade Name	Targeted Enzyme	Approved Clinical Indication	Reference
1.	Romidepsin	Istodax®	Inhibits class I and II HDACs	CTCL monotherapy	[60]
2.	Belinostat	Baleodac®	Pan-HDAC inhibitor	PTCL monotherapy	[61]
3.	Tucidinostat	Epizada®	Inhibits class I and IIb HDACs	PTCL monotherapy	[62]
4.	Panobinostat	Farydak®	Inhibits class I, IIb, and IV HDACs	Multiple myeloma AML	[63]
5.	Vorinostat	Zolinza®	Inhibits class I and II HDACs	CTCL monotherapy Triple negative breast cancer	[64]
6.	Azacitidine	Vidaza®	Inhibits DNMT1	AML Chronic myelomonocytic leukemia	[65]
7.	Decitabine	Dacogen®	Inhibits DNMT1	CMML	[66]
9.	Cladribine	Mavenclad®	Inhibits DNMT1	Brain and central nervous system cancer Lymphoma AML	[67]
10.	Fludarabine	Fludara®	Inhibits DNMT1	Head and neck cancer Myeloma Metastatic melanoma	[68,69]
12.	CPI-0209 (Tulmimetostat)		EZH2 histone methyltransferase	Solid tumor	[70]
13.	DS-3201 (Valemetostat)	Ezharmia®	EZH1 and EZH2 histone methyltransferase	AML SCLC Adult T-cell lymphoma	[71]
14.	Tazematostat	Tazveric®	EZH2 histone methyltransferase	Malignant mesothelioma Synovial sarcoma	[72]
15.	Azacytidine	Vidaza®	Inhibits DNMT	SLE, SS	[73]
16.	Procainamide	Pronestyl®	Inhibits DNMT1	SLE	[74]
17.	Trichostatin A	—	Inhibits HDAC	SLE, T1D	[75]
18.	Vorinostat	Zolinza®	Inhibits HDAC	SLE	[76]
19.	Givinostat	Gavinostat®	Inhibits class I HDAC	RA	[77]
20.	Largazole	—	Inhibits class I HDAC	RA	[37]
21.	Romidepsin	Istodax®	Inhibits HDAC 2/3	RA	[30]

5.6.1.2 DNA Demethylases

Along with DNMTs, DNA demethylase is crucial in drug resistance. Ten-eleven translocations (TETs) from the DNA demethylase family in association with DNMTs regulate DNA methylation [89]. In one study, TET1 enhanced reactive oxygen species (ROS) and epithelial-to-mesenchymal transition (EMT) by binding to the promoter DUOX2 through methylation-regulated transcription [90]. Along with this, high expression of TET1 has been linked to oxaliplatin resistance in gastric cancer, while to the contrary, TET1 silencing induces gefitinib resistance in lung cancer [91,92]. Moreover, reduced expression of TET2 caused Erα suppression, leading to the development of MDR in breast cancer [93].

Multiple DNA methylations that inhibit drugs have shown potency in resensitizing the resistance in multiple cancers; some of them have gone for phase trials (Table 5.2). Azacitidine, a DNMT inhibitor, has been under phase I trial for the reversion of resistance in oesophageal cancer (clinical trial no. NCT01386346) and head and neck cancer (no. NCT02178072); in 2018, it entered phase III trial testing its safety and efficacy in advanced leukaemia (no. NCT00887068). Another DNMT1 inhibitor drug, decitabine, is under phase III trial for advanced ovarian cancer (no. NCT02159820) and phase IV trial in relapse and refractory diffuse large B-cell lymphoma (no. NCT03579082).

5.7 TARGETING HISTONE MODIFICATIONS FOR OVERCOMING DRUG RESISTANCE

Histone modification is another important characteristic of resistance that includes histone deacetylases, histone acetyltransferases, histone demethylases, and histone methyltransferases.

5.7.1 Histone Deacetylases

Studies have revealed that many lysine remains at the end of histones can undergo acetylation and deacetylation, which can alter chromatin organization and gene transcription, promoting tumour resistance [94]. Histone acetylation is regulated by two opposing catalytic complexes: HDACs and HATs [95]. HDACs mainly eliminate the acetyl groups from histone and non-histone proteins with lysine residues; studies have reported the vital role of HDACs in cancer progression and resistance. HDAC6 has been the key player in regulating resistance in multiple cancers [96]. Additionally, deacetylation of specific protein1 (Sp1) by HDAC2 and HDAC6 induces temozolomide-resistant glioblastoma [97]. Thus, targeting HDACs can suppress the progression of cancer and mainly drug resistance by inhibiting their expression. By now, four HDACs inhibitors have been approved by the US Food and Drug Administration (FDA) for clinical use, which are panobinostat, belinostat, vorinostat, and romidepsin [98].

5.7.2 Histone Acetyltransferases

HATs work opposite to HDACs by addition of acetyl group to histone and non-histone proteins with lysine remains. HATs have been classified into a number of protein acetylation enzymes like MYST, CREB-binding protein (CBP), GCN5-related N-acetyltransferases family (GNAT) and Hat1 [99]. From the MYST family, hMOF shows atypical expression in multiple cancers: MDM2 expression regulated by hMOF induced cisplatin resistance in ovarian cancer in one study [100]. Another enzyme, TIP60, acetylates a number of histone and non-histone proteins, leading to resistance [101]. Similarly, phosphorylation induced by CREB1 uplifts CBP-p300 complex role leading to resistance in prostate cancer [102], and high levels of GNAT elevate A1B1 expression, promoting cancer progression and resistance [103].

Another acetylation protein, HAT1, is associated with the progression of multiple cancers [104]. High levels of HAT1 induce resistance against enzalutamide in castration-resistant prostate cancer [105]. Thus, targeting HATs by developing nano inhibitors can be a promising tool for treating cancer and overcoming resistance in multiple cancers. In recent years, many HAT inhibitors have been in clinical and preclinical trials for reverting of the drug resistance (Table 5.2). EP31670, an inhibitor of CBP-p300, entered phase I trial for safety and efficacy in metastatic solid tumours (no. NCT05488548). In another phase I/II trial, CCS1477, an inhibitor of CBP-p300, was used in a regimen for reverting multiple metastatic cancers (nos. NCT03568656 and NCT04068597).

5.7.3 Histone Demethylases

Histone demethylases (HDMs) have been popularly known for their potent role in tumour progression. Multiple studies have revealed that HDMs are associated with MDR by inducing autophagy, metabolic reprogramming, and EMT [106]. HDMs are classified into three groups: Jumonji C (JmjC)-domain-containing demethylase (JMJD), lysine-specific demethylase (LSD), and histone arginine demethylases [88]. Targeting several upstream and downstream genes of the JMJD family, the LSD family, and proteins of histone arginine demethylases would provide better therapeutic opportunities in developing epigenetic modification drugs for treatment of advanced MDR cancers.

5.7.4 Histone Methyltransferases

HMTs are histone-modifying enzymes that catalyse the addition of methyl groups to lysine and arginine remains of histone proteins. Methyl groups catalysed to lysine residues are known as histone lysine methyltransferases (HKMTs), while the catalysis of methyl groups onto arginine residues is carried out by histone arginine methyltransferases [107]. HMTs have been closely linked with drug resistance in cancer. Tumourogenesis caused by EZH2 due to HEK27 methylation can lead to overexpression of EZH2, leading to cisplatin-resistant ovarian cancer [108]. Additionally, SETDB1 targeting H3K9 for methylation interacts with PELP1 to activate the Akt pathway, leading to tamoxifen resistance in breast cancer [109]. HMTs hold greater promise for inhibitory drugs for cancer therapy in reverting the resistance back to drug sensitivity.

5.8 NONCODING RNA FOR OVERCOMING DRUG RESISTANCE

NcRNAs play a major role in gene expression, acting as either an oncogene or a TSG [110]. A number of ncRNAs have been used as biomarkers for detecting drug resistance including miR-103a-3p, miR-214, miR-221, and miR-155-5p [111]. Studies have shown that ncRNAs have a deep association with EMT for development of resistance; for instance, miR-1280 inhibits EMT and CSCs in colorectal cancer by disrupting the Notch pathway [112]. The long ncRNAs from the family of large ncRNAs interact with DNA and motif to exert oncogenic or TSG effects in cancer. The long ncRNA-MEG3 inhibits the tumour growth by upregulating p53 and binding it to the promoter region of growth differentiation factor 15 [113]. On the basis of characteristics of ncRNAs, directly targeting ncRNAs could be an operative strategy for treatment of metastatic cancers. Several clinical trials are underway which might give confirmed regimen in the near future.

TABLE 5.2
Clinical Trials of Epigenetic Drugs in Different Cancer Models

Sl No.	Cancer Model	Intervention	Description	Study Type	Study Phase	Sample size	Clinical Trial No.
1.	Advanced malignancy	• *Grifola Frondosa* • Azacitidine • Lenalidomide	Epigenetic immunomodulation in patients with advanced malignancy using azacitidine, lenalidomide, and *Grifola frondosa*	Int.	Phase 1	42	NCT01200004
2.	Advanced breast cancer	• Tamoxifen • Vorinostat • Pembrolizumab	Reversing therapy resistance with epigenetic-immune modification	Int.	Phase II	38	NCT02395627
3.	Advanced neuroendocrine tumours	• ASTX727 • Lutathera	Epigenetic modification of somatostatin receptor-2 to improve therapeutic outcome with Lutathera in patients with metastatic neuroendocrine tumours	Int.	Phase I	27	NCT05178693
4.	Advanced ovarian cancer	• Decitabine • Paclitaxel and carboplatin	Lower-dose decitabine-Primed TC (carboplatin-paclitaxel) regimen in ovary cancer	Int.	Phase II & III	500	NCT02159820
5.	Multiple cancers	• Durvalumab • Tazemetostat	Combining epigenetic and immune therapy to beat cancer	Int.	Phase II	173	NCT04705818
6.	Refractory solid tumours	• Hydralazine • Magnesium valproate	Study of epigenetic therapy to overcome chemotherapy resistance in refractory solid tumours	Int.	Phase II	15	NCT00404508
7.	Gastrointestinal cancer	• Domatinostat • Avelumab	Epigenetic modulation of the immune response in gastrointestinal cancers	Int.	Phase II	75	NCT03812796
8.	Gastrointestinal cancer	• Terelizumab • CapeOx • Trastuzumab • Radiotherapy	Efficacy prediction model based on epigenetic multi-omics sequencing in neoadjuvant immunotherapy of gastrointestinal tumours	Int.	Phase II	200	NCT05515796
9.	Oesophageal cancer	• Azacitidine • Oxaliplatin • Epirubicin • Capecitabine	Epigenetic priming using azacitidine with neoadjuvant chemotherapy in patients with resectable oesophageal cancer	Int.	Phase I	12	NCT01386346
10.	Advanced breast cancer	• CC-486 • Fulvestrant	Epigenetic modifying effects of CC-486 (oral azacitidine) in combination with fulvestrant	Int.	Phase II	92	NCT02374099
11.	Leukaemia	• Azacitidine	Post-transplant azacitidine to prevent acute myelogenous leukaemia and myelodysplastic syndrome relapse	Int.	Phase III	246	NCT00887068

(Continued)

TABLE 5.2 (*Continued*)
Clinical Trials of Epigenetic Drugs in Different Cancer Models

Sl No.	Cancer Model	Intervention	Description	Study Type	Study Phase	Sample size	Clinical Trial No.
12.	Head and neck cancer	• 5-Azacitadine	Demethylation therapy in patients with advanced head and neck small cell carcinoma	Int.	Phase I	18	NCT02178072
13.	Advanced prostate cancer	• Azacitidine	Using Vidaza™ to restore responsiveness of patients' prostate cancers to hormone therapy	Int.	Phase II	36	NCT00384839
14.	Leukaemia	• Busulfan • Cytarabine • Daunorubicin Hydrochloride • Decitabine • Etoposide • Filgrastim	Maintenance therapy with decitabine in leukaemia	Int.	Phase II	546	NCT00416598
15.	Advanced lymphoma	• Decitabine	A clinical trial of decitabine in relapse and refractory diffuse large B-cell lymphoma	Int.	Phase IV	60	NCT03579082
16.	Advanced thyroid cancer	• Decitabine • Iodine I 131 • Recombinant thyrotropin alfa • Fludeoxyglucose F 18	Decitabine in patients with metastatic papillary thyroid cancer	Int.	Phase II	12	NCT00085293
17.	Advanced germ cell tumour	• Guadecitabine • Cisplatin	Guadecitabine (SGI-110) plus cisplatin in relapsed refractory germ cell tumours	Int.	Phase I	14	NCT02429466
18.	Metastatic solid tumours	• EP31670	EP31670 in patients with targeted advanced solid tumours	Int.	Phase I	50	NCT05488548
19.	Haematological malignancy	• CCS1477 • Pomalidomide • Dexamethasone	CCS1477 in haematological malignancies	Int.	Phase I & II	250	NCT04068597
20.	Advanced multiple cancers	• CCS1477 • Abiraterone acetate • Enzalutamide	CCS1477 as monotherapy and combination in advanced solid metastatic tumours.	Int.	Phase I & II	350	NCT03568656

Int.: Intervention.

5.9 TARGETING EPIGENETIC MODIFICATIONS FOR DRUG DISCOVERY IN AUTOIMMUNE DISEASES

Currently, there are over 100 different autoimmune disorders recognized, and the majority of them are chronic in nature. Autoimmunity is known to be caused by a variety of agents, including immunologic, environmental, genetic, epigenetic, and hormonal factors. The need of the hour is to find a better treatment regimen because the existing therapies for autoimmune disorders are linked to low efficacy, a poor safety profile, and high cost [114].

Researchers have discovered that variations in the epigenetic states of diseased and healthy tissues can act as useful disease biomarkers. Although the category of approved drugs for epigenetic inhibitions is still inadequate, scientists are investigating novel strategies to improve the efficiency and lessen the adverse effects of these drugs, such as antibody-drug conjugates and protein degraders [115]. Methods such as CRISPR/Cas9 and multitarget assembly designing are being explored to rectify genetic mutations and offer improved anticancer benefits. Various epigenetic modifications have been explored as stable therapeutic strategies for treating autoimmune diseases. Epigenetic target includes chromatin modifications, histone acetylation and methylation enzymes, DNA methylation and demethylation enzymes. By now, a number of HDAC inhibitors and DNMT inhibitors have been targeted for epigenetic modulations in therapeutics for autoimmune diseases. There is hope for the foreseeable future of epigenetic drug discovery, as new agents may be developed further and put through clinical trials [116].

5.9.1 Chromatin Modifications

During T-cell maturation, the expression of cytokines is programmed by chromatin modifications. Thus chromatin-modifying enzymes can be a potential target for developing therapy against autoimmune disease. Reversibility is a key component of the epigenetic alterations with maintenance of equilibrium between the enzymes being the key component in all histone modifications. Therefore, using specific antagonists to inhibit modification enzymes, chromatin dynamics can be shifted in either direction. Since blocking a broad histone-modifying enzyme would only influence specific classes of genes and not have a general effect on gene expression, the requirement for targeting has significant pharmacological consequences. The environment in which the targeting proteins are found in a particular cell will determine the gene-specific effects. For instance, hyperacetylation would only occur in the genes that HATs target if HDACs were inhibited.

5.9.2 Histone Acetylation and Methylation Enzymes

HATs are a popular histone modification being explored as therapeutic targets in autoimmune disease. Studies have reported that HATs are the hallmark of active genes. HATs are a potent therapeutic target in autoimmune diseases due to their role in dysregulated gene expression in T-cell development [117]. HAT inhibitors such as benzamides, anacardic acid, and CBP-p300 inhibited 4-hydroxyquinolines have been under several clinical and preclinical trials [118]. Several classes of histone acetylase could be potentially targeted as cellular inhibitors of HAT like SET/TAF-Ibeta/INHAT complex [119].

5.9.3 DNA Methylation and Demethylation Enzymes

DNA methylation and demethylation play crucial roles in the regulation of gene expression, and their dysregulation has been implicated in autoimmune diseases. Hypermethylation leads to increased DNA methylation in the promoter regions of certain genes and can lead to gene silencing, affecting immune-related genes. This results in the suppression of regulatory T cells (Tregs) or the overexpression of pro-inflammatory genes. Similarly, hypomethylation of specific genes can lead to their overexpression, contributing to autoimmune responses. Inhibitors of DNMTs provide a wide range opportunity for targeting the discovery of epigenetic drugs in autoimmune disease [120].

5.10 CHALLENGES AND OPPORTUNITIES IN EPIGENETIC DRUG DISCOVERY IN MULTIRESISTANT CANCER AND AUTOIMMUNE DISEASES

Epigenetics holds immense promise for revolutionizing medicine, especially in cancer treatment. The road to successful epigenetic drug discovery is fraught with challenges yet also brimming with exciting opportunities.

5.10.1 Challenges

Challenges associated with epigenetic drug discovery stem from the complexity of targeting complex processes rather than individual proteins, as is the case with traditional drugs. This makes it difficult to identify specific and effective targets. Additionally, epigenetic modifications are involved in many cellular processes, making it difficult to develop drugs with specific effects and avoid unwanted side effects. Limited understanding is another challenge for the field of epigenetics, which is still evolving; much remains unknown about the specific roles of different epigenetic modifications in disease.

Delivery of some epigenetic drugs to the target site within the cell can be challenging. Designing clinical trials for epigenetic drugs is also a big challenge as it requires careful consideration due to their unique mechanisms of action and effects. To overcome the challenges with epigenetic drug discovery, current research is increasingly focusing on phytotherapy, which utilizes natural plant-based compounds as therapeutic agents because of their potent pharmacological activities and reduced side effects. These phytocompounds also play a role in epigenetic changes, which helps in treating multiresistant cancer and autoimmune diseases, particularly in combination with epigenetic and chemotherapeutic drugs rather than alone.

5.10.2 Opportunities

The booming opportunities in the discovery of epigenetic drugs relate to the novel and specific epigenetic targets that are being identified, opening doors for the development of more effective drugs. Combined epigenetic drugs with other targeted therapies or immunotherapy, potent phytocompounds could lead to improved treatment outcomes. Biomarkers that predict response to epigenetic therapy can also help personalize treatment and improve patient outcomes. Advancements in gene editing and other technologies are providing new tools for studying and manipulating epigenetic modifications. Existing drugs that have shown epigenetic activity can also be used in the treatment of specific cancer and autoimmune diseases.

5.11 PHYTOCOMPOUNDS TARGETING EPIGENETIC ALTERATION IN CANCER AND AUTOIMMUNE DISEASE

Epigenetic modifications have emerged as a new area of research for cancer treatment using dietary phytocompounds in the last three decades. DNA modification, histone modifications, and ncRNA expression are the major alterations that occur in epigenetic mechanisms and are drivers of both cancer and autoimmune disorders. Phytocompounds are naturally occurring compounds found in plants that have numerous pharmaceutical effects and are used as traditional medicine as well as in research. Several phytocompounds have been shown to be directly involved in epigenetic alterations, offering promising avenues for treating these diseases. In cancer, epigenetic dysregulation plays a vital role in tumorigenesis and progression, and in autoimmune disorders, it deregulates immune response. Dietary phytocompounds either alone or combined with conventional therapies such as chemo- and radiotherapy show efficacy and reduced side effects in patients. This section explores the significance of phytocompounds used in epigenetic alterations in cancer and autoimmune disease, which may help to unlock potential of these compounds in personalized medicine.

5.11.1 Epigenetic Changes in Cancer and Autoimmune Disease

Epigenetic modifications are vital in shaping the emergence and advancement of cancer as well as autoimmune disorders. These diseases share many common epigenetic mechanisms but also display distinct features in their epigenetic scenarios. Both cancer and autoimmune diseases involve alterations in DNA methylation, histone modification, and ncRNA expressions. However, the specific modifications differ between these diseases. For example, DNA hypermethylation is more prevalent in cancer, while DNA hypomethylation is more common in autoimmune diseases. In cancer, promoter regions of tumor suppressor genes often become hypermethylated, leading to their silencing and uncontrolled cell division. Cancer cells often exhibit aberrant histone modifications such as increased H3K27me3 and decreased H3K9 acetylation, promoting tumorigenesis [121]. Additionally, some miRNAs are often downregulated in cancer, leading to oncogene expression, and ncRNAs can promote tumorigenesis by various mechanisms [122]. In autoimmune diseases, hypomethylation of repetitive elements and immune-related genes can modulate inflammatory pathways and aberrant histone modifications in immune cells can lead to TH17 cell differentiation and proinflammatory cytokine production [14]. Some miRNAs and long ncRNAs regulate immune cell function, and their dysregulation can contribute to autoimmune disease development [123].

5.11.2 Phytocompounds as Epigenetic Modulators

As previously discussed, phytocompounds are bioactive compounds found in plants that have potential to modulate gene expression through epigenetic mechanisms; that is, they alter gene expression without changing the underlying DNA sequence. Several studies have shown that phytocompounds can target epigenetic mechanisms and exert beneficial effects on treating cancer and autoimmune disease [124,125]. For instance, curcumin, resveratrol, and quercetin have shown involvement in epigenetic modifications targeting multiple pathways in various cancers and autoimmune diseases [126].

5.11.3 Epigenetic Targets of Phytocompounds in Cancer and Autoimmune Diseases

DNMTs, HMTs, HDMs, HATs, and histone deacetylases are the most common epigenetic modifications occurs in cancer and autoimmune diseases [14,127]. For example, curcumin inhibits DNA methylation and increases the expression of tumour genes [128]. It also inhibits HDACs, leading to anti-inflammatory effects in RA [129,130]. Similarly, resveratrol found in red wine and grapes inhibits HDACs, which increases histone acetylation and activates genes involved in apoptosis and cell cycle arrest in cancer [131]. It also inhibits SIRT1, promoting anti-inflammatory effects and reducing T-cell activation in rat models of autoimmune diseases [132]. In addition, quercetin regulates the expressions of various miRNAs and can inhibit the expression of miRNAs that promote cancer proliferation and metastasis [133,134]

Quercetin also regulates miRNA expression involved in T-cell function and inflammatory pathways, possibly offering therapeutic potential in autoimmune diseases like psoriasis and RA [135]. Epigallocatechin-3-gallate, found in green tea, inhibits DNMTs, leading to DNA hypomethylation. It also modulates the activities of HATs and HDACs, contributing to changes in histone acetylation patterns and affecting gene expression in cancer, lupus, IBD, and RA [136].

5.11.4 Therapeutic Potential and Challenges

Despite the potential of phytocompounds in targeting epigenetic alterations for both cancer and autoimmune diseases, significant challenges remain unresolved. For example, many phytocompounds have low bioavailability, meaning that they are poorly absorbed and rapidly metabolized in the body, leading to limited therapeutic efficacy. These compounds also have complex pharmacokinetics, so understanding the absorption, distribution, metabolism and excretion is crucial for optimizing their therapeutic potential.

Furthermore, as epigenetic pathways are complex and involve multiple enzymes, targeting these pathways with single compounds may not be sufficient to achieve therapeutic efficacy, and many phytocompounds have multiple biological activities, which can lead to unwanted side effects; more research is needed to identify and develop phytocompounds with greater specificity for desired epigenetic targets. In addition, as these compounds have limited delivery potential to target tissues, developing nanovesicles to carry these compounds to targeted tissue would be helpful. Most phytocompounds lack or have limited clinical data, so there must be large-scale clinical trials to validate their efficacy and safety. The development and clinical trial costs can be high, potentially hindering commercialization and broad accessibility. Despite these challenges, research on the therapeutic potential of phytocompounds continues to increase rapidly. Advances in delivery systems, target identification, and clinical trial design have the potential to overcome these hurdles and pave the way for the development of safe and effective epigenetic therapies for cancer and autoimmune diseases.

5.11.5 Preclinical and Clinical Evidence

Several phytocompounds like curcumin and resveratrol show antiproliferative activities by inhibiting cell growth and promoting apoptosis in cancer cells. Ongoing clinical trials investigate the efficacy of curcumin in treating various cancers like breast, colorectal, and pancreatic [137]. Epigallocatechin gallate (EGCG) and lycopene improved the survival rate of patients with head and neck squamous cell carcinoma in clinical trials [138], and resveratrol has shown the potential to prevent and treat various cancers, although more research must be implemented to confirm the effectiveness of these trials [139].

Similarly, gingerol and boswellic acid can modulate the immune system by reducing inflammation and promoting immune tolerance in autoimmune diseases [140]. Ginger has been shown to be efficacious in reducing pain and inflammation in patients with RA in clinical settings [141]. Some trials have also demonstrated the efficacy of boswellic acid in treating ulcerative colitis and Crohn's disease [142]. Vitamin C and beta carotene, which have anti-oxidant properties, have been shown to protect cells from damage caused by free radicals, which are implicated in the development of autoimmune diseases. High doses of vitamin C improved symptoms in some patients with lupus and RA [143]. Apart from these compounds discussed in preclinical and clinical trials, more in the research community should be encouraged to identify their effectiveness, to advance the knowledge about phytocompounds which can used as drugs (Tables 5.3 and 5.4).

5.12 COMBINAING THERAPIES TO SENSITIZE MULTIRESISTANT CANCER AND AUTOIMMUNE DISEASES

Combining epigenetic drugs, phytocompounds, and traditional chemotherapy or radiotherapy for the treatment of multiresistant cancer and autoimmune disease is an area of ongoing research, and it holds potential for improving therapeutic outcomes. This multitherapy approach offers several advantages over traditional therapies such as overcoming resistance, enhancing efficacy, reducing side effects, and immunomodulation.

TABLE 5.3
Phytocompounds in Cancer Studies

Compound Name	Source	Therapeutic Potential	Cancer Type	Epigenetic Targets	Clinical Evidence	Ref.
Curcumin	Turmeric	Anti-inflammatory, antioxidant	Various cancers	HDAC, DNMT	Preclinical studies, phase I/II trials	[144]
Resveratrol	Grapes, red wine	Antioxidant, anti-inflammatory	Hepatic, pancreatic, breast, prostate, and colorectal cancer	SIRT1, DNMT	Preclinical studies, animal models	[139]
Quercetin	Apples, onions, citrus fruits	Anticancer, anti-inflammatory	Breast, lung, and nasopharyngeal cancer	HDAC, DNMT	In vitro studies, animal models	[145]
Sulforaphane	Broccoli, cruciferous vegetables	Anticancer, antioxidant	Breast, prostate, and lung cancer, melanoma	HDAC, DNMT	Phase I/II trials	[146]
EGCG	Green tea	Anticancer, antioxidant	Colon, kidney, breast, and brain cancer, leukemia	DNMT, HDAC	Preclinical studies, phase I trials	[147]
Genistein	Soy products	Anticancer, anti-angiogenic	Breast and prostate cancer	HDAC, DNMT	In vitro studies, animal models	[148]
Lycopene	Tomatoes, watermelon	Antioxidant, anti-angiogenic	Prostate and breast cancer	GSTP1	Observational trials	[149]
Gingerol	Ginger	Anti-inflammatory, antioxidant	Various cancers	HDAC, DNMT	Preclinical studies, phase I trials	[150, 151]
Betulinic Acid	Birch bark, berries	Apoptosis induction, antiangiogenic	Melanoma	ER-alpha, PIWIL1	Preclinical studies, phase I trials	[152]
Berberine	Goldenseal, barberry	Anticancer, anti-inflammatory	Liver and colon cancers	HDAC, DNMT	Preclinical studies, phase I trials	[153]
Allicin	Garlic	Anticancer, anti-inflammatory	Various cancers	HDAC, DNMT	In vitro studies, animal models	[154]
Pterostilbene	Blueberries, grapes	Antioxidant, anti-inflammatory	Various cancers	SIRT1	Preclinical studies, phase I trials	[155]

(Continued)

TABLE 5.3 (*Continued*)
Phytocompounds in Cancer Studies

Compound Name	Source	Therapeutic Potential	Cancer Type	Epigenetic Targets	Clinical Evidence	Ref.
Hesperidin	Citrus fruits	Antiangiogenic, anti-inflammatory	Breast and colorectal cancer	DNMTs	Preclinical, clinical	[153]
Theaflavins	Black tea	Antioxidant, antiangiogenic	Colon and prostate cancer	DNMTs	Preclinical,clinical trials	[13]
Carnosol	Rosemary	Antioxidant, anti-inflammatory	Breast and prostate cancer	HATs, HDACs	Preclinical studies, phase I trials	[156]
Betanin	Beets	Anticancer, antiangiogenic	Ovary and prostate cancer	HDACs, DNMTs	Preclinical and clinical studies	[157]
Nimbolide	Neem (*Azadirachta indica*)	Apoptosis induction, antiangiogenic	Skin, liver, lung, and esophageal cancer	DNMTs	Preclinical studies, phase I/II trials	[158]
Plumbagin	*Plumbago indica*	Apoptosis induction, antiangiogenic	Breast, pancreatic, prostate, and colon cancer	HATs, HDACs	Preclinical studies, phase I/II trials	[159]
Camptothecin	*Camptotheca acuminata*	Apoptosis induction, antiangiogenic	Colon cancer	HMGB1, APEX1, and POLE3,	Preclinical studies, phase I/II trials	[160]
Artemisinin	Sweet wormwood (*Artemisia annua*)	Antiangiogenic, anticancer	Various cancers	DNMT1	Phase I/II trials	[161]

TABLE 5.4
Phytocompounds in Autoimmune Disease

Compound Name	Source	Therapeutic Potential	Autoimmune Disease	Epigenetic Targets	Clinical Evidence	Ref.
Curcumin	Turmeric	Anti-inflammatory, immunomodulatory	Rheumatoid arthritis	HDAC, DNMT	Preclinical studies, clinical trials	[162]
Resveratrol	Grapes, red wine	Anti-inflammatory, antioxidant	Multiple sclerosis	SIRT1, DNMT	Preclinical studies, clinical trials	[163]
Quercetin	Fruits, vegetables	Immunomodulatory, anti-inflammatory	Lupus	HDAC, DNMT	In vitro studies, animal models	[164]
Polyphenols	Green tea	Immunomodulatory, antioxidant	Inflammatory bowel disease	DNMT	Preclinical studies, clinical trials	[164]
Berberine	Goldenseal, barberry	Immunomodulatory, anti-inflammatory	Inflammatory bowel disease, myasthenia gravis	HDAC	Preclinical studies, clinical trials	[165]
Artemisinin	Sweet wormwood (*Artemisia annua*)	Immunomodulatory, anti-inflammatory	Rheumatoid arthritis	HAT	Preclinical studies, clinical trials	[166]
Ashwagandha Extract	*Withania somnifera*	Immunomodulatory, anti-inflammatory	Psoriasis	DNMT, HMT	Preclinical studies, clinical trials	[167]
Sophoridine	*Sophora flavescens*	Immunomodulatory, anti-inflammatory	Lupus	HMT, HAT	Preclinical studies, clinical trials	[168]
Celastrol	Thunder god vine (*Tripterygium*)	Immunomodulatory, anti-inflammatory	Rheumatoid arthritis	HDAC, DNMT	Preclinical studies, clinical trials	[169]
Rutin	Buckwheat, citrus fruits	Immunomodulatory, antioxidant	Psoriasis		Preclinical studies, clinical trials	[170]
Cannabidiol	*Cannabis sativa*	Anti-inflammatory, immunomodulatory	Multiple sclerosis	HMT	Preclinical studies, clinical trials	[171]
Andrographolide	*Andrographis paniculata*	Anti-inflammatory, immunomodulatory	Rheumatoid arthritis		Preclinical studies, clinical trials	[172]
Salidroside	*Rhodiola rosea*	Immunomodulatory, anti-inflammatory	Autoimmune encephalomyelitis.	SIRT1	Preclinical studies, clinical trials	[173]

For example, epigenetic drugs can resensitize resistance cancer cells and autoimmune cells to conventional therapies by modifying chromatin structure and gene expressions. A study combined the HDAC inhibitor vorinostat with curcumin and 5-fluorouracil (5-FU) chemotherapy in a colorectal cancer cell line. The combination treatment significantly reduced cell viability and induced apoptosis, in contrast with 5-FU alone [174]. Similarly, resveratrol was combined with the HDACI entinostat and cisplatin chemotherapy in NSCLC cells. The combination treatment displayed increased antitumor activity and reduced tumor growth versus cisplatin alone [175]. A clinical trial combined panobinostat with curcumin and docetaxel chemotherapy in patients with advanced breast cancer. The combination therapy demonstrated improved overall survival and progression-free survival compared to docetaxel alone [176]. A study combined 5-azacytidine with curcumin in a rat model of RA significantly reduced inflammation and joint damage, in contrast with either treatment alone [177]. Similarly, vorinostat was combined with the phytocompound quercetin and mesalamine (5-aminosalicylic acid) in a mouse model of IBD. The combination therapy significantly reduced inflammation and improved disease activity compared to mesalamine alone [178]. A clinical trial combined decitabine with the phytocompound resveratrol helped with relapsing-remitting multiple sclerosis. The combination therapy demonstrated a good safety profile and showed potential for reducing disease activity and improving clinical outcomes [179].

Phytocompounds, with their varied mechanisms of actions, can enhance this effect by targeting multiple signaling pathways involved in resistance and autoimmune disease. Combination therapy also often shows synergistic effects, leading to improved efficacy compared with single treatment, allowing lower doses of each drug and reducing potential side effects and toxicity. As phytocompounds are generally considered safer than conventional therapies with little to no side effects, their inclusion in combinatorial treatment can help to minimize toxicity and burden on patients. In addition, certain phytocompounds exhibit immunomodulatory properties, potentially boosting the immune system's ability to sensitize resistant cancer cells and autoimmune diseases. These procedures can further enhance the effectiveness of combined therapy.

5.12.1 Challenges and Considerations

There are multiple potential challenges with using combination therapy; for example, optimizing dosage and schedule for each patient is a hectic process, and it is necessary to identify the most effective combinations based on specific stages of cancer and resistance and in autoimmune disease. Furthermore, addressing side effects and interactions between different agents and conducting further research in clinical trials to validate the effectiveness of this approach is essential. The most important point will be that the individual responses to these treatments can vary, demanding a personal approach based on the patient's genetic and epigenetic profile.

5.13 CONCLUSION AND FUTURE DIRECTIONS

Epigenetic alterations are increasingly recognized as key drivers in the development and progression of multiresistant cancer and autoimmune diseases. As a result, the field has garnered significant attention, highlighting the need for novel drugs targeting epigenetic alterations. While a lot of challenges are there in this field like the process of epigenetics mechanism and its target specificity is complex. Drugs that overcome drug resistance triggered by epigenetic alterations target the molecular mechanisms that drive the changes and improve patient outcomes. Epigenetic drugs can target distinct pathways in the immune system and provide a more accurate and potentially less toxic approach in the treatment of autoimmune diseases.

Identifying the epigenetic profiles of individual patients can pave the way for personalized treatment modalities, optimizing the efficacy and reducing side effects. As research is advancing, we can hope for more effective epigenetic therapies going to clinical trials, offering hope to millions tolerating the pain of multiresistant cancers and autoimmune diseases. Naturally occurring

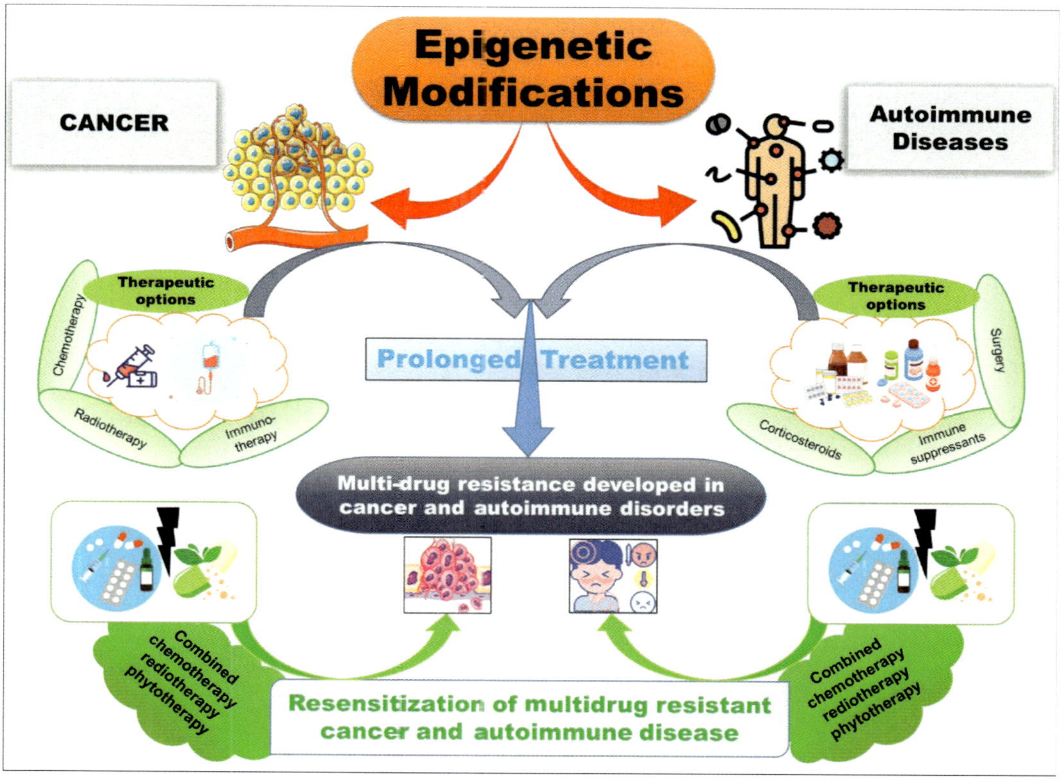

FIGURE 5.1 The contributions of epigenetics in drug discovery and development.

phytocompounds are also proving effective at targeting epigenetic alterations in these diseases and also overcoming therapy resistance, particularly when they are combined with epigenetic drugs and chemotherapeutic agents. In conclusion, epigenetic drug discovery holds immense promise for revolutionizing the treatment of these challenging and complex diseases. By addressing the remaining challenges and capitalizing on the exciting opportunities, we can move closer to a future where epigenetic therapies become a mainstay in the fight against multiresistant cancers and autoimmune diseases (Figure 5.1).

5.14 ACKNOWLEDGMENT

We would like to acknowledge the National Institute of Technology Rourkela, Odisha, India, for providing a facility to write this book chapter, and we also thank the editors and reviewers for their helpful suggestions on this manuscript.

REFERENCES

[1] Hausman, D. M., What is cancer?" *Perspectives in Biology Medicine* 2019, 62 (4), 778–784.
[2] Yasunaga, M., "Antibody therapeutics and immunoregulation in cancer and autoimmune disease." *Seminars in Cancer Biology* 2020, 64, 1–12.
[3] Giat, E.; Ehrenfeld, M.; Shoenfeld, Y., "Cancer and autoimmune diseases." *Autoimmunity Reviews* 2017, 16 (10), 1049–1057.

[4] Koustas, E.; Trifylli, E.-M.; Sarantis, P., et al., "Immunotherapy as a therapeutic strategy for gastrointestinal cancer—current treatment options and future perspectives." *International Journal of Molecular Sciences* 2022, 23 (12), 6664.

[5] Sadeqi Nezhad, M.; Seifalian, A.; Bagheri, N., et al., "Chimeric antigen receptor based therapy as a potential approach in autoimmune diseases: How close are we to the treatment?" *Frontiers in Immunology* 2020, 11, 603237.

[6] Garcia-Martinez, L.; Zhang, Y.; Nakata, Y.; Chan, H. L.; Morey, L., "Epigenetic mechanisms in breast cancer therapy and resistance." *Nature Communications* 2021, 12 (1), 1786.

[7] Karagianni, P.; Tzioufas, A. G., "Epigenetic perspectives on systemic autoimmune disease." *Journal of Autoimmunity* 2019, 104, 102315.

[8] Romero-Garcia, S.; Prado-Garcia, H.; Carlos-Reyes, A., "Role of DNA methylation in the resistance to therapy in solid tumors." *Frontiers in Oncology* 2020, 10, 1152.

[9] Gómez-Miragaya, J.; Moran, S.; Calleja-Cervantes, M. E., et al., "The altered transcriptome and DNA methylation profiles of docetaxel resistance in breast cancer PDX models." *Molecular Cancer Research* 2019, 17 (10), 2063–2076.

[10] Li, W.; Wu, H.; Sui, S., et al., "Targeting histone modifications in breast cancer: A precise weapon on the way." *Frontiers in Cell & Developmental Biology* 2021, 9, 736935.

[11] Luo, M.; Yang, X.; Chen, H.-N.; Nice, E. C.; Huang, C., "Drug resistance in colorectal cancer: An epigenetic overview." *Biochimica et Biophysica Acta-Reviews on Cancer* 2021, 1876 (2), 188623.

[12] Rong, D.; Sun, G.; Wu, F., et al., "Epigenetics: Roles and therapeutic implications of non-coding RNA modifications in human cancers." *Molecular Therapy-Nucleic Acids* 2021, 25, 67–82.

[13] Rubatto, M.; Borriello, S.; Sciamarrelli, N., et al., "Exploring the role of epigenetic alterations and non-coding RNAs in melanoma pathogenesis and therapeutic strategies." *Melanoma Research* 2023, 33 (6), 462–474.

[14] Surace, A. E. A.; Hedrich, C. M., "The role of epigenetics in autoimmune/inflammatory disease." *Frontiers in Immunology* 2019, 10, 1525.

[15] Khan, A. Q.; Ahmad, F.; Raza, S. S., et al. "Role of non-coding RNAs in the progression and resistance of cutaneous malignancies and autoimmune diseases." *Seminars in Cancer Biology* 2022, 83, 208–226.

[16] Cooles, F. A.; Tarn, J.; Lendrem, D. W., et al., "Interferon-α-mediated therapeutic resistance in early rheumatoid arthritis implicates epigenetic reprogramming." *Annals of the Rheumatic Diseases* 2022, 81 (9), 1214–1223.

[17] Wu, H.; Chang, C.; Lu, Q., "The epigenetics of lupus erythematosus." *Epigenetics in Allergy Autoimmunity* 2020, 185–207.

[18] Krishnachaitanya, S. S.; Liu, M.; Fujise, K.; Li, Q., "MicroRNAs in inflammatory bowel disease and its complications." *International Journal of Molecular Sciences* 2022, 23 (15), 8751.

[19] Pathak, A.; Tomar, S.; Pathak, S., "Epigenetics and cancer: A comprehensive review." *Asian Pacific Journal of Cancer Biology* 2023, 8 (1), 75–89.

[20] Mohammad, H. P.; Barbash, O.; Creasy, C., "Targeting epigenetic modifications in cancer therapy: Erasing the roadmap to cancer." *Journal of Nature Medicine* 2019, 25 (3), 403–418.

[21] Liu, Z.; Ren, Y.; Weng, S., et al., "A new trend in cancer treatment: The combination of epigenetics and immunotherapy." *Frontiers in Immunology* 2022, 13, 809761.

[22] Clarke, J., "DNA methylation inhibitor resets tolerance in autoimmune arthritis." *Nature Reviews Rheumatology* 2021, 17 (7), 379.

[23] Man, X.; Li, Q.; Wang, B., et al., "DNMT3A and DNMT3B in breast tumorigenesis and potential therapy." *Frontiers in Cell Developmental Biology* 2022, 10, 1020.

[24] Zhang, T.-J.; Zhang, L.-C.; Xu, Z.-J.; Zhou, J.-D., "Expression and prognosis analysis of DNMT family in acute myeloid leukemia." *Journal of Aging and Health* 2020, 12 (14), 14677.

[25] Li, J.; Li, L.; Wang, Y., et al., "Insights into the role of DNA methylation in immune cell development and autoimmune disease." *Frontiers in Cell and Developmental Biology* 2021, 9, 757318.

[26] Li, G.; Tian, Y.; Zhu, W.-G., "The roles of histone deacetylases and their inhibitors in cancer therapy." *Frontiers in Cell Developmental Biology* 2020, 8, 576946.

[27] Ramaiah, M. J.; Tangutur, A. D.; Manyam, R. R., "Epigenetic modulation and understanding of HDAC inhibitors in cancer therapy." *Journal of Life Sciences* 2021, 277, 119504.

[28] Bubna, A. K., "Vorinostat—an overview." *Indian Journal of Dermatology* 2015, 60 (4), 419.

[29] Pojani, E.; Barlocco, D., "Romidepsin (FK228), A histone deacetylase inhibitor and its analogues in cancer chemotherapy." *Current Medicinal Chemistry* 2021, 28 (7), 1290–1303.

[30] El Omari, N.; Lee, L.-H.; Bakrim, S., et al., "Molecular mechanistic pathways underlying the anticancer therapeutic efficiency of romidepsin." *Biomedicine Pharmacotherapy* 2023, 164, 114774.

[31] El Omari, N.; Bakrim, S.; Khalid, A., et al., "Anticancer clinical efficiency and stochastic mechanisms of belinostat." *Biomedicine Pharmacotherapy* 2023, 165, 115212.

[32] Maksimova, V.; Makus, J.; Popova, V., et al., "Histone methyltransferases as a new target for epigenetic action of vorinostat." *Biochemistry* 2023, 88 (7), 968–978.

[33] Xiao, C.; Fan, T.; Tian, H., et al. "H3K36 trimethylation-mediated biological functions in cancer." *Clinical Epigenetics* 2021, 13, 1–19.

[34] Das, P.; Taube, J. H., "Regulating methylation at H3K27: A trick or treat for cancer cell plasticity." *Cancers* 2020, 12 (10), 2792.

[35] Wood, K.; Tellier, M.; Murphy, S., "DOT1L and H3K79 methylation in transcription and genomic stability." *Biomolecules* 2018, 8 (1), 11.

[36] Chen, Y.; Ren, B.; Yang, J., et al., "The role of histone methylation in the development of digestive cancers: A potential direction for cancer management." *Signal Transduction Targeted Therap* 2020, 5 (1), 143.

[37] Mazzone, R.; Zwergel, C.; Artico, M., et al., "The emerging role of epigenetics in human autoimmune disorders." *Clinical Epigenetics* 2019, 11 (1), 1–15.

[38] Moufarrij, S.; Srivastava, A.; Gomez, S., et al., "Combining DNMT and HDAC6 inhibitors increases anti-tumor immune signaling and decreases tumor burden in ovarian cancer." *Scientific Reports* 2020, 10 (1), 3470.

[39] Hamminger, P.; Rica, R.; Ellmeier, W., "Histone deacetylases as targets in autoimmune and autoinflammatory diseases." *Advances in Immunology* 2020, 147, 1–59.

[40] Bondarev, A. D.; Attwood, M. M.; Jonsson, J., et al., "Recent developments of HDAC inhibitors: Emerging indications and novel molecules." *British Journal of Clinical Pharmacology* 2021, 87 (12), 4577–4597.

[41] Moran, B.; Davern, M.; Reynolds, J. V.; Donlon, N. E.; Lysaght, J., "The impact of histone deacetylase inhibitors on immune cells and implications for cancer therapy." *Cancer Letters* 2023, 216121.

[42] Gatla, H. R.; Muniraj, N.; Thevkar, P., et al., "Regulation of chemokines and cytokines by histone deacetylases and an update on histone decetylase inhibitors in human diseases." *International Journal of Molecular Sciences* 2019, 20 (5), 1110.

[43] Clarke, J., "DNA methylation inhibitor resets tolerance in autoimmune arthritis." *Nature Reviews Rheumatology* 2021, 17 (7), 379–379.

[44] Li, J.; Li, L.; Wang, Y., et al., "Insights into the role of DNA methylation in immune cell development and autoimmune disease." *Frontiers in Cell Developmental Biology* 2021, 9, 757318.

[45] Hurtado, C.; Acevedo Saenz, L. Y.; Vasquez Trespalacios, E. M., et al., "DNA methylation changes on immune cells in systemic lupus erythematosus." *Autoimmunity* 2020, 53 (3), 114–121.

[46] Pitaksalee, R.; Parmar, R.; Hotgett, R.; Emery, P.; Ponchel, F., "DNA-hypomethylation in the TNF-alpha gene predicts rheumatoid arthritis classification in patients with early inflammatory symptoms." *Cells* 2023, 12 (19), 2376. doi: 10.3390/cells12192376.

[47] Dashti, M.; Nizam, R.; Hebbar, P., et al., "Differentially methylated and expressed genes in familial type 1 diabetes." *Scientific Reports* 2022, 12 (1), 11045.

[48] El Hadad, J.; Schreiner, P.; Vavricka, S. R.; Greuter, T., "The genetics of inflammatory bowel disease." *Molecular Diagnosis Therapy* 2023, 1–9.

[49] Kmiołek, T.; Paradowska-Gorycka, A., "miRNAs as biomarkers and possible therapeutic strategies in rheumatoid arthritis." *Cells* 2022, 11 (3), 452.

[50] Domingo, S.; Solé, C.; Moliné, T.; Ferrer, B.; Cortés-Hernández, J., "MicroRNAs in several cutaneous autoimmune diseases: Psoriasis, cutaneous lupus erythematosus and atopic dermatitis." *Cells* 2020, 9 (12), 2656.

[51] Martínez-Hernández, R.; Marazuela, M., "MicroRNAs in autoimmune thyroid diseases and their role as biomarkers." *Best Practice Research Clinical Endocrinology Metabolism* 2023, 37 (2), 101741.

[52] Prinz, F.; Jonas, K.; Balihodzic, A., et al., "MicroRNA mimics can distort physiological microRNA effects on immune checkpoints by triggering an antiviral interferon response." *RNA Biology* 2022, 19 (1), 1305–1315.

[53] Pashangzadeh, S.; Motallebnezhad, M.; Vafashoar, F.; Khalvandi, A.; Mojtabavi, N., "Implications the role of miR-155 in the pathogenesis of autoimmune diseases." *Frontiers in Immunology* 2021, 12, 669382.

[54] Abdallah, F.; Henriet, E.; Suet, A., et al., "miR-21-3p/IL-22 axes are major drivers of psoriasis pathogenesis by modulating keratinocytes proliferation-survival balance and inflammatory response." *Cells* 2021, 10 (10), 2547.

[55] Basak, J.; Piotrzkowska, D.; Kucharska-Lusina, A.; Majsterek, I., "Association between the cytosine hydroxymethylation and the expression of microRNA in multiple sclerosis in Polish population." *International Journal of Molecular Sciences* 2023, 24 (18), 13923.

[56] Jung, S. M.; Kim, W.-U., "Targeted immunotherapy for autoimmune disease." *Immune Network* 2022, 22 (1).
[57] Patil, S.; Gs, V.; Sarode, G. S., et al., "Exploring the role of immunotherapeutic drugs in autoimmune diseases: A comprehensive review." *Journal of Oral Biology Craniofacial Research* 2021, 11 (2), 291–296.
[58] Bacalao, M. A.; Satterthwaite, A. B., "Recent advances in lupus B cell biology: PI3K, IFNγ, and chromatin." *Frontiers in Immunology* 2021, 11, 615673.
[59] Rosás-Umbert, M.; Ruiz-Riol, M.; Fernández, M. A., et al., "In vivo effects of romidepsin on T-cell activation, apoptosis and function in the BCN02 HIV-1 Kick&Kill clinical trial." *Frontiers in Immunology* 2020, 11, 418.
[60] Ma, H.; Cheng, B.; Falchi, L., et al., "Survival benefit in patients with peripheral T-cell lymphomas after treatments with novel therapies and clinical trials." *Hematological Oncology* 2020, 38 (1), 51–58.
[61] Stuver, R.; Epstein-Peterson, Z. D.; Johnson, W. T., et al., "Current treatment of peripheral T-cell lymphoma." *Oncology* 2022, 36 (5).
[62] Sun, Y.; Hong, J. H.; Ning, Z., et al., "Therapeutic potential of tucidinostat, a subtype-selective HDAC inhibitor, in cancer treatment." *Frontiers in Pharmacology* 2022, 13, 932914.
[63] San José-Enériz, E.; Gimenez-Camino, N.; Agirre, X.; Prosper, F., "HDAC inhibitors in acute myeloid leukemia." *Cancers* 2019, 11 (11), 1794.
[64] Palczewski, M. B.; Kuschman, H. P.; Bovee, R.; Hickok, J. R.; Thomas, D. D., "Vorinostat exhibits anticancer effects in triple-negative breast cancer cells by preventing nitric oxide-driven histone deacetylation." *Biological Chemistry* 2021, 402 (4), 501–512.
[65] DiNardo, C. D.; Jonas, B. A.; Pullarkat, V., et al., "Azacitidine and venetoclax in previously untreated acute myeloid leukemia." *New England Journal of Medicine* 2020, 383 (7), 617–629.
[66] Alexandraki, A.; Strati, K., "Decitabine treatment induces a viral mimicry response in cervical cancer cells and further sensitizes cells to chemotherapy." *International Journal of Molecular Sciences* 2022, 23 (22), 14042.
[67] Li, M.; Zhang, D., "DNA methyltransferase-1 in acute myeloid leukaemia: Beyond the maintenance of DNA methylation." *Annals of Medicine* 2022, 54 (1), 2011–2023.
[68] Behbahani, T. E.; Rosenthal, E. L.; Parker, W. B.; Sorscher, E. J., "Intratumoral generation of 2-fluoroadenine to treat solid malignancies of the head and neck." *Head Neck* 2019, 41 (6), 1979–1983.
[69] Sidana, S.; Hosoya, H.; Jensen, A., et al., "Bendamustine vs. fludarabine/cyclophosphamide lymphodepletion prior to BCMA CAR-T cell therapy in multiple myeloma." *Blood Cancer Journal* 2023, 13 (1), 158.
[70] Drescher, C.; Walter, H.; Pons-Tostivint, E., et al., PR088/# 198 EZH2/EZH1 inhibitor tulmimetostat (CPI-0209): Preliminary phase II results and first biomarker findings in patients with arid1a-mutant ovarian clear cell or endometrial carcinomas (OCCC/EC). *BMJ Specialist Journals* 2023, 33 (Suppl 4), A80.1–A80. Doi:10.1136/ijgc-2023-IGCS.124.
[71] Dou, F.; Tian, Z.; Yang, X., et al., "Valemetostat: First approval as a dual inhibitor of EZH1/2 to treat adult T-cell leukemia/lymphoma." *Drug Discoveries Therapeutics* 2022, 16 (6), 297–299.
[72] Italiano, A.; Isambert, N.; Metges, J.-P., et al., CAIRE: A basket multicenter open-label phase 2 study evaluating the EZH2 inhibitor tazemetostat in combination with durvalumab in patients with advanced solid tumors. *American Society of Clinical Oncology* 2022, 40 (16), TPS2703.
[73] Landman, S.; van der Horst, C.; van Erp, P. E., et al., "Immune responses to azacytidine in animal models of inflammatory disorders: A systematic review." *Journal of Translational Medicine* 2021, 19 (1), 1–17.
[74] Solhjoo, M.; Goyal, A.; Chauhan, K., "Drug-induced lupus erythematosus." StatPearls. 2023 [cited 2025 May 29]. Available at https://www.ncbi.nlm.nih.gov/books/NBK441889/
[75] Li, Y.; Zhao, Y.; Peng, H., et al., "Histone deacetylase inhibitor Trichostatin a reduces endothelial cell proliferation by suppressing STAT5A-related gene transcription." *Frontiers in Oncology* 2021, 11, 746266.
[76] Samuelov, L.; Bochner, R.; Magal, L., et al., "Vorinostat, a histone deacetylase inhibitor, as a potential novel treatment for psoriasis." *Experimental Dermatology* 2022, 31 (4), 567–576.
[77] Montoya, T.; Castejón, M. L.; Muñoz-García, R.; Alarcón-de-la-Lastra, C., "Epigenetic linkage of systemic lupus erythematosus and nutrition." *Nutrition Research Reviews* 2023, 36 (1), 39–59.
[78] Castro-Muñoz, L. J.; Ulloa, E. V.; Sahlgren, C., et al., "Modulating epigenetic modifications for cancer therapy." *Oncology Reports* 2023, 49 (3), 1–23.
[79] Sahafnejad, Z.; Ramazi, S.; Allahverdi, A., "An update of epigenetic drugs for the treatment of cancers and brain diseases: A comprehensive review." *Genes* 2023, 14 (4), 873.
[80] Wajapeyee, N.; Gupta, R., "Epigenetic alterations and mechanisms that drive resistance to targeted cancer therapies." *Cancer Research* 2021, 81 (22), 5589–5595.
[81] Khan, F.; Agarwal, P.; Gupta, S., et al., "BRCA1 promoter methylation & its immunohistochemical correlation in sporadic breast cancer." *Indian Journal of Medical Research* 2023, 158 (1), 47–54.

[82] Liu, P.; Yang, F.; Zhang, L., et al., "Emerging role of different DNA methyltransferases in the pathogenesis of cancer." *Frontiers in Pharmacology* 2022, 13, 958146.
[83] Effendi, W. I.; Nagano, T., "Epigenetics approaches toward precision medicine for idiopathic pulmonary fibrosis: Focus on DNA methylation." *Biomedicines* 2023, 11 (4), 1047.
[84] Yu, Y. C.-Y.; Hui, T. Z.; Kao, T.-H., et al., "Transient DNMT3L expression reinforces chromatin surveillance to halt senescence progression in mouse embryonic fibroblast." *Frontiers in Cell and Developmental Biology* 2020, 8, 103.
[85] Liang, N.; Yang, T.; Huang, Q., et al., "Mechanism of cancer stemness maintenance in human liver cancer." *Cell Death & Disease* 2022, 13 (4), 394.
[86] Park, D. D.; Phoomak, C.; Xu, G., et al., "Metastasis of cholangiocarcinoma is promoted by extended high-mannose glycans." *Proceedings of the National Academy of Sciences* 2020, 117 (14), 7633–7644.
[87] Camero, S.; Vitali, G.; Pontecorvi, P., et al., "DNMT3A and DNMT3B targeting as an effective radiosensitizing strategy in embryonal rhabdomyosarcoma." *Cells* 2021, 10 (11), 2956.
[88] Wang, N.; Ma, T.; Yu, B., "Targeting epigenetic regulators to overcome drug resistance in cancers." *Signal Transduction and Targeted Therapy* 2023, 8 (1), 69.
[89] Zhang, X.; Zhang, Y.; Wang, C.; Wang, X., "TET (Ten-eleven translocation) family proteins: Structure, biological functions and applications." *Signal Transduction and Targeted Therapy* 2023, 8 (1), 297.
[90] Meitzler, J. L.; Konaté, M. M.; Doroshow, J. H., "Hydrogen peroxide-producing NADPH oxidases and the promotion of migratory phenotypes in cancer." *Archives of Biochemistry and Biophysics* 2019, 675, 108076.
[91] Forloni, M.; Gupta, R.; Nagarajan, A., et al., "Oncogenic EGFR represses the TET1 DNA demethylase to induce silencing of tumor suppressors in cancer cells." *Cell Reports* 2016, 16 (2), 457–471.
[92] Tang, X.; Liang, Y.; Sun, G., et al., "Upregulation of CRABP2 by TET1-mediated DNA hydroxymethylation attenuates mitochondrial apoptosis and promotes oxaliplatin resistance in gastric cancer." *Cell Death and Disease* 2022, 13 (10), 848.
[93] Lyu, R.; Zhu, X.; Shen, Y., et al., "Tumour suppressor TET2 safeguards enhancers from aberrant DNA methylation and epigenetic reprogramming in ERα-positive breast cancer cells." *Epigenetics* 2022, 17 (10), 1180–1194.
[94] Han, D.; Huang, M.; Wang, T., et al., "Lysine methylation of transcription factors in cancer." *Cell Death and Disease* 2019, 10 (4), 290.
[95] Park, S.-Y.; Kim, J.-S., "A short guide to histone deacetylases including recent progress on class II enzymes." *Experimental and Molecular Medicine* 2020, 52 (2), 204–212.
[96] Ramaiah, M. J.; Tangutur, A. D.; Manyam, R. R., "Epigenetic modulation and understanding of HDAC inhibitors in cancer therapy." *Life Sciences* 2021, 277, 119504.
[97] Yang, W.-B.; Hsu, C.-C.; Hsu, T.-I., et al., "Increased activation of HDAC1/2/6 and Sp1 underlies therapeutic resistance and tumor growth in glioblastoma." *Neuro-Oncology* 2020, 22 (10), 1439–1451.
[98] Parveen, R.; Harihar, D.; Chatterji, B. P., "Recent histone deacetylase inhibitors in cancer therapy." *Cancer* 2023, 129 (21), 3372–3380.
[99] Wu, D.; Qiu, Y.; Jiao, Y.; Qiu, Z.; Liu, D., "Small molecules targeting HATs, HDACs, and BRDs in cancer therapy." *Frontiers in Oncology* 2020, 10, 560487.
[100] Cai, M.; Xu, S.; Jin, Y., et al., "hMOF induces cisplatin resistance of ovarian cancer by regulating the stability and expression of MDM2." *Cell Death Discovery* 2023, 9 (1), 179.
[101] Tan, K. N.; Avery, V. M.; Carrasco-Pozo, C., "Metabolic roles of androgen receptor and Tip60 in androgen-dependent prostate cancer." *International Journal of Molecular Sciences* 2020, 21 (18), 6622.
[102] Pan, W.; Zhang, Z.; Kimball, H., et al., "Abiraterone acetate induces CREB1 phosphorylation and enhances the function of the CBP-p300 complex, leading to resistance in prostate cancer cells." *Clinical Cancer Research* 2021, 27 (7), 2087–2099.
[103] Xiao, H.-T.; Jin, J.; Zheng, Z.-G., "Emerging role of GCN5 in human diseases and its therapeutic potential." *Biomedicine and Pharmacotherapy* 2023, 165, 114835.
[104] Fiorentino, F.; Mai, A.; Rotili, D., HAT inhibitors in cancer therapy. In *Histone Modifications in Therapy*, Elsevier; 2020, pp. 51–80.
[105] Hong, Z.; Xiang, Z.; Zhang, P., et al., "Histone acetyltransferase 1 upregulates androgen receptor expression to modulate CRPC cell resistance to enzalutamide." *Clinical and Translational Medicine* 2021, 11 (7), e495.
[106] Loras, A.; Segovia, C.; Ruiz-Cerdá, J. L., "Epigenomic and metabolomic integration reveals dynamic metabolic regulation in bladder cancer." *Cancers* 2021, 13 (11), 2719.
[107] Liao, X.; Wu, C.; Shao, Z., et al., "SETD4 in the proliferation, migration, angiogenesis, myogenic differentiation and genomic methylation of bone marrow mesenchymal stem cells." *Stem Cell Reviews and Reports* 2021, 1–16.

[108] Duan, R.; Du, W.; Guo, W., "EZH2: A novel target for cancer treatment." *Journal of Hematology and Oncology Reports* 2020, 13 (1), 1–12.
[109] Liu, Z.; Liu, J.; Ebrahimi, B., et al., "SETDB1 interactions with PELP1 contributes to breast cancer endocrine therapy resistance." *Breast Cancer Research* 2022, 24 (1), 26.
[110] Verhoeff, T. J.; Holloway, A. F.; Dickinson, J. L., "Non-coding RNA regulation of integrins and their potential as therapeutic targets in cancer." *Cellular Oncology* 2023, 46 (2), 239–250.
[111] He, B.; Zhao, Z.; Cai, Q., et al., "miRNA-based biomarkers, therapies, and resistance in Cancer." *International Journal of Biological Sciences* 2020, 16 (14), 2628.
[112] Huang, B.; Yang, H.; Cheng, X., et al., "tRF/miR-1280 suppresses stem cell–like cells and metastasis in colorectal cancer." *Cancer Research* 2017, 77 (12), 3194–3206.
[113] Al-Rugeebah, A.; Alanazi, M.; Parine, N. R., "MEG3: An oncogenic long non-coding RNA in different cancers." *Pathology and Oncology Research* 2019, 25, 859–874.
[114] Ullah, H.; Khan, H., Epigenetic drug development for autoimmune and inflammatory diseases. In *Histone Modifications in Therapy*, Elsevier; 2020, pp. 395–413.
[115] Zhang, L.; Lu, Q.; Chang, C., "Epigenetics in health and disease." *Epigenetics in Allergy and Autoimmunity* 2020, 3–55.
[116] Nepali, K.; Liou, J.-P., "Recent developments in epigenetic cancer therapeutics: Clinical advancement and emerging trends." *Journal of Biomedical Science* 2021, 28 (1), 1–58.
[117] Lv, X.; Zhou, M.; Zhang, Q., et al., "Abnormal histones acetylation in patients with primary Sjögren's syndrome." *Clinical Rheumatology* 2022, 41 (5), 1465–1472.
[118] Sharma, R.; Sharma, S.; Thakur, A., et al., "The role of epigenetic mechanisms in autoimmune, neurodegenerative, cardiovascular, and imprinting disorders." *Mini Reviews in Medicinal Chemistry* 2022, 22 (15), 1977–2011.
[119] Di Mambro, A.; Esposito, M. T., "Thirty years of SET/TAF1β/I2PP2A: From the identification of the biological functions to its implications in cancer and Alzheimer's disease." *Bioscience Reports* 2022, 42 (11), BSR20221280.
[120] Ballestar, E.; Sawalha, A. H.; Lu, Q., "Clinical value of DNA methylation markers in autoimmune rheumatic diseases." *Nature Reviews Rheumatology* 2020, 16 (9), 514–524.
[121] Yang, Y.; Zhang, M.; Wang, Y., "The roles of histone modifications in tumorigenesis and associated inhibitors in cancer therapy." *Journal of the National Cancer Center* 2022, 2 (4), 277–290.
[122] Si, W.; Shen, J.; Zheng, H.; Fan, W., "The role and mechanisms of action of microRNAs in cancer drug resistance." *Clinical Epigenetics* 2019, 11 (1), 1–24.
[123] Yang, J.; Li, Z.; Wang, L., et al., "The role of non-coding RNAs (miRNA and lncRNA) in the clinical management of rheumatoid arthritis." *Pharmacological Research* 2022, 106549.
[124] George, B. P.; Chandran, R.; Abrahamse, H., "Role of phytochemicals in cancer chemoprevention: Insights." *Antioxidants* 2021, 10 (9), 1455.
[125] Fan, X.-X.; Pan, H.-D.; Li, Y., et al., "Novel therapeutic strategy for cancer and autoimmune conditions: Modulating cell metabolism and redox capacity." *Pharmacology & Therapeutics* 2018, 191, 148–161.
[126] Carlos-Reyes, Á.; López-González, J. S.; Meneses-Flores, M., et al., "Dietary compounds as epigenetic modulating agents in cancer." *Frontiers in Genetics* 2019, 10, 79.
[127] Mohammad, H. P.; Barbash, O.; Creasy, C. L., "Targeting epigenetic modifications in cancer therapy: Erasing the roadmap to cancer." *Nature Medicine* 2019, 25 (3), 403–418.
[128] Hassan, F.-U.; Rehman, M. S.-U.; Khan, M. S., et al., "Curcumin as an alternative epigenetic modulator: Mechanism of action and potential effects." *Frontiers in Genetics* 2019, 10, 514.
[129] Makuch, S.; Więcek, K.; Woźniak, M., "The immunomodulatory and anti-inflammatory effect of curcumin on immune cell populations, cytokines, and in vivo models of rheumatoid arthritis." *Pharmaceuticals* 2021, 14 (4), 309.
[130] Kou, H.; Huang, L.; Jin, M., et al., "Effect of curcumin on rheumatoid arthritis: A systematic review and meta-analysis." *Frontiers in Immunology* 2023, 14, 1121655.
[131] Akone, S. H.; Ntie-Kang, F.; Stuhldreier, F., et al., "Natural products impacting DNA methyltransferases and histone deacetylases." *Frontiers in Pharmacology* 2020, 11, 992.
[132] Shen, P.; Deng, X.; Chen, Z., et al., "SIRT1: A potential therapeutic target in autoimmune diseases." *Frontiers in Immunology* 2021, 12, 779177.
[133] Murata, M.; Komatsu, S.; Miyamoto, E., et al., "Quercetin up-regulates the expression of tumor-suppressive microRNAs in human cervical cancer." *Bioscience of Microbiota, Food and Health* 2023, 42 (1), 87–93.
[134] Kim, D. H.; Khan, H.; Ullah, H., et al., "MicroRNA targeting by quercetin in cancer treatment and chemoprotection." *Pharmacological Research* 2019, 147, 104346.

[135] Benameur, T.; Soleti, R.; Porro, C., "The potential neuroprotective role of free and encapsulated quercetin mediated by miRNA against neurological diseases." *Nutrients* 2021, 13 (4), 1318.
[136] Ciesielski, O.; Biesiekierska, M.; Balcerczyk, A., "Epigallocatechin-3-gallate (EGCG) alters histone acetylation and methylation and impacts chromatin architecture profile in human endothelial cells." *Molecules* 2020, 25 (10), 2326.
[137] Li, C.; Xu, Y.; Zhang, J., et al., "The effect of resveratrol, curcumin and quercetin combination on immuno-suppression of tumor microenvironment for breast tumor-bearing mice." *Scientific Reports* 2023, 13 (1), 13278.
[138] Biernacka, K. M.; Holly, J. M.; Martin, R. M., et al., "Effect of green tea and lycopene on the insulin-like growth factor system: The ProDiet randomized controlled trial." *European Journal of Cancer Prevention* 2019, 28 (6), 569.
[139] Ren, B.; Kwah, M. X.-Y.; Liu, C., et al., "Resveratrol for cancer therapy: Challenges and future perspectives." *Cancer Letters* 2021, 515, 63–72.
[140] Yu, G.; Xiang, W.; Zhang, T., et al., "Effectiveness of Boswellia and Boswellia extract for osteoarthritis patients: A systematic review and meta-analysis." *BMC Complementary Medicine and Therapies* 2020, 20 (1), 1–16.
[141] Aryaeian, N.; Mahmoudi, M.; Shahram, F., et al., "The effect of ginger supplementation on IL2, TNFα, and IL1β cytokines gene expression levels in patients with active rheumatoid arthritis: A randomized controlled trial." *Medical Journal of the Islamic Republic of Iran* 2019, 33, 154.
[142] Roy, N. K.; Parama, D.; Banik, K., et al., 'An update on pharmacological potential of boswellic acids against chronic diseases." *International Journal of Molecular Sciences* 2019, 20 (17), 4101.
[143] Constantin, M. M.; Nita, I. E.; Olteanu, R., et al., "Significance and impact of dietary factors on systemic lupus erythematosus pathogenesis." *Experimental and Therapeutic Medicine* 2019, 17 (2), 1085–1090.
[144] Mansouri, K.; Rasoulpoor, S.; Daneshkhah, A., et al., "Clinical effects of curcumin in enhancing cancer therapy: A systematic review." *BMC Cancer* 2020, 20, 1–11.
[145] Asgharian, P.; Tazekand, A. P.; Hosseini, K., et al., "Potential mechanisms of quercetin in cancer prevention: Focus on cellular and molecular targets." *Cancer Cell International* 2022, 22 (1), 257.
[146] Mitsiogianni, M.; Trafalis, D. T.; Franco, R., et al., "Sulforaphane and iberin are potent epigenetic modulators of histone acetylation and methylation in malignant melanoma." *European Journal of Nutrition* 2021, 60, 147–158.
[147] Li, F.; Qasim, S.; Li, D.; Dou, Q. P., "Updated review on green tea polyphenol epigallocatechin-3-gallate as a cancer epigenetic regulator." *Seminars in Cancer Biology* 2022, 335–352.
[148] Tuli, H. S.; Tuorkey, M. J.; Thakral, F., et al., "Molecular mechanisms of action of genistein in cancer: Recent advances." *Frontiers in Pharmacology* 2019, 10, 1336.
[149] van Steenwijk, H. P.; Bast, A.; de Boer, A., "The role of circulating lycopene in low-grade chronic inflammation: A systematic review of the literature." *Molecules* 2020, 25 (19), 4378.
[150] Kiyama, R., "Nutritional implications of ginger: Chemistry, biological activities and signaling pathways." *The Journal of Nutritional Biochemistry* 2020, 86, 108486.
[151] Nafees, S.; Zafaryab, M.; Mehdi, S. H., et al., "Anti-cancer effect of gingerol in cancer prevention and treatment." *Anti-Cancer Agents in Medicinal Chemistry (Formerly Current Medicinal Chemistry-Anti-Cancer Agents)* 2021, 21 (4), 428–432.
[152] Jiang, W.; Li, X.; Dong, S.; Zhou, W., "Betulinic acid in the treatment of tumour diseases: Application and research progress." *Biomedicine & Pharmacotherapy* 2021, 142, 111990.
[153] Liu, D.; Meng, X.; Wu, D.; Qiu, Z.; Luo, H., "A natural isoquinoline alkaloid with antitumor activity: Studies of the biological activities of berberine." *Frontiers in Pharmacology* 2019, 10, 9.
[154] Xu, S.; Liao, Y.; Wang, Q.; Liu, L.; Yang, W., "Current studies and potential future research directions on biological effects and related mechanisms of allicin." *Critical Reviews in Food Science and Nutrition* 2023, 63 (25), 7722–7748.
[155] Lin, W.-S.; Leland, J. V.; Ho, C.-T.; Pan, M.-H., "Occurrence, bioavailability, anti-inflammatory, and anticancer effects of pterostilbene." *Journal of Agricultural and Food Chemistry* 2020, 68 (46), 12788–12799.
[156] O'Neill, E. J.; Den Hartogh, D. J.; Azizi, K.; Tsiani, E., "Anticancer properties of carnosol: A summary of in vitro and in vivo evidence." *Antioxidants* 2020, 9 (10), 961.
[157] Lechner, J. F.; Stoner, G. D., "Red beetroot and betalains as cancer chemopreventative agents." *Molecules* 2019, 24 (8), 1602.
[158] Nagini, S.; Nivetha, R.; Palrasu, M.; Mishra, R., "Nimbolide, a neem limonoid, is a promising candidate for the anticancer drug arsenal." *Journal of Medicinal Chemistry* 2021, 64 (7), 3560–3577.

[159] Petrocelli, G.; Marrazzo, P.; Bonsi, L., et al., "Plumbagin, a natural compound with several biological effects and anti-inflammatory properties." *Life* 2023, 13 (6), 1303.

[160] Farhana, A.; Koh, A. E.-H.; Tong, J. B., et al., "Nanoparticle-encapsulated camptothecin: Epigenetic modulation in DNA repair mechanisms in colon cancer cells." *Molecules* 2021, 26 (17), 5414.

[161] Li, Y.; Zhou, X.; Liu, J.; Yuan, X.; He, Q., "Therapeutic potentials and mechanisms of artemisinin and its derivatives for tumorigenesis and metastasis." *Anti-Cancer Agents in Medicinal Chemistry (Formerly Current Medicinal Chemistry-Anti-Cancer Agents)* 2020, 20 (5), 520–535.

[162] Marton, L. T.; Barbalho, S. M.; Sloan, K. P., et al., "Curcumin, autoimmune and inflammatory diseases: Going beyond conventional therapy–a systematic review." *Critical Reviews in Food Science and Nutrition* 2022, 62 (8), 2140–2157.

[163] Gianchecchi, E.; Fierabracci, A., "Insights on the effects of resveratrol and some of its derivatives in cancer and autoimmunity: A molecule with a dual activity." *Antioxidants* 2020, 9 (2), 91.

[164] Khan, H.; Sureda, A.; Belwal, T., et al., "Polyphenols in the treatment of autoimmune diseases." *Autoimmunity Reviews* 2019, 18 (7), 647–657.

[165] Song, J.; Yang, J.; Jing, S., et al., "Berberine attenuates experimental autoimmune myasthenia gravis via rebalancing the T cell subsets." *Journal of Neuroimmunology* 2022, 362, 577787.

[166] Gao, X.; Lin, X.; Wang, Q.; Chen, J., "Artemisinins: Promising drug candidates for the treatment of autoimmune diseases." *Medicinal Research Reviews* 2024, 44 (2), 867–891.

[167] Nowak-Perlak, M.; Szpadel, K.; Jabłońska, I.; Pizon, M.; Woźniak, M., "Promising strategies in plant-derived treatments of psoriasis-update of in vitro, in vivo, and clinical trials studies." *Molecules* 2022, 27 (3), 591.

[168] Li, Y.; Wang, G.; Liu, J.; Ouyang, L., "Quinolizidine alkaloids derivatives from Sophora alopecuroides Linn: Bioactivities, structure-activity relationships and preliminary molecular mechanisms." *European Journal of Medicinal Chemistry* 2020, 188, 111972.

[169] Xinqiang, S.; Erqin, D.; Yu, Z., et al., "Potential mechanisms of action of celastrol against rheumatoid arthritis: Transcriptomic and proteomic analysis." *PLoS ONE* 2020, 15 (7), e0233814.

[170] Pandey, P.; Khan, F.; Qari, H. A.; Oves, M., "Rutin (Bioflavonoid) as cell signaling pathway modulator: Prospects in treatment and chemoprevention." *Pharmaceuticals* 2021, 14 (11), 1069.

[171] Holloman, B. L.; Nagarkatti, M.; Nagarkatti, P., "Epigenetic regulation of cannabinoid-mediated attenuation of inflammation and its impact on the use of cannabinoids to treat autoimmune diseases." *International Journal of Molecular Sciences* 2021, 22 (14), 7302.

[172] Burgos, R. A.; Alarcón, P.; Quiroga, J.; Manosalva, C.; Hancke, J., "Andrographolide, an anti-inflammatory multitarget drug: All roads lead to cellular metabolism." *Molecules* 2020, 26 (1), 5.

[173] Jin, M.; Wang, C.; Xu, Y., et al., "Pharmacological effects of salidroside on central nervous system diseases." *Biomedicine & Pharmacotherapy* 2022, 156, 113746.

[174] Toden, S.; Okugawa, Y.; Jascur, T., et al., "Curcumin mediates chemosensitization to 5-fluorouracil through miRNA-induced suppression of epithelial-to-mesenchymal transition in chemoresistant colorectal cancer." *Carcinogenesis* 2015, 36 (3), 355–367.

[175] Mokhtari, R. B.; Homayouni, T. S.; Baluch, N., et al., "Combination therapy in combating cancer." *Oncotarget* 2017, 8 (23), 38022.

[176] Tan, B. L.; Norhaizan, M. E., "Curcumin combination chemotherapy: The implication and efficacy in cancer." *Molecules* 2019, 24 (14), 2527.

[177] Dai, Q.; Zhou, D.; Xu, L.; Song, X., "Curcumin alleviates rheumatoid arthritis-induced inflammation and synovial hyperplasia by targeting mTOR pathway in rats." *Drug Design, Development and Therapy* 2018, 4095–4105.

[178] Lyu, Y.-L.; Zhou, H.-F.; Yang, J., et al., "Biological activities underlying the therapeutic effect of quercetin on inflammatory bowel disease." *Mediators of Inflammation* 2022, 2022.

[179] van Brummelen, R.; van Brummelen, A. C., "The potential role of resveratrol as supportive antiviral in treating conditions such as COVID-19–a formulator's perspective." *Biomedicine & Pharmacotherapy* 2022, 148, 112767.

6 Epigenetic Drug Discovery in Inflammatory Diseases

Silpa R S,[1] Akey Krishna Swaroop,[1] Esakkimuthukumar M,[1] Saranya R,[1] Prabha T,[2] Suresh S,[3] and Jubie S[1]

[1]JSS Academy of Higher Education and Research, Ooty, Tamilnadu, India

[2]Nandha College of Pharmacy, Erode, Tamilnadu, India

[3]Health & Social Science Cluster, Singapore Institute of Technology, Singapore

6.1 INTRODUCTION

In 1942, the term "epigenetics" was introduced by developmental biologist Conrad H. Waddington to describe heritable alterations in gene expression patterns that transpire independent modifications in the primary DNA sequence but are adequate to govern the gene expression profile (Shanmugam and Sethi 2013). Within the framework of the provided definition, epigenetics encompasses all molecular pathways that regulate the transformation of a genotype into a specific phenotype.

Epigenetics is a firmly established scientific field focused on the study of modifications in gene function that can be transmitted across cell divisions (mitotically) or generations (meiotically) without requiring changes to the underlying DNA sequence (Dupont, Armant, and Brenner 2009). Holliday characterized epigenetics as "the study of the mechanisms governing the temporal and spatial regulation of gene activity during the development of a complex organism," and Russo et al. defined it as "the exploration of mitotically and/or meiotically inheritable modifications in gene function that cannot be elucidated by alterations in the DNA sequence." The Greek prefix *epi* translates as "on top of" or "in addition to." Another widely accepted definition of epigenetics is, "the enduringly inheritable phenotype arising from chromosome changes without modifications to the DNA sequence" (Seo et al. 2014). Epigenetics transmits inheritable information through cell division distinct from the primary DNA sequence (Feinberg 2007), involving the chemical modification of DNA and associated proteins. This in turn restructures chromatin and selectively activates or deactivates genes, thereby shaping their expression throughout development (Gomez, Dutra, and Moreira 2009). Epigenetics encompasses a diverse range of biochemical mechanisms that regulate DNA function without directly involving alterations to adenine, guanine, thymine and cytosine nucleotides, the DNA sequences. This distinguishes epigenetic modifications from genetic mutations. Importantly, epigenetic alterations persist through cell division, rendering them heritable in nature (Huber et al. 2007).

Inflammation, derived from the Latin *inflammatio*, is an intricate biological reaction exhibited by bodily tissues in response to stimuli such as pathogens, compromised cells, and irritants (Linlin Chen et al. 2017). It is a protective response involving immune cells, blood vessels, and molecular mediators, and its primary purposes are to remove the initial cause of cellular injury, clear away necrotic cells and tissues damaged during the initial stage and in subsequent inflammatory process, and initiate tissue repair. This response is characterized by five hallmark signs: increased heat, pain, redness, swelling, and impaired activity (expressed in Latin as *calor, dolor, rubor, tumor*, and *functio laesa*). Inflammation is categorized as a mechanism of innate immunity, distinguishing it from adaptive immunity, which is pathogen specific (Abbas, Lichtman, and Pillai 2019).

DOI: 10.1201/9781032706511-6

Pathological conditions associated with inflammation are typically identified by the suffix "-itis," as in bronchitis, esophagitis, gastritis, colitis, pancreatitis, prostatitis, cervicitis, and hepatitis, just to name a few. However, it's worth noting that certain inflammatory conditions, such as asthma and pneumonia, deviate from this naming system (Aggarwal and Gehlot 2009). The interplay between epigenetic modifications and genetic abnormalities plays a pivotal role in driving chronic inflammatory diseases (Egger et al. 2004). The emergence of epigenetics holds promise for its utility as preventive, diagnostic, and therapeutic markers. As drugs targeting precise epigenetic mechanisms responsible for gene expression regulation continue to advance, epigenetic tools are a suitable and effective approach for clinical intervention in the treatment of diverse diseases (Tong et al. 2022) (see Figure 6.1).

6.1.1 Epigenetic Mechanisms

6.1.1.1 DNA Methylation

DNA methylation entails the shift of covalent methyl group from S-adenosyl-L-methionine (SAM) to cytosine residue on the carbon atom in the fifth position within cytosine-phosphodiester-guanine (CpG) dinucleotides. This process primarily occurs happens within promoter gene regions characterized by a high expression of dinucleotides of CpG, referred to as "CpG islands," and suppresses gene activity. DNMT1, DNMT3, and DNMT3b catalyse DNA methylation. Methyl groups cause duplicate repression by hindering the connection of transcription factors to DNA, commonly referred as gene silencing. Methyl-CpG-binding domain proteins (MBDs) interact with these exposed methylation sites (Srinivasan n.d.).

DNMT1 primarily methylates already methylated profiles, which preserves DNA methylation. Simultaneously, DNMT3A and DNMT3B contribute to the de novo methylation of unmethylated substrates. DNA methylation has a significant role in tissue-specific gene expression and regulation. DNA methylation analysis in cells is informed by the methylome, which is specific to a particular cell state. Each cell state consists of a diverse mix of cell types, each with its unique methylome.

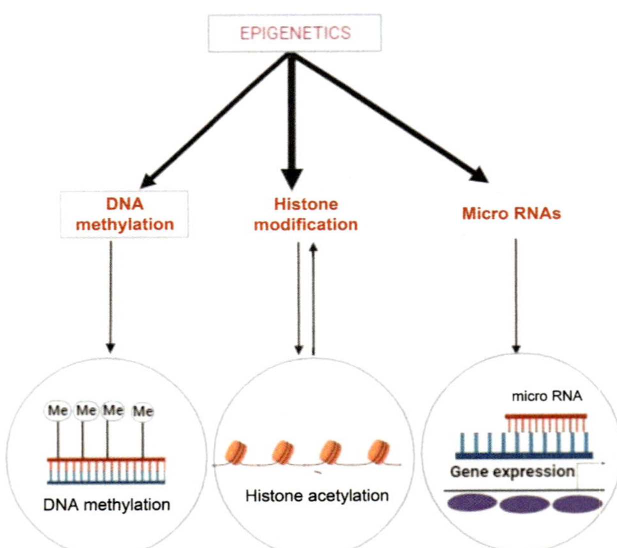

FIGURE 6.1 Epigenetic mechanisms: DNA methylation, histone modifications, and microRNA (miRNA) processes.

In such cases, observed alterations in DNA methylation may simply show shifts in the fraction of different cell types (Fogel, Richard-Miceli, and Tost 2017).
DNA methylation serves additional crucial functions, including inactivating the X chromosome and cutting transposable elements, which constitute a significant portion—approximately 42%—of the human genome. Suppressing the activity of these transposable elements, also known as retro transposons, is essential to preserve genomic stability (Ospelt, Gay, and Klein 2017). The diversity of epigenetic changes in mammalian DNA has expanded with the identifying 5-hydroxymethylcytosine (5hmC) and its oxidative derivatives, specifically 5-formylcytosine and 5-carboxylcytosine, within mammalian neurons and embryonic stem cells. These modifications arise through catalytic oxidation mediated by TET proteins, originating from 5-methylcytosines (Kriaucionis and Heintz 2009).

6.1.1.2 Factors Affecting DNA Methylation

6.1.1.2.1 Age

Embryonic development is pivotal, and prenatal exposure to stressful environmental factors can lead to modifications in DNA methylation patterns during adulthood, with some exhibiting sex-specific effects (Murphy et al. 2012). These alterations manifest as age-related transformations in all tissue types, reflecting biological rather than chronological aging. These modifications are intricately linked to both intrinsic cell factors and alterations in the cell composition of whole blood (Jaffe and Irizarry 2014).

6.1.1.2.2 Sex

While disparities in DNA methylation of sex chromosomes and sex-related genes have been thoroughly documented, the identification of variations in DNA methylation at other autosomal loci between the two sexes remains inconsistent (McCarthy et al. 2014). More extensive research is imperative for validating these observations and ascertaining whether these distinctions persist into adulthood.

6.1.1.2.3 Smoking

Smoking significantly impacts DNA methylation, with alterations in DNA methylation patterns being associated with the intensity of smoking. Some CpG sites return to DNA methylation levels typically observed in non-smokers within 12 weeks after smoking cessation, while others retain these alterations even after a decade (Tsaprouni et al. 2014). These modifications are specific to the cell types they occur in (Su et al. 2016). Additionally, passive exposure to smoking induces DNA methylation changes in the same way as direct exposure (Kim 2019). Recent data suggests that e-cigarettes (vapes) also have the capacity to modify epigenetic and transcriptional regulatory mechanisms (Martin et al. 2016).

6.1.1.2.4 Medication

Numerous medications employed in the management of inflammatory conditions exert an impact on DNA methylation, encompassing both prescribed and complementary drugs. Specifically, agents such as glucocorticoids, non-steroidal anti-inflammatory drugs (NSAIDs), sulfasalazine, methotrexate, and tumor necrosis factor-α inhibitors induce alterations in DNA methylation (Wilson et al. 2015).

6.1.1.2.5 Alcohol and Diet

Alcohol is recognized for its capacity to induce DNA methylation changes even at low concentrations, but its influence remains inadequately comprehended (Philibert et al. 2012). Dietary elements also influence DNA methylation, although prior investigations have predominantly been obesity and cancer intervention studies. Long interspersed nuclear element (LINE), serving as a surrogate for overall DNA methylation levels, was identified as a universal marker for dietary response: The investigation revealed a correlation between LINE methylation levels and serum glucose concentrations as well as associations between hydroxymethylation levels and BMI, waist circumference, and total cholesterol. Importantly, it is imperative to recognize that BMI can modulate DNA methylation levels as a result of adiposity (Nicoletti et al., 2016).

6.1.1.2.6 Sample Cell Type

Each cell type possesses distinct DNA methylation profiles and variations in the same patterns within tissues are notably driven by the unique composition of cell types present. Furthermore, the impacts of environmental factors and diseases on DNA methylation patterns are specific to particular cell types. In silico deconvolution can determine cell proportions, with numerous methods extensively outlined in a recent review, offering guidance on their appropriate application. However, it's important to note that these methods do not allow for the precise identification of cell-type-specific alterations in DNA methylation or the mitigation of the preservation of distinct cell-specific signals in complex samples (Whyte et al. 2019).

6.1.1.3 Chromatin Remodeling and Histone Modification

Numerous histone modifications have been extensively studied, with acetylation, methylation, phosphorylation, and ubiquitination standing out as the most prominent and crucial in chromatin structure regulation and its impact on transcriptional activity (Alaskhar Alhamwe et al. 2018) (Figure 6.2).

6.1.1.3.1 Histone Acetylation

Histone acetylation regulates two distinct classes of enzymes that exert opposite effects: histone acetyltransferases (HATs) and histone deacetylases (HDACs). The transfer of an acetyl group from acetyl-CoA to specific amino acid residues within the histone tails is facilitated by HATs, resulting in the changing of a positive charge on the histones. This process weakens the interactivity between histones and the negatively charged phosphate groups of DNA. Conversely, HDACs are responsible for the removal of acetyl groups from lysine residues in the histone tails, thus act as repressors of gene expression (Alaskhar Alhamwe et al. 2018). HATs can be categorized into five (or sometimes six) classes. The GCN5-related N-acetyltransferase (GNAT) family includes enzymes like KAT2A and KAT2B, which are associated in histone and transcription factor acetylation. This, in turn, plays a crucial role in cell cycle regulation, DNA replication and repair, and has recently been implicated in centrosome function (Nagy and Tora 2007), (Fournier and Tora 2017). Lysine acetyltransferase (KAT) enzymes catalyzed the acetylation reaction. One of the major sets of human KATs are; the MYST (*Moz, Ybf2/Sas3, Sas2, Tip60*) proteins form the largest family of histone acetyltransferases and contribute to transcription regulation and DNA repair (Klein et al. 2014). The Table 6.1. represents the KAT family of histone acetyltransferases.

Other HAT groups are smaller, with KAT3A and KAT3B enzymes belonging to the p300/CBP family, and KAT4/TAF1/TBP and KAT12/TIFIIIC90 being part of the general transcriptional factor-related HAT family (Allis et al. 2007). Furthermore, the steroid receptor co-activators

TABLE 6.1
The KAT Family of Histone Acetyltransferases

Sl. No	Protein
1.	KAT6A, known as MOZ (monocytic leukemic zinc-finger protein) or MYST3
2.	KAT6B, known as MORF (MOZ-related factor) or MYST4
3.	KAT7, known as HBO1 or MYST2
4.	KAT8, known as hMOF or MYST1
5.	KAT5, or Tip60

family encompasses KAT13A/SRC1, KAT13B/SCR3/AIB1/ACTR, KAT13C/p600, and KAT13D/CLOCK. Finally, the cytoplasmic HATs include KAT1/HAT1 and HAT4/NAA60 (Wapenaar and Dekker 2016). On the other hand, the HDAC superfamily consists of 18 enzymes categorized into four groups (Table 6.2) (Hull, Montgomery, and Leyva 2016).

6.1.1.3.2 Histone Methylation

Histone methylation is orchestrated by histone methyltransferases (HMTs), encompassing lysine methyltransferases (KMTs) and arginine methyltransferases (PRMTs), while histone demethylation is facilitated by histone demethylases (HDMs). Unlike histone lysine acetylation, which modifies the electrical charge of histones and thereby their activity with DNA, methylation of histone lysine or arginine does not directly impact this electrostatic bond; rather, it indirectly modulates chromatin by recruiting and binding distinct regulatory proteins to (Morera, Lübbert, and Jung 2016).

HMTs can shift up to three methyl groups from cofactor S-adenosyl-L-methionine (SAM) to lysine or arginine residues on histones (Kaniskan, Martini, and Jin 2018). KMTs exhibit higher specificity than HATs, typically targeting specific lysine residues. For example, methylation of the H3K4 residue, responsible for dictating histone modifications, their location, characteristics, and transcriptional effects in mammals, is catalyzed by KMT2A/MLL1, KMT2A/MLL2, KMT2G/hSET1B, or KMT2H/ASH1 and KMT2F/hSET1A (Hyun et al. 2017). HDMs are divided into two groups based on their catalytic mechanism and sequence homology. The first comprises amine-oxidase type lysine-specific demethylases (LSDs or KDM1s) and consists of KDM1B/LSD2/AOF1 and KDM1A/LSD1/AOF2, which detach methyl groups from mono- and dimethylated H3K4. The second class involves JumonjiC (JMJC) domain-containing HDMs, which demethylate mono-, di-, and trimethylated lysine residues at numerous histone amino acids (Bennani-Baiti 2012).

6.1.1.3.3 Histone Ubiquitination

Protein ubiquitination, a pivotal post-translational modification, exerts regulatory control over nearly every facet of cellular functionality within various eukaryotic cell signaling pathways. Histone ubiquitination, specifically facilitated by histone ubiquitin ligases and subject to reversal by ubiquitin-specific peptidases, referred to as deubiquitinating enzymes (DUBs) holds a central role in this intricate regulatory network (Ravid and Hochstrasser 2008). Monoubiquitination, in particular, plays a crucial role in governing processes such as protein translocation, transcriptional regulation and DNA damage signalling. Notably, histone 2A monoubiquitination (H2Aub) predominantly associates with gene silencing, while monoubiquitination of histone 2B (H2Bub) is typically linked to transcriptional triggering. Furthermore, the process of polyubiquitination serves as a dual mechanism, marking proteins for either degradation or activation within specific signaling pathways (Weake and Workman 2008).

TABLE 6.2
The HDAC Superfamily

Group	Enzyme
Class I	HDAC1, HDAC2, HDAC3, and HDAC8
Class IIa	HDAC4, HDAC5, HDAC7, and HDAC9
Class IIb	HDAC6 and HDAC10
Class III	SIRTs; SIRT 1-7
Class IV	HDAC11

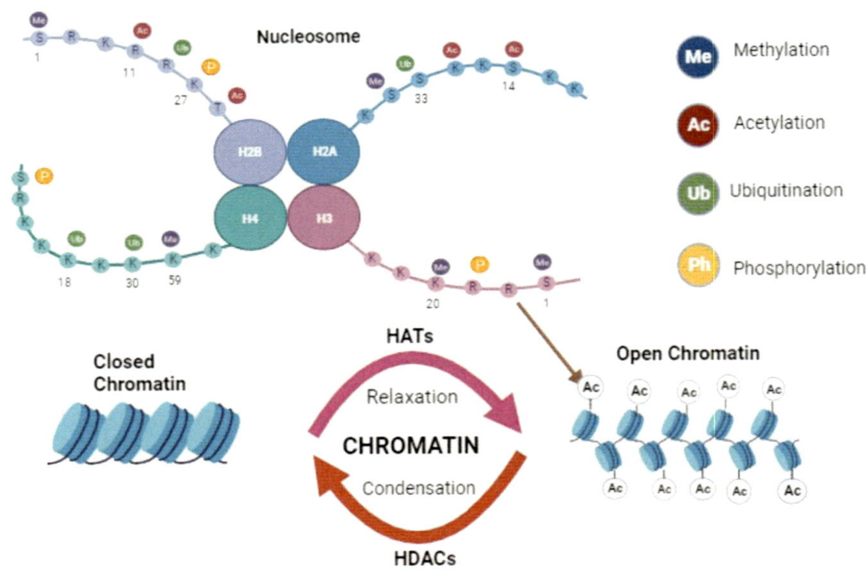

FIGURE 6.2 Epigenetic processes.

6.1.1.4 MicroRNAs

Epigenetic regulation through RNA mechanisms are less studied than DNA methylation and histone modification. RNA plays a role in regulating monoallelic expression of imprinted genes and in X-chromosome inactivation is established with RNA involvement, and recent researchers found RNA-based mechanisms in epigenetic regulation. Non-coding infrastructural RNAs (ncRNAs) have long been understood to play prominent roles in epigenetics (Gibney and Nolan 2010); ncRNAs such as microRNAs (miRNAs), short interfering RNAs (siRNAs), and long ncRNAs originate from introns and intergenic regions. Within intergenic regions during transcription, non-coding transcripts play a critical role in modulating chromatin accessibility, an open chromatin structure maintained thereby (Surace and Hedrich 2019).

MiRNAs are small ncRNAs molecules spanning a length of 17 to 25 nucleotides that are endogenously expressed. They exert their regulatory effects by causing the degradation of target mRNA or by suppressing mRNA translation through imperfect attachment to the 3′-untranslated portion of their targets. MiRNA genes are transcribed by RNA polymerase II, producing pri-miRNA in the nucleus as the primary transcript. The pri-miRNA is processed by microprocessing complex DGCR8, forming a hairpin stem-loop structure that is then transformed into a precursor miRNA (pre-miRNA) having a length of 70 nucleotides. The pre-miRNA is subsequently transported to the cytoplasm, where it undergoes secondary processing mediated by Dicer. During this process, one strand of the hairpin structure becomes part of the ribonucleoprotein complex referred to as miRNA-encouraged cutting complex (Boyanapalli and Kong 2015). Long ncRNAs (lncRNAs; Table 6.3) are greater than 200 nt and are 'loose' or 'noisy' transcription factors. Most eukaryotes contain large amounts of these transcription factors.

LncRNAs exert distinct control over gene expression, either through the functionality of the primary transcript or during the process of transcription itself. These molecules exhibit tissue-specific patterns, undergo modulation throughout development, and consolidate to particular cell compartments. Importantly, their association with human diseases and evidence of evolutionary selection underscore their significance in biological processes.

TABLE 6.3
The Five Classes of LncRNAs

Class	Function
Sense	Joining one or more exons of same transcript on the same strand
Anti-sense	Joining one or more exons of another transcript on the same strand or the opposite strand
Bidirectional	Originating from genomic proximity to a neighboring coding transcript on the opposite strand
Intronic	Originating entirely from within an intron of a second transcript
Intergenic	Occuring within the genomic interval between two genes

In specific instances, the mere transcription of an lncRNA can either enhance or diminish the transcription from a downstream promoter. This effect may occur by altering the recruitment of RNA polymerase II or by influencing chromatin configuration. Alternatively, the hybridization of an antisense transcript with a sense transcript can impact the splicing of the sense transcript or induce the generation of endogenous siRNAs mediated by Dicer. Additionally, lncRNAs can connect with protein partners to alter or constrain the functions of the given proteins within the cellular environment. Moreover, these RNAs undergo refinement that leads to various types of small regulatory RNAs (Gibney and Nolan 2010).

6.1.2 Inflammatory Diseases

Epigenetics in diagnostics and therapeutics application are recently under scrutiny. For diagnostic purpose it involve the utilization of biomarkers for confirming diagnosed condition and predicting treatment responses. Epigenetic plays a pivotal role in gene regulation within various inflammation mediators. So here we are discussing epigenetics in chronic inflammatory diseases. Extensive research has been conducted on epigenetic alterations that predispose individuals to cancer development during the period of inflammatory diseases (Fleisher et al. 2000).

A broad spectrum of disorders with both genetic and epigenetic risk factors can result from improper inflammation, even when temporary inflammation protects tissues from pathogenic invasions. It is difficult to comprehend the idea of epigenetic variability in relation to the healthy and pathological functioning of immune cells. There is a significant difference between genetically complicated and monogenic disorders when it comes to the effects of epigenetic modifications in inflammatory diseases. For instance, a single genetic flaw may be the cause of monogenic disorders like CAPS, either directly or indirectly through certain epigenetic modifications that occur later (Shanmugam and Sethi 2013).

These changes are also linked to a range of clinical symptoms. Genetic susceptibility variants in different diseases are linked to diverse downstream epigenetic modifications. Understanding how epigenetic modifications influence disease requires thoroughly investigating each disease's biology. Studies conducted in the last five years have shown that epigenetic changes are common in inflammatory rheumatic diseases that are monogenic and genetically complicated such as systemic lupus erythematous (SLE) and RA. Tissue destruction—whether caused by organ-specific or systemic inflammation—is a hallmark of autoimmune and inflammatory disorders (Shanmugam and Sethi 2013).

Autoimmune disorders are distinguished by the pathophysiological involvement of autoantibodies and/or auto-responding lymphocytes, driven by the adaptive immune system. Recent discoveries, however, indicate that the situation is more complex, and only a small number of monogenic disorders, such as primary type I interferonopathies, cryopyrin-associated periodic syndromes (CAPS), and other "monogenic" disorders, may be categorized as mostly autoinflammatory, although as the illness progresses, this classification may alter. Adaptive immune cells are presented with intracellular and nuclear components when tissue injury frequently leads to cell death. This ultimately results in their activation, which in turn sets off the generation of autoantibodies and/or self-directed lymphocyte (Dupont, Armant, and Brenner 2009).

In conditions such as psoriasis, an initial heterogeneous immunological pattern drives inflammation, leading to the concept of an 'inflammatory spectrum' has been put forth (Seo et al. 2014). On this spectrum of systemic autoimmune and inflammatory disorders, "classical" autoimmune conditions are at one end, and monogenic autoinflammatory conditions are at the other. By seeing autoimmune and inflammatory diseases as dynamic processes with changing molecular phenotypes, the inflammatory spectrum also makes it possible to define changeable clinical stereotypes, outcomes, and treatment responses (Feinberg 2007).

Our understanding of autoimmune and inflammatory diseases is constantly evolving. The observed inter-individual diversity in symptoms and outcomes—sometimes even within families or among individuals with the same diagnosis (Gomez, Dutra, and Moreira 2009)—reflects that conditions that were previously different can change over time along the inflammatory spectrum. For example, adult Still's illness and systemic JIA can transition from a predominantly autoinflammatory to an autoimmune phenotype (Huber et al. 2007).

Autoimmune and inflammatory diseases have complex and little-understood molecular pathophysiology. Genetically predisposed (more common) and monogenic (rare) illnesses can be linked to autoimmune and inflammatory processes influenced by both environmental and personal factors that can change an individual's phenotype and course of illness. (Egger et al. 2004). This complexity is seen in "classical" autoimmune illnesses, in which genetically identical monozygotic twins develop different inflammatory and autoimmune symptoms that also differ for different diseases: 13% for multiple sclerosis, 20% for psoriasis, 24% for type 1 diabetes, and 14.3%–40% for SLE (Murphy et al. 2012).

These studies have led to two key hypotheses: (i) In genetically complex disorders, genomic variants define disease susceptibility, but individual phenotypes and outcomes are influenced by additional factors, including epigenetics; and (ii) in rare auto inflammatory disorders, disease-causing single-gene mutations trigger illness onset, although they do not always predict individual (Jaffe and Irizarry 2014). By regulating chromatin approachability to transcriptional regulatory factors, epigenetic processes regulate and fine tune gene expression without altering the DNA's fundamental sequence (McCarthy et al. 2014). Despite a shared genotype among all diploid cells, environment-driven, dynamic, and heritable processes shape cell and tissue differences within an organism (Tsaprouni et al. 2014).

6.1.3 Ankylosing Spondylitis

Spondyloarthropathy encompasses a diverse range of rheumatic conditions that share common pathological and genetic characteristics (Table 6.4). They are categorized as either peripheral or axial based on the primary body regions involved; for instance, ankylosing spondylitis (AS; Figure 6.3) primarily affects the axial portion (Zhu et al. 2019).

TABLE 6.4
Seronegative Spondyloarthropathies

Sl. No	Disease
1	Ankylosing spondylitis
2	Reactive arthritis or Reiter's syndrome
3	Psoriatic arthropathy
4	Spondylitis associated with inflammatory bowel diseases (IBD) such as Crohn's disease and ulcerative colitis
5	Undifferentiated spondyloarthropathies

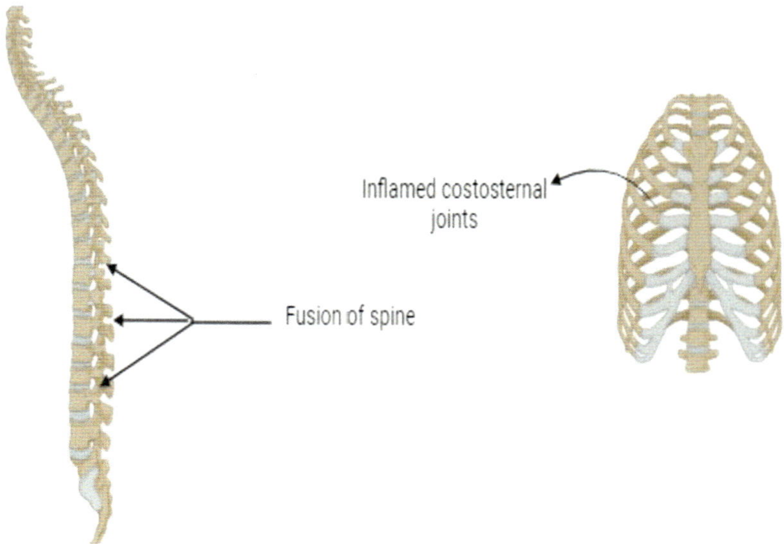

FIGURE 6.3 Ankylosing spondolytis.

These diseases are described as seronegative because of their antinuclear antibodies and negative rheumatoid factors. AS is the most prevalent seronegative spondyloarthropathy, with estimated prevalence ranging from 0.03% to 1.8%, a figure influenced by the frequency of human leucocyte antigen (HLA)-B27 within the population (Ebrahimiadib et al. 2021).

AS is a chronic autoimmune inflammatory condition primarily impacting the spine, sacroiliac joint, and peripheral joints. It gives rise to considerable pain, restricted mobility, and spinal deformities (Yang et al. 2021). In advanced cases, fibrosis and calcification occur due to persistent inflammation that leads to loss of movement and fusion of the spine—called 'bamboo spine'—that causes immobility. Clinical presentations include back pain, continuous spinal sternness, and inflammation in shoulders, hips, peripheral joints, and fingers and toes. Furthermore, extra-articular manifestations such as inflammatory bowel disease (IBD) and acute anterior uveitis are noted (Zhu et al. 2019).

The pathogenesis of AS remains incompletely elucidated due to the intricate nature of the disease. Research has indicated that AS is hereditary, with over 90% of the risk attributed to genetic factors. HLA-B27 is the primary genetic predisposing element for AS, encoded in the short arm of chromosome six as a class I major histocompatibility complex allele, which is present in approximately 95% of AS patients. Nevertheless, HLA-B27 accounts for less than 50% of overall risk, highlighting the involvement of additional genes in AS pathogenesis and susceptibility (Yang et al. 2021). HLA-B60, HLA-B61, HLA-DR8, HLA-DRB1m and MICA (MHC class I chain-related gene A) are the additional genes thought to play a role in the disease (Ebrahimiadib et al. 2021). Genome-wide association studies have identified non-HLA susceptibility genes, including IL6R, NOTCH1, IL1O, CXCR2, and TNFRSF1A, as also playing a prominent role in AS progression. However, they do not fully account for the estimated heritability, suggesting that epigenetic modifications play a role in AS susceptibility (Yang et al. 2021).

6.1.3.1 Epigenetics in AS

6.1.3.1.1 DNA Methylation

Karami et al. compared AS patients with a healthy control group and found lower B-cell chronic lymphocytic leukemia/lymphoma 11B (BCL11B) gene transcription and higher promoter methylation.

Their findings showed that messenger RNA (mRNA) overexpression and promoter hyper methylation are correlated, but not with clinical symptoms (Karami et al. 2017).

In earlier research, HLA-B27-positive AS patients' serum contained a suppressor of cytokine signalling 1 (SOCS-1) methylation, but HLA-B27-positive controls' sera did not. Furthermore, the authors found a strong correlation between the level of spondylopathy, sacroiliitis, and critical phase reactant C-reactive protein (CRP) and SOCS-1 methylation (Lai et al. 2014). In a study in Iran, AS patients exhibited greater DNMT1 promoter methylation and less DNMT1 expression than in the control group, and high DNMT1 promoter methylation was associated with the decreased DNMT1 expression. The authors proposed that DNMT1 dysregulation affected AS development by converting other target genes' methylation levels (Aslani et al. 2016).

Although HLA-B27 has been closely linked to the development of AS in numerous studies, the precise process by which the allele causes the risk of AS is still unknown. Relevant research indicates that HLA-B27 affects DNA methylation at CpG sites surrounding HLA-B sites, potentially having distinct pathophysiological implications. Moreover, aberrant methylation of the IFN regulatory factor 8 gene's promoter and hyper methylation of IL12B may contribute to the onset, progression, and pathophysiology of AS. (44). AS has been linked to DNMT3A, DNMT3B, and DNMT3L, indicating the relevance of methylation variation in the disease's pathogenesis (Whyte et al. 2019).

6.1.3.1.2 MiRNAs in AS

MiRNAs are endogenous ncRNAs that are intricately involved in the regulation of gene expression across various stages of physiological and pathological processes. Extensive evidence underscores the substantial contribution of miRNAs to autoimmune diseases. They are essential for both innate and adaptive immunity, affecting growth, specialization, and the operation of myeloid and lymphoid cells. In AS, miRNAs specifically target pathways that may be involved in the pathophysiology of the disease. Authors of 47 studies have examined the function of miRNAs in AS to date, with an emphasis on their potential as novel biomarkers and their ability to influence downstream targets linked to the illness's progression. Numerous processes such as ossification, osteoblast development, inflammatory responses, and the control of immune cells like lymphocytes and macrophages, have been linked to these miRNAs (Motta et al. 2020).

6.1.3.1.2.1 MiRNAs in New Bone Formation Fibroblasts, the most common cells in connective tissue, have a particular miRNA signature after being exposed to an osteoclast-like medium in osteogenesis and ectopic ossification in a number of illnesses, including AS. Notably, after fibroblasts were cultured with a pro-osteoclast medium, a miRNA array study revealed considerable downregulation of miR-20a, miR-300, miR-185, miR-30d, miR-320a, miR-130b, miR-33a, miR-155, and miR-222. These discovered miRNAs are anticipated to target genes including BMP2 (bone morphogenetic protein 2), osteocalcin and Runx2 (RUNT-related transcription factor 2), according to analysis performed using target gene prediction tools. These genes are essential for osteogenic distinction and have major involvement in the pathogenesis of AS. This finding implies that the interaction between fibroblasts and osteoclasts, which is controlled by miRNAs, important for the development of AS (Kumar and Roger, 2019). In addition, peripheral blood mononuclear cells from AS patients had higher expression of miR-29a, a miRNA linked to osteoblast development, than did cells from rheumatoid arthritis patients and healthy controls patients. This suggests that miR-29a could be a beneficial diagnostic marker for the development of new in AS (Huang et al. 2014). In the human osteoblast-like cell line hFOB1.19, miR-29a also increases invasion and migration, suppresses apoptosis, and encourages osteoblast proliferation. On the other hand, the effects of DKK1, one of miR-29a's targets, suggest a pathogenic role for it in bone deposition, a characteristic of AS. Notably, the bones of AS patients show higher expression of miR-29a (F. Zhang et al. 2019).

6.1.3.1.2.2 MiRNAs as Modulators of Inflammation AS patients' plasma showed different mRNA levels of let-7i, miR-16, and miR-221, which modulated Bach1 and controlled the production of pro-inflammatory IL-1β and MCP-1. Notably, among these, let-7i emerged as a potential diagnostic biomarker for AS, exhibiting higher levels in AS patients than in controls. A comprehensive analysis of the functions of noncoding genes in AS that included pathway finding, gene prediction, and network design indicated the critical participation of two microRNAs, namely miR-17–5p and miR-27b-3p that controlled inflammation and promoted osteogenic differentiation in ligament fibroblasts (C. Zhang et al. 2017).

6.1.3.1.2.3 MiRNAs as T-Cell Modulators AS patients had higher levels of let-7i, miR-16, and miR-221 than controls, with significant correlations with clinical parameters found only for miR-221 and Let-7i. Let-7i affects T-cell expression of Toll-like receptor 4 (TLR4), which produced IFNγ in vitro and contributed to the immunopathogenesis of AS (N-S Lai et al. 2013). Subsequent research revealed that Let-7i directly targets the insulin-like growth factor-1 receptor (IGF1R). Leukemia \ and AS T-cells showed down-\regulated IGF1R in the presence of Let-7i. The resulting IGF1R inhibition affected the formation of the downstream genes mTOR, Akt, PARP, and BCL-2, which resulted in self-digestion of miRNA-mediated defence to prevent T-helper cells from dying (Hou et al. 2014). In another investigation involving over 600 Chinese individuals with AS and 600 healthy controls, three different SNPs (rs2910164, rs2431697, and rs57095329) of miR-146, a well-known immune system and inflammation regulator, were analysed. The researchers observed no correlation with the disease: The frequencies of these three SNPs were consistent between cases and controls (Niu et al. 2015). Additionally, a unique miRNA profile was found for Th17 cells that were isolated from the blood of AS patients. Notably, miR-10b was upregulated and demonstrated regulatory effects on cytokine production in Th17 cells. Furthermore, miR-10b may prevent CD4+ T-cells from expressing MAP3K7, which limited the production of the pro-inflammatory cytokine IL-17A (L. Chen et al. 2017).

6.2 CONCLUSION

Epigenetic regulation, specifically through DNA methylation and microRNA modification, plays a significant role in ankylosing spondylitis and can potentially serve as diagnostic and therapeutic biomarkers. AS is the most prevalent subtype among seronegative spondyloarthropathies, and its prevalence is influenced by the frequency of HLA-B27 within the population. LncRNAs are transcription factors that can interact with protein partners and modulate their activity within the cell environment. They can also give rise to small regulatory RNAs through refinement processes. AS primarily affects the axial portion of the body, specifically the spine, sacroiliac joint, and peripheral joints. The biology and pathways of AS are not fully understood, but studies have shown a close association between HLA-B27 and AS progression.

REFERENCES

Abbas, Abul K., Andrew H. Lichtman, and Shiv Pillai. 2019. *Basic Immunology E-Book: Basic Immunology E-Book*. Elsevier Health Sciences.

Aggarwal, Bharat B., and Prashasnika Gehlot. 2009. "Inflammation and Cancer: How Friendly Is the Relationship for Cancer Patients?" *Current Opinion in Pharmacology, Cancer/Immunomodulation*, 9 (4): 351–369. https://doi.org/10.1016/j.coph.2009.06.020

Alaskhar Alhamwe, Bilal, Razi Khalaila, Johanna Wolf, Verena von Bülow, Hani Harb, Fahd Alhamdan, Charles S. Hii, et al. 2018. "Histone Modifications and Their Role in Epigenetics of Atopy and Allergic Diseases." *Allergy, Asthma & Clinical Immunology* 14 (1): 39. https://doi.org/10.1186/s13223-018-0259-4

Allis, C. David, Shelley L. Berger, Jacques Cote, Sharon Dent, Thomas Jenuwien, Tony Kouzarides, Lorraine Pillus, et al. 2007. "New Nomenclature for Chromatin-Modifying Enzymes." *Cell* 131 (4): 633–636. https://doi.org/10.1016/j.cell.2007.10.039

Aslani, Saeed, Mahdi Mahmoudi, Masoud Garshasbi, Ahmad Reza Jamshidi, Jafar Karami, and Mohammad Hossein Nicknam. 2016. "Evaluation of DNMT1 Gene Expression Profile and Methylation of Its Promoter Region in Patients with Ankylosing Spondylitis." *Clinical Rheumatology* 35 (11): 2723–2731. https://doi.org/10.1007/s10067-016-3403-x

Bennani-Baiti, Idriss M. 2012. "Integration of ERα-PELP1-HER2 Signaling by LSD1 (KDM1A/AOF2) Offers Combinatorial Therapeutic Opportunities to Circumventing Hormone Resistance in Breast Cancer." *Breast Cancer Research* 14 (5): 112. https://doi.org/10.1186/bcr3249

Boyanapalli, Sarandeep S. S., and Ah-Ng Tony Kong. 2015. "'Curcumin, the King of Spices': Epigenetic Regulatory Mechanisms in the Prevention of Cancer, Neurological, and Inflammatory Diseases." *Current Pharmacology Reports* 1 (2): 129–139. https://doi.org/10.1007/s40495-015-0018-x

Chen, L., M. H. Al-Mossawi, A. Ridley, T. Sekine, A. Hammitzsch, J. de Wit, D. Simone, et al. 2017. "miR-10b-5p Is a Novel Th17 Regulator Present in Th17 Cells from Ankylosing Spondylitis." *Annals of the Rheumatic Diseases* 76 (3): 620–625. https://doi.org/10.1136/annrheumdis-2016-210175

Chen, Linlin, Huidan Deng, Hengmin Cui, Jing Fang, Zhicai Zuo, Junliang Deng, Yinglun Li, Xun Wang, and Ling Zhao. 2017. "Inflammatory Responses and Inflammation-Associated Diseases in Organs." *Oncotarget* 9 (6): 7204–7218. https://doi.org/10.18632/oncotarget.23208

Dupont, Cathérine, D. Randall Armant, and Carol A. Brenner. 2009. "Epigenetics: Definition, Mechanisms and Clinical Perspective." *Seminars in Reproductive Medicine* 27 (5): 351–357. https://doi.org/10.1055/s-0029-1237423

Ebrahimiadib, Nazanin, Sahar Berijani, Mohammadreza Ghahari, and Fatemeh Golsoorat Pahlaviani. 2021. "Ankylosing Spondylitis." *Journal of Ophthalmic & Vision Research* 16 (3): 462–469. https://doi.org/10.18502/jovr.v16i3.9440

Egger, Gerda, Gangning Liang, Ana Aparicio, and Peter A. Jones. 2004. "Epigenetics in Human Disease and Prospects for Epigenetic Therapy." *Nature* 429 (6990): 457–463. https://doi.org/10.1038/nature02625

Feinberg, Andrew P. 2007. "Phenotypic Plasticity and the Epigenetics of Human Disease." *Nature* 447 (7143): 433–440. https://doi.org/10.1038/nature05919

Fleisher, A. S., M. Esteller, N. Harpaz, A. Leytin, A. Rashid, Y. Xu, J. Liang, et al. 2000. "Microsatellite Instability in Inflammatory Bowel Disease-Associated Neoplastic Lesions Is Associated with Hypermethylation and Diminished Expression of the DNA Mismatch Repair Gene, hMLH1." *Cancer Research* 60 (17): 4864–4868.

Fogel, O., C. Richard-Miceli, and J. Tost. 2017. "Chapter Six—Epigenetic Changes in Chronic Inflammatory Diseases." In *Advances in Protein Chemistry and Structural Biology*, edited by Rossen Donev, 106:139–189. Chromatin Remodelling and Immunity. Academic Press. https://doi.org/10.1016/bs.apcsb.2016.09.003

Fournier, Marjorie, and László Tora. 2017. "KAT2-Mediated PLK4 Acetylation Contributes to Genomic Stability by Preserving Centrosome Number." *Molecular & Cellular Oncology* 4 (2): e1270391. https://doi.org/10.1080/23723556.2016.1270391

Gibney, E. R., and C. M. Nolan. 2010. "Epigenetics and Gene Expression." *Heredity* 105 (1): 4–13. https://doi.org/10.1038/hdy.2010.54

Gomez, Ricardo Santiago, Walderez Ornelas Dutra, and Paula Rocha Moreira. 2009. "Epigenetics and Periodontal Disease: Future Perspectives." *Inflammation Research* 58 (10): 625–629. https://doi.org/10.1007/s00011-009-0041-7

Hou, Chunfeng, Mengzhu Zhu, Min Sun, and Yanliang Lin. 2014. "MicroRNA Let-7i Induced Autophagy to Protect T Cell from Apoptosis by Targeting IGF1R." *Biochemical and Biophysical Research Communications* 453 (4): 728–734. https://doi.org/10.1016/j.bbrc.2014.10.002

Huang, Jinxian, Guoxiang Song, Zhihua Yin, Xiuxia Luo, and Zhizhong Ye. 2014. "Elevated miR-29a Expression Is Not Correlated with Disease Activity Index in PBMCs of Patients with Ankylosing Spondylitis." *Modern Rheumatology* 24 (2): 331–334. https://doi.org/10.3109/14397595.2013.854077

Huber, Lars, Joanna Stanczyk, Astrid Jüngel, and Steffen Gay. 2007. "Epigenetics in Inflammatory Rheumatic Diseases." *Arthritis and Rheumatism* 56 (November): 3523–3531. https://doi.org/10.1002/art.22948

Hull, Elizabeth E., McKale R. Montgomery, and Kathryn J. Leyva. 2016. "HDAC Inhibitors as Epigenetic Regulators of the Immune System: Impacts on Cancer Therapy and Inflammatory Diseases." *BioMed Research International* 2016 (July): e8797206. https://doi.org/10.1155/2016/8797206

Hyun, Kwangbeom, Jongcheol Jeon, Kihyun Park, and Jaehoon Kim. 2017. "Writing, Erasing and Reading Histone Lysine Methylations." *Experimental & Molecular Medicine* 49 (4): e324–e324. https://doi.org/10.1038/emm.2017.11

"IJMS | Free Full-Text | From Crosstalk between Immune and Bone Cells to Bone Erosion in Infection." n.d. Accessed October 20, 2023. https://www.mdpi.com/1422-0067/20/20/5154

Jaffe, Andrew E., and Rafael A. Irizarry. 2014. "Accounting for Cellular Heterogeneity Is Critical in Epigenome-Wide Association Studies." *Genome Biology* 15 (2): R31. https://doi.org/10.1186/gb-2014-15-2-r31

Kaniskan, H. Ümit, Michael L. Martini, and Jian Jin. 2018. "Inhibitors of Protein Methyltransferases and Demethylases." *Chemical Reviews* 118 (3): 989–1068. https://doi.org/10.1021/acs.chemrev.6b00801

Karami, J., M. Mahmoudi, A. Amirzargar, M. Gharshasbi, A. Jamshidi, S. Aslani, and M. H. Nicknam. 2017. "Promoter Hypermethylation of BCL11B Gene Correlates with Downregulation of Gene Transcription in Ankylosing Spondylitis Patients | Genes & Immunity." n.d. Accessed September 29, 2023. https://www.nature.com/articles/gene201717

Kim, Sungroul. 2019. "Chapter 51—Overview of Cotinine Cutoff Values for Smoking Status Classification." In *Neuroscience of Nicotine*, edited by Victor R. Preedy, 419–431. Academic Press. https://doi.org/10.1016/B978-0-12-813035-3.00051-4

Klein, Brianna J., Marie-Eve Lalonde, Jacques Côté, Xiang-Jiao Yang, and Tatiana G. Kutateladze. 2014. "Crosstalk between Epigenetic Readers Regulates the MOZ/MORF HAT Complexes." *Epigenetics* 9 (2): 186–193. https://doi.org/10.4161/epi.26792

Kriaucionis, Skirmantas, and Nathaniel Heintz. 2009. "The Nuclear DNA Base 5-Hydroxymethylcytosine Is Present in Purkinje Neurons and the Brain." *Science* 324 (5929): 929–930. https://doi.org/10.1126/science.1169786

Lai, Ning-Sheng, Jian-Liang Chou, Gary C. W. Chen, Su-Qin Liu, Ming-Chi Lu, and Michael W. Y. Chan. 2014. "Association between Cytokines and Methylation of SOCS-1 in Serum of Patients with Ankylosing Spondylitis." *Molecular Biology Reports* 41 (6): 3773–3780. https://doi.org/10.1007/s11033-014-3242-2

Lai, N.-S., H.-C. Yu, H.-C. Chen, C.-L. Yu, H.-B. Huang, and M.-C. Lu. 2013. "Aberrant Expression of MicroRNAs in T Cells from Patients with Ankylosing Spondylitis Contributes to the Immunopathogenesis." *Clinical and Experimental Immunology* 173 (1): 47–57. https://doi.org/10.1111/cei.12089

Martin, Elizabeth M., Phillip W. Clapp, Meghan E. Rebuli, Erica A. Pawlak, Ellen Glista-Baker, Neal L. Benowitz, Rebecca C. Fry, and Ilona Jaspers. 2016. "E-Cigarette Use Results in Suppression of Immune and Inflammatory-Response Genes in Nasal Epithelial Cells Similar to Cigarette Smoke." *APSselect* 3 (8): L135–L144. https://doi.org/10.1152/ajplung.00170.2016@apsselect.2016.3.issue-8

McCarthy, Nina S., Phillip E. Melton, Gemma Cadby, Seyhan Yazar, Maria Franchina, Eric K. Moses, David A. Mackey, and Alex W. Hewitt. 2014. "Meta-Analysis of Human Methylation Data for Evidence of Sex-Specific Autosomal Patterns." *BMC Genomics* 15 (1): 981. https://doi.org/10.1186/1471-2164-15-981

Morera, Ludovica, Michael Lübbert, and Manfred Jung. 2016. "Targeting Histone Methyltransferases and Demethylases in Clinical Trials for Cancer Therapy." *Clinical Epigenetics* 8 (1): 57. https://doi.org/10.1186/s13148-016-0223-4

Motta, Francesca, Maria Cristina Carena, Carlo Selmi, and Matteo Vecellio. 2020. "MicroRNAs in Ankylosing Spondylitis: Function, Potential and Challenges." *Journal of Translational Autoimmunity* 3 (January): 100050. https://doi.org/10.1016/j.jtauto.2020.100050

Murphy, Susan K., Abayomi Adigun, Zhiqing Huang, Francine Overcash, Frances Wang, Randy L. Jirtle, Joellen M. Schildkraut, Amy P. Murtha, Edwin S. Iversen, and Cathrine Hoyo. 2012. "Gender-Specific Methylation Differences in Relation to Prenatal Exposure to Cigarette Smoke." *Gene* 494 (1): 36–43. https://doi.org/10.1016/j.gene.2011.11.062

Nagy, Z., and L. Tora. 2007. "Distinct GCN5/PCAF-Containing Complexes Function as Co-Activators and Are Involved in Transcription Factor and Global Histone Acetylation." *Oncogene* 26 (37): 5341–5357. https://doi.org/10.1038/sj.onc.1210604

Nicoletti, Carolina Ferreira, Carla Barbosa Nonino, Bruno Affonso Parenti de Oliveira, Marcela Augusta de Souza Pinhel, Maria Luisa Mansego, Fermin Ignacio Milagro, Maria Angeles Zulet, and José Alfredo Martinez. 2016. "DNA Methylation and Hydroxymethylation Levels in Relation to Two Weight Loss Strategies: Energy-Restricted Diet or Bariatric Surgery | SpringerLink." n.d. Accessed September 29, 2023. https://link.springer.com/article/10.1007/s11695-015-1802-8

Niu, Zhenmin, Jiucun Wang, Hejian Zou, Chengde Yang, Wei Huang, and Li Jin. 2015. "Common MIR146A Polymorphisms in Chinese Ankylosing Spondylitis Subjects and Controls." *PLoS ONE* 10 (9): e0137770. https://doi.org/10.1371/journal.pone.0137770

Ospelt, Caroline, Steffen Gay, and Kerstin Klein. 2017. "Epigenetics in the Pathogenesis of RA." *Seminars in Immunopathology* 39 (4): 409–419. https://doi.org/10.1007/s00281-017-0621-5

Philibert, Robert, Jeffrey Plume, Frederick Gibbons, Gene Brody, and Steven Beach. 2012. "The Impact of Recent Alcohol Use on Genome Wide DNA Methylation Signatures." *Frontiers in Genetics* 3. https://www.frontiersin.org/articles/10.3389/fgene.2012.00054

Ravid, Tommer, and Mark Hochstrasser. 2008. "Diversity of Degradation Signals in the Ubiquitin–Proteasome System." *Nature Reviews Molecular Cell Biology* 9 (9): 679–689. https://doi.org/10.1038/nrm2468

Seo, Ji-Yun, Yoon-Jung Park, Young-Ah Yi, Ji-Yun Hwang, In-Bog Lee, Byeong-Hoon Cho, Ho-Hyun Son, and Deog-Gyu Seo. 2014. "Epigenetics: General Characteristics and Implications for Oral Health." *Restorative Dentistry & Endodontics* 40 (1): 14–22. https://doi.org/10.5395/rde.2015.40.1.14

Shanmugam, Muthu K., and Gautam Sethi. 2013. "Role of Epigenetics in Inflammation-Associated Diseases." *Sub-Cellular Biochemistry* 61: 627–657. https://doi.org/10.1007/978-94-007-4525-4_27

Srinivasan, Prathiba. n.d. "The Emerging Role of Epigenetics in the Pathogenesis of Periodontal Disease a Review." Accessed September 8, 2023. https://www.academia.edu/24940860/The_emerging_role_of_epigenetics_in_the_pathogenesis_of_periodontal_disease_A_review

Su, Dan, Xuting Wang, Michelle R. Campbell, Devin K. Porter, Gary S. Pittman, Brian D. Bennett, Ma Wan, et al. 2016. "Distinct Epigenetic Effects of Tobacco Smoking in Whole Blood and among Leukocyte Subtypes." *PLoS ONE* 11 (12): e0166486. https://doi.org/10.1371/journal.pone.0166486

Surace, Anna Elisa Andrea, and Christian M. Hedrich. 2019. "The Role of Epigenetics in Autoimmune/Inflammatory Disease." *Frontiers in Immunology* 10. https://www.frontiersin.org/articles/10.3389/fimmu.2019.01525

Tong, Liping, Huan Yu, Xingyun Huang, Jie Shen, Guozhi Xiao, Lin Chen, Huaiyu Wang, Lianping Xing, and Di Chen. 2022. "Current Understanding of Osteoarthritis Pathogenesis and Relevant New Approaches." *Bone Research* 10 (1): 1–17. https://doi.org/10.1038/s41413-022-00226-9

Tsaprouni, Loukia G., Tsun-Po Yang, Jordana Bell, Katherine J. Dick, Stavroula Kanoni, James Nisbet, Ana Viñuela, et al. 2014. "Cigarette Smoking Reduces DNA Methylation Levels at Multiple Genomic Loci But the Effect Is Partially Reversible Upon Cessation." *Epigenetics* 9 (10): 1382–1396. https://doi.org/10.4161/15592294.2014.969637

Wapenaar, Hannah, and Frank J. Dekker. 2016. "Histone Acetyltransferases: Challenges in Targeting Bi-Substrate Enzymes." *Clinical Epigenetics* 8 (1): 59. https://doi.org/10.1186/s13148-016-0225-2

Weake, Vikki M., and Jerry L. Workman. 2008. "Histone Ubiquitination: Triggering Gene Activity." *Molecular Cell* 29 (6): 653–663. https://doi.org/10.1016/j.molcel.2008.02.014

Whyte, Jessica M., Jonathan J. Ellis, Matthew A. Brown, and Tony J. Kenna. 2019. "Best Practices in DNA Methylation: Lessons from Inflammatory Bowel Disease, Psoriasis and Ankylosing Spondylitis." *Arthritis Research & Therapy* 21 (1): 133. https://doi.org/10.1186/s13075-019-1922-y

Wilson, Lauren E., Sangmi Kim, Zongli Xu, Sophia Harlid, Dale P. Sandler, and Jack A. Taylor. 2015. "Non-Steroidal Anti-Inflammatory Drug Use and Genomic DNA Methylation in Blood." *PLoS ONE* 10 (9): e0138920. https://doi.org/10.1371/journal.pone.0138920

Yang, Hui, Yuting Chen, Wei Xu, Ming Shao, Jixiang Deng, Shanshan Xu, Xing Gao, et al. 2021. "Epigenetics of Ankylosing Spondylitis: Recent Developments." *International Journal of Rheumatic Diseases* 24 (4): 487–493. https://doi.org/10.1111/1756-185X.14080

Zhang, Chen, Chen Wang, Zhenyu Jia, Wenwen Tong, Delin Liu, Chongru He, Xuan Huang, and Weidong Xu. 2017. "Differentially Expressed mRNAs, lncRNAs, and miRNAs with Associated Co-Expression and ceRNA Networks in Ankylosing Spondylitis." *Oncotarget* 8 (69): 113543–113557. https://doi.org/10.18632/oncotarget.22708

Zhang, Fuwen, Kun Cao, Gongwen Du, Qi Zhang, and Zongsheng Yin. 2019. "miR-29a Promotes Osteoblast Proliferation by Downregulating DKK-1 Expression and Activating Wnt/β-Catenin Signaling Pathway." *Advances in Clinical and Experimental Medicine* 28 (10): 1293–1300. https://doi.org/10.17219/acem/104533

Zhu, Wei, Xuxia He, Kaiyuan Cheng, Linjie Zhang, Di Chen, Xiao Wang, Guixing Qiu, Xu Cao, and Xisheng Weng. 2019. "Ankylosing Spondylitis: Etiology, Pathogenesis, and Treatments." *Bone Research* 7 (1): 1–16. https://doi.org/10.1038/s41413-019-0057-8

7 Epigenetic Drug Discovery in Neurodegenerative Diseases

Madhavi Patel,[1] *Priya Shah,*[2] *and M V N L Chaitanya*[3]
[1]Bhaikaka University, Gokalnagar, Karamsad, Gujarat, India
[2]Department of Pharmacognosy, L.J. Institute of Pharmacy, L J University, Ahmedabad, India
[3]School of Pharmaceutical Sciences, Lovely Professional University, Phagwara, Punjab, India

7.1 INTRODUCTION

Neurodegenerative illnesses pose significant difficulties and are a serious health concern for contemporary nations worldwide, particularly due to the rising ageing populations. These disorders are characterised by the ongoing degeneration and death of nerve cells, which is irreversible and leads to a worsening of symptoms over time. The prevalence of these ailments often rises with advancing age, particularly Parkinson's disease (PD) and Alzheimer's disease (AD). The prevalence of these neurodegenerative illnesses is resulting in immeasurable economic and human consequences. The International Burden of Disease Study, conducted by the World Health Organization, the World Bank, and the Harvard School of Public Health, predicted that dementia and other neurological disorders would be the eighth most significant contributor to disease burdens in industrialized countries by 2020 and projected that by 2050, they will surpass cancer to become the second most prevalent cause of death worldwide (GBD 2019 Dementia Forecasting Collaborators, 2022).

These projections are merely approximations, but when combined with the existing status, they undeniably reinforce the notion that neurodegenerative illnesses are becoming a growing public concern. The complex interplay of environmental and genetic factors is responsible for most of these problems, with only a small fraction being attributed to well-defined hereditary variables such as Huntington's disease (HD). Today, identifying precise risk factors, effective biomarkers, promising targeted treatment, and even definitive finding continues to pose challenges. The pathological features observed in the brain throughout the progression of neurodegeneration exhibit significant similarities across all types of neurodegenerative disorders that affect cognitive and motor functions (Forlenza et al., 2010).

Many of the mechanisms that cause neurodegenerative disorders remains a mystery. The aggregation of intracellular proteins is a hallmark of most of these diseases, but it remains uncertain whether this is the primary cause or a consequence of another cellular dysfunction. Neurodegeneration can occur through a myriad of different pathways some of which include disruptions in protein degradation or homeostasis, changes in gene expression or transcriptional control, and mitochondrial malfunction. Improving early diagnosis and developing disease-modifying medicines requires a deeper understanding of disease pathophysiology, an area that needs attention immediately (Prohovnik et al., 2006) (Figure 7.1).

Epigenetics refers to modifications in gene expression that are not determined by changes in the DNA sequence but instead have internal and environmental influences. In the past ten years, researchers have identified changes in epigenetic playing a role in the onset and progression of neurodegenerative diseases. The epigenetic configuration provides a platform for the interaction between an individual's genetic composition and environmental influences (Hwang et al., 2017).

DOI: 10.1201/9781032706511-7

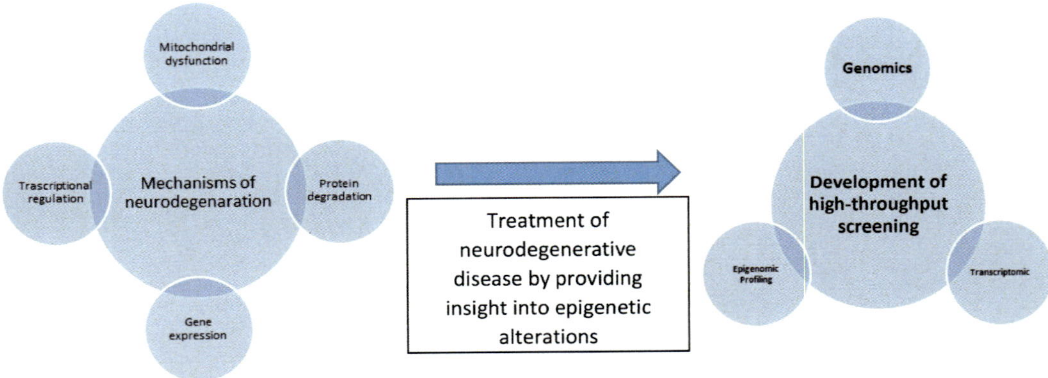

FIGURE 7.1 Mechanisms of neurodegeneration and the latest genomics treatments.

Furthermore, advancements in high-throughput methods for analysing the genome, transcriptome, and epigenome offer vital knowledge on the changes in gene expression patterns associated with dementia. Eepigenetic changes occur by histone modifications, DNA methylation, changes in coding of RNAs and noncoding RNAs (ncRNAs) that regulate gene expression part. Above all, these mechanisms regulate changes in gene and protein expression without changing DNA sequences (Xylaki et al., 2019).

Epigenetic mechanisms associated with neurons are referred to as neuroepigenetics, a field with the potential to disrupt CNS maturation and function. Hence, it is unsurprising that epigenetic alterations play a role, at least in part, in the development and progression of neurodegenerative disorders. This has led scientists to propose that targeting the epigenome using compounds that possess epigenetic properties could offer an effective strategy for preventing and treating neurodegeneration (Harrison and Dexter, 2013).

7.2 EPIGENETIC REGULATION

The growing interest in epigenetics, a rapidly advancing field in biomedical research, has led to significant advancements, including the discovery of histone variants or modifications, the identification of CpG island shores, and most importantly, the ability to analyze epigenetic marks throughout the entire genome at the level of individual nucleotides, known as epigenomic analysis. The significance of epigenetics in human disease was initially addressed in the field of oncology and subsequently expanded to include development of neurodegenerative diseases.

Heritable expression of gene alterations can result from DNA sequence modifications and epigenetic processes. This happens, for instance, in monozygotic twins whose DNA sequences are identical but whose patterns of histone modifications and DNA methylation are distinct. Cancer, neurological disorders, autoimmune diseases, and cyclic vomiting syndrome diseases are only a few complex illnesses that are affected by these variations in susceptibility. In brief, the study of different epigenetic mechanisms has enabled the mapping of crucial epigenetic marks that regulate gene expression.

7.2.1 DNA Methylation

One epigenetic mechanism is DNA methylation, which downregulates gene expression by adding a methyl group to the fifth position of cytosine monitored by guanines and formation of 5-methylcytosine in cytosine-phosphate-guanine (CpG) dinucleotide clusters (Mastroeni et al., 2010, Moore et al., 2013). Ageing, environmental changes, and disease conditions, influence methylation occurs at CpG islands and non-CpG (CpH) sites, leading to mutations or alterations in gene expression (Pi et al., 2022).

Centromeric chains, alternatively called repeating sequences, also include CpG dinucleotides; methylation of these is essential for stabilizing the chromosome and preventing translocations (Zampieri et al., 2015). Hydroxyl methylcytosine (5-hmC) is another epigenetic alteration of cytosine residue that occurs through its oxidation. Widely present throughout the CNS, 5-hmC is recognized as being linked to numerous neurodegenerative illnesses and neurodevelopmental problems. In neuronal genes, more methylation occurs because of inhibition of transcription by direct and indirect way (Fournier et al., 2012) DNA methyl transferases (DNMTs) such as DNMT1, DNMT3A, and DNMT3B (de novo pathway) induce DNA methylation. Gene expression changes and mutations of DNMT3A and 3B are associated with the development of dementia and other neurodegenerative disorders. DNMT1 maintained the de novo pathway (DNMT3A/DNMT3B) initiated by DNA methylation. Mutations in or promoter variations of the APP gene have been reported in neurodegenerative diseases (Figure 7.2).

7.2.2 Histone Acetylation

Histones, proteins that are plentiful in the nucleus, encase DNA, stabilizing the double-stranded molecule and serving as a location for gene transcription. The nucleosome is formed when histones combine to create an octamer that is then encased in the DNA. The N-terminal of histone proteins can undergo a number of post-translational modifications, such as methylation (Figure 7.2), acetylation, ADP ribosylation, and ubiquitination (Khan et al., 2023; Gräff et al., 2012; Narayan et al., 2015; Zhang et al., 2012). The most important histone alterations are acetylation and methylation (Harr et al., 2016). Histones undergo acetylation at lysine residues through the action of histone acetyl

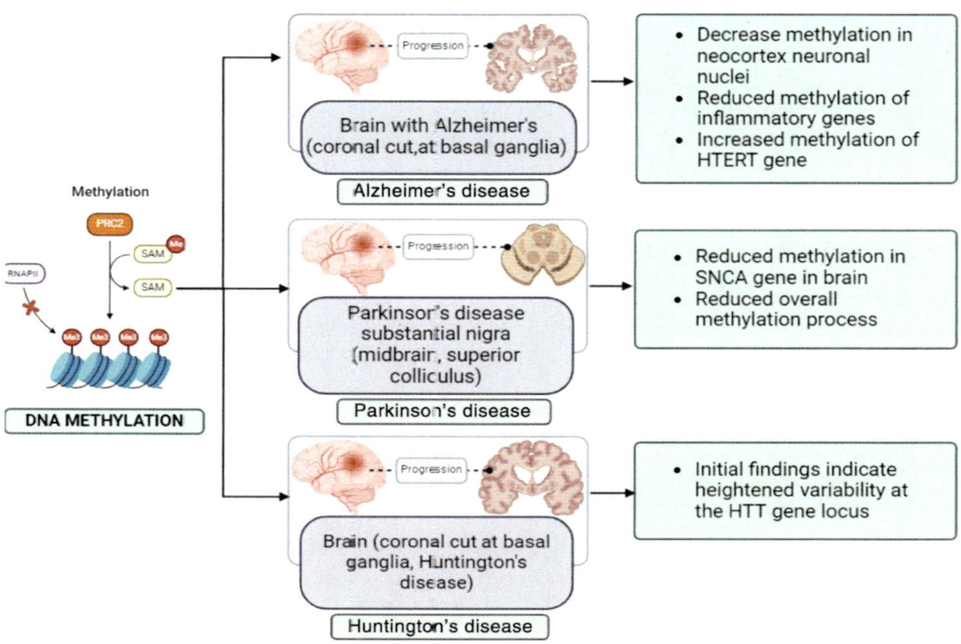

FIGURE 7.2 DNA methylation in neurodegenerative diseases.

transferases (HATs), which activates transcription by promoting the formation of a euchromatin structure. On the other hand, histone deacetylase (HDAC) induces deacetylation, resulting in transcriptional repression by promoting the formation of a heterochromatin structure. Both enzymes are essential for memory processing and synaptic plasticity. Elevated levels of histone acetylation lead to the formation of less compact chromatin regions, resulting in enhanced DNA access to chromatin-binding proteins and transcription enzymes (Harr et al., 2016).

Conversely, low acetylation has the reverse effect. Neurodegenerative disease is caused by low histone acetylation. HDAC inhibitors enhance the strength of CA1 synapses, and HDAC inhibitors have demonstrated memory augmentation. The administration of HDAC inhibitors in vitro, which increased acetylation at H3K9 and H3K14 and methylation at H3K4, restored lost memory following brain shrinkage and neuronal death. Different histone modifications affected each other and interacted DNA methylation (Harr et al., 2016).

7.2.3 Noncoding RNA-Mediated Modulation of Gene Expression

NcRNA is intricately associated with histone modifications and DNA methylation, with each exerting separate affects. Histone alterations and DNA methylation, essential epigenetic markers, are interconnected (Wei et al., 2017). NcRNAs, a diverse group comprising piRNAs, endogenous siRNAs, miRNAs, and long ncRNAs (lncRNAs) regulate gene expression after transcription by affecting mRNA stability, translation, and compartmentalization (Yang et al., 2020).

Gene expression modulates nucleosomes, which impede transcription by impeding the access of activators and transcription factors. Significantly, nucleosome was displaced at a distance of around 30 base pairs from the transcription start site (TSS), which affected the activity of RNA polymerase II. Gene activation is closely linked to the removal of nucleosomes jsut before the transcription start site (TSS), whereas the presence of nucleosomes at the TSS is connected with gene suppression (Portela and Esteller, 2010).

Chromatin remodeling complexes, such as the SWI/SNF, ISWI, CHD, and INO80 families, have a crucial function in destabilizing or removing nucleosomes via ATP hydrolysis. This dynamic mechanism controls the activation of certain genes, impacting important biological processes such as cell growth, development of form, programmed cell death, and specialization (Ho and Crabtree, 2010). In recent research, short RNAs selectively interacted with promoter regions, influencing chromatin changes that resulted in gene silencing and regulating transcription (Kim et al., 2006). In brief, the complex interaction among ncRNAs, histone modifications, DNA methylation, and nucleosomes is vital for governing gene expression in different important biological processes.

7.3 NEURODEGENERATIVE DISEASES AND EPIGENETICS

Neurodegenerative disorders such as AD, PD, HD, and prion disease are significantly impacted by environmental factors including cigarette smoking, dietary habits, and exercise. These environmental influences lead to changes in the epigenome, regulating gene expression and contributing to neuronal cell death, ultimately resulting in various neurodegenerative conditions (Tsankova et al., 2007).

7.3.1 AD

AD is a common neurological condition that occurs with age, manifests as cognitive impairment and progressive dementia, and is characterized by memory loss. AD is a complex ailment influenced by a range of factors, including genetics, metabolism, nutrition, environment, and social aspects, all playing roles in its onset and progression. The histological characteristics of AD consist of the presence of senile plaques made up of amyloid beta (Aβ) fibrils, neurofibrillary tangles formed of tau protein coupled with microtubules, and substantial loss of cholinergic neurons, especially in the neocortical association areas and hippocampus (Hardy, 2006; Ballatore et al., 2007). Additionally, aging and lifestyle-related disorders contribute to memory loss and induce epigenetic changes (Marques et al., 2013). Presenilin 1, amyloid precursor protein, and presenilin 2 (PSEN1 and PSEN2) are the genes that mutate and trigger AD.

7.3.1.1 DNA Methylation in AD

West et al. (1995) found the first alteration in DNA methylation in the APP promoter region in the temporal lobe. Studies conducted both in vivo and in vitro showed that modifications to environmental factors, such as exposure to heavy metals and vitamin B deficiency, result in methylation changes in the genes linked to the production of Aβ peptide (Eid et al., 2016; Fuso et al., 2005; Wu et al., 2008). Researchers discovered that the different stages of Alzheimer's disease and dementia can be linked to substantial alterations in 5-mC or 5-hmC levels (Bradley-Whitman and Lovell, 2013; Ellison et al., 2017).

Studies confirmed that the hypo-/hypermethylation of various genes (PRRT1, PPT2, PPT2-EGFL8, PRDM16, LOC100507547, CDH23, RNF39, and C10orf105) responsible for the involvement of epigenetics through the regulation of several cellular pathways, for instance synaptic plasticity and intracellular communication (Piaceri et al., 2015; Tannorella et al., 2015). In a study on the role of genes in AD progression via DNA methylation in neuronal and blood DNA, methylation of BDNF led to cognitive impairment, followed by changes in 5 hmC as AD progressed (Xie et al., 2017a; Xie et al., 2017b; Irier and Jin, 2012; Song et al., 2011). Mladenova et al. (2018) studied and identified the roles of epigenetic mutants of the ADARB2 gene in the hippocampus responsible for symptoms learning dysfunction, impaired synaptic plasticity, and diminishing memory in monozygotic and dizygotic twin pairs of AD patients (Mladenova et al., 2018; Konki et al., 2019). It was also observed that the total 5-hmC levels in the mitochondrial DNA in the different parts of brain samples were higher in AD patients (Mposhi et al., 2017).

Even though it is now clear that differential methylation plays a role in the etiology and course of AD, there are contradicting results that raise questions about the specific role of DNA in AD pathogenesis. The enzyme NEP breaks down β-amyloid proteins, and AD pathophysiology is characterized by a dearth of NEP that results in the buildup of amyloid plaques (Nagata et al., 2018; Turner et al., 2004). While there is ongoing debate regarding the connection between NEP and AD, there is compelling evidence for a significant link between epigenetics and AD, emphasizing the importance of further investigating epigenetic interventions for AD.

7.3.1.2 Histone Modifications in AD

According to research on transgenic mouse models of AD, the link between histone changes and the disease progression is significant. Histone H4 undergoes hypermethylation and activates acetylation at lysine 12 (H4K12ac) when amyloid proteins begin to build up in the brain. At every stage of AD, alterations in H3K27me3 and H3K4me3 have been found (Watson et al., 2016).

7.3.1.3 NcRNAs in AD

NcRNAs has a major role in regulating the structure of chromatin and gene expression. They have been associated with AD due to their involvement in crucial processes related to brain development and cognition. Distinct variations in miRNA expression have been observed between individuals with sporadic AD and individuals of the same age without the disease. However, it is still uncertain if these alterations are a result or a consequence of the disease progression. BACE1 is primarily responsible for the production of Aβ peptide in both the bloodstream and the brain. BACE1 is controlled by miRNA-107 and long ncRNAs and produces brain Aβ peptide (Watson et al., 2016). The dysregulation of miRNA-107 and the lncRNAs miR-33, involved in lipid metabolism, and miR-34a and miR-155, involved in neuroinflammation, plays a role in the epigenetic development of AD (Kim et al., 2015; Schipper et al., 2007; Guedes et al., 2014).

7.3.2 PD

PD is a degenerative condition that affects 1%–3% of those people 65 and older. It is marked by involuntary shaking at rest, muscle stiffness, slowed movement, and difficulty maintaining balance while walking (Zhang et al., 2022; Pringsheim et al., 2021; Livingston and Monroe-Duprey, 2023). While levodopa substitute and deep neurological stimulation help alleviate PD symptoms, there is currently no medication available that can halt the advancement of the condition. An estimated 6.3 million individuals worldwide are affected by this degenerative condition, and this figure is projected to rise to 8.3 million by 2030 (Dorsey et al., 2007).

Loss of dopaminergic neurons from the substantia nigra pars compacta (SNCA) and the buildup of intracellular protein inclusions dubbed Lewy bodies (LBs), predominantly made of alpha-synuclein (aSyn), are the pathological hallmarks of PD (Dickson et al., 2009; Spillantini et al., 1997). PD is attributed to genetic mutations in the SNCA and PD genes, which are responsible for both sporadic cases of PD with unknown etiology and familial cases of PD. Moreover, mutations in various genes such as PD loci of PARK1-15, which include Parkin, SNCA, LRRK2, Dj-1, and related genes like MAPT, SMPD1, GBA, and LRP10, have been reported as interacting the cellular molecules in mitochondria and inducing neurodegeneration (Coppedè, 2012; Yu et al., 2015; Song et al., 2022; Panicker et al., 2021).

Research has shown that epigenetic controls, in addition to differences in the PD genes themselves, can alter the production of proteins and messenger RNAs linked to disease pathophysiology. Epigenetic changes change gene and chromosome expression while leaving the gene's DNA sequence unchanged. So along with genes, other factors such as nutrition and environment could be responsible for the development of PD.

7.3.2.1 DNA Methylation in PD

PD patients display substantial dysregulation of CpG island methylation. Researchers have observed accumulations of LBs with high expression of α-synuclein (Espay et al., 2020) and hypermethylation of the SNCA promoter, resulting in metabolic changes in the middle brain (Frieling et al., 2007). Higher expression of SNCA was found when SNCA intron 1 was inhibited by methylation. Additionally, the SNs, putamens, and cortices of patients with PD showed noticeably less methylation than in normal individuals (Jowaed et al., 2010; Cai et al., 2011; Sohrabi et al., 2023; Panicker et al., 2021).

7.3.2.2 Histone Modifications in PD

Research has demonstrated that increased acetylation of the histones H2A, H3, and H4 is associated with the pathogenesis of PD, and chromatin remodelling is a significant factor primarily due to histone modification. The alterations in histone modification can be induced by exposure to toxins such as the pesticides dieldrin and paraquat. These neurotoxins are implicated in the heightened histone acetylation in mesencephalic dopaminergic neurons, attributed to proteasomal dysfunction. Furthermore, the administration of MPTP results in dopamine depletion accompanied by reduced histone H3 trimethylation at Lys4 or by H3 acetylation. This played a role in the development and persistence of levodopa-induced dyskinesia (LDID) in PD, possibly by activating the D1 receptor (Nicholas et al., 2008; Darmopil et al., 2009; Södersten et al., 2014).

Moreover, deposition of an aSyn co-localized with different types of H1 and H3 decreases the acetylation and reduces the overexpression of dHDAC6 in various animal models. The reduced dHDAC6 improved the damaging properties of aSyn overexpression, such as the loss of or damage to dopaminergic neurons and locomotor dysfunction (Du et al., 2010). Analytical techniques such as immunoblotting, chromatin immunoprecipitation sequencing (ChIP-seq), and RNA sequencing were performed and showed acetylation in various histones during the clinical studies of PD patients (Toker et al., 2021). The results showed significant hyperacetylation of H2BK15, H3K9/14, H3K27, H3K56, and H4K12 followed by upregulation of SIRT1 and SIRT3 in PD patients (Figure 7.3).

Increased acetylation of H3K27 is associated with PRKN, SNCA, MAPT, and PARK7 genes (Dickson, 2018; Lim et al., 2020).

7.3.2.3 NcRNA in PD

Both ncRNAs and miRNAs are actively involved in PD for gene expression regulation. MiR-133b is responsible for maturation regulation and functioning of dopaminergic neurons in the midbrain. PD patients' midbrains exhibited deficiency of miR-133b (Kim et al., 2007). Lower expression of Pitx3- transcription factor is responsible for low expression of miR-133b; in the healthy brain, it promotes gene expression of dopaminergic neurons and increased miR-133b transcription. In contrast, PD brains exhibited overexpression of miR-133b, which decreases Pitx3 protein level via negative feedback (Martinat et al., 2006). Meanwhile, concentrations of α-synuclein are regulated by miR-7 and miR-153. These miRNAs bind to 3′UTR of α-synuclein and are responsible for the downregulation of translation and transcription (Coppedè, 2012), and in a study on the expression of miRNAs in *Caenorhabditis elegans* in a PD model, some miRNAs were underexpressed (Asikainen et al., 2010).

7.3.3 HD

HD, a neurodegenerative genetic disease, affects medium spiny nerves present in striatum of brain. Unsteady expansion of cytosine-adenine-guanine (CAG) repeats in the Huntington gene (HTT) produces mutant huntingtin (mHtt) with a tract of toxic polyQ (polyglutamine) at N-terminus (Landles and Bates, 2004; MacDonald et al., 1993). Mutant Htt expression initiates a series of detrimental events in a cell. Neuronal functions disrupted by this protein have been documented in numerous in vitro and in vivo animal models, revealing many cellular events such as mitochondrial and synaptic dysfunction, transcriptional dysregulation, vesicular transport deficit, and proteostasis impairment and ultimately resulting in neuronal dysfunction and death (Jimenez-Sanchez et al., 2017). The severity of HD depends on the expansion of CAG repeats, where chronic neuronal

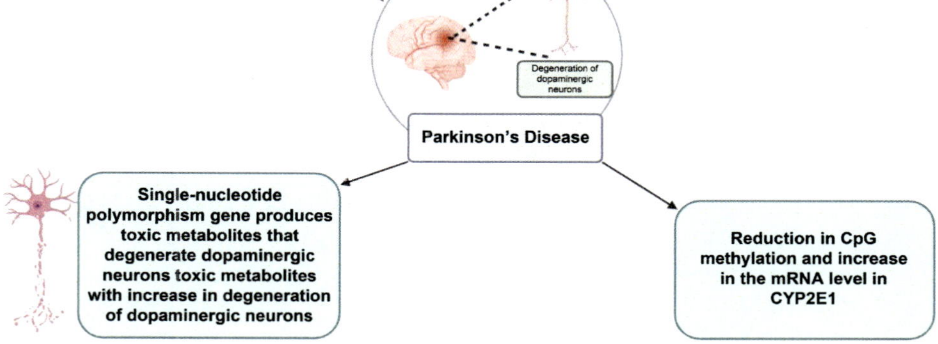

FIGURE 7.3 Epigenetic targets involved in PD pathogenesis.

dysfunction leads to neuronal death, ultimately contributing to the development of mania, depression, apathy, suicide, motor skill learning deficiency, and dystonia.

7.3.3.1 Histone Modifications in HD

Along with DNA methylation, histone modifications control the structure of chromatin and aids in defining gene expression. Many reports have summarised histone acetylation and deacetylation (Sharma and Taliyan, 2015). Overexpression of mutant protein Htt changes histone acetylation levels. Reduced H3 acetylation is observed in the striatum of HD mice at the promoter of the genes Drd2, Actb, Penk1, or Grin1, which are typically underexpressed for the function of the striatal neurons. Due to acetylation, these genes are upregulated (Sadri-Vakili et al., 2007). Studies also reports augmented H3K9me2 and H3K9me3 that causes heterochromatin structure in HD patients' striata (Zsindely and Bodai, 2018).

Methylation of various lysine residues demonstrated different transcriptional outcomes (Kimura, 2013). A mouse HD model and human HD patients displayed reduced H3K4me3 at promoters of specific downregulated genes (Vashishtha et al., 2013). A study using neural progenitor cells and embryonic stem cells demonstrates CAG repeat alterations in H3K4me3 due to mutant Htt, which correlated with gene expression changes (Biagioli et al., 2015). In a Drosophila model of HD, a systematic genetic interaction study identified H3K27me2/3 demethylase UTX as a promising therapeutic target, where UTX inhibition was shown to be neuroprotective (Song et al., 2018). The frontal cortex and blood of HD patients showed overexpressed H2AFY, a member Y of the H2A histone family (Hu et al., 2011). The striata of HD patients showed increased monoubiquitylation at uH2A, which affected transcription. Studies on mice model demonstrated hypoacetylation in H3K9,14 and H4K12 (Valor et al., 2013).

7.3.3.2 DNA Methylation in HD

Despite the significant research that has been conducted so far, no substantial points toward a function for DNA methylation in HD. Researchers did find that alterations in DNA methylation occurred due to mutant Htt at both promoter proximal and distal regulatory regions (Francelle et al., 2017). Future researchers should explore how mutant Htt promotes abnormal DNA methylation to specific gene loci, which is still unclear. Potential pathways include DNA methylation in particular genomic areas and enhanced DNMT enzyme activity by mutant Htt expression. In the context of mutant Htt, these pathways could be mediated by abnormal protein–protein interactions or by posttranslational changes of DNMTs (Hyeon et al., 2021).

7.3.4 Prion Disease

Prion diseases are infectious and lethal neuroconditions, with Creutzfeldt–Jakob Disease) being one of the most common; it is found in humans but closely resemble spongiform encephalopathy in cattle. Infectious prion protein results from specific conformational changes or misfolded forms in the protein. Sporadic CJD results in rapid neurodegeneration associated with myoclonus, cerebellar ataxia, and motor and sensory abnormality (Viré and Mead, 2023).

Another inherited prion disease is fatal familial insomnia, which is associated with protein folding, revealing unusual epigenetics. Any sudden change in temperature, pH, and intracellular metabolites cause environmental stress. The incidence of prion switching increases under extreme stress, followed by epigenetic modifications that occur in extreme conditions (Halfmann and Lindquist, 2010).

Alterations in DNA methylation have been much less explored in human prion diseases. Some studies have documented miRNA profiling in prion diseases. Mice infected with prion disease showed a unique time course of deregulation of hsa-miR-16-5p as the disease advanced. Specifically, expression is increased in key neurons cultured during the pre-symptomatic phase but downregulated throughout the clinically developed phase of the disease (Norsworthy et al., 2020). Another study observed elevated hsa-miR-16-5p levels in the sCJD frontal cortex in humans (Llorens et al., 2018). Table 7.1 presents gene mutations that commonly cause neurodegenerative diseases.

TABLE 7.1
Gene Mutations Involved in Neurodegenerative Diseases

Disease	Mutation	Reference
AD	APP promoter hypomethylation	(Guedes et al., 2014)
	Elevated PSEN1 in the cortex	(Monti et al., 2020)
	PSEN1 hypomethylation	(Monti et al., 2020)
	Hypomethylated in CREB-regulated transcription factor 1	(Mendioroz et al., 2016)
	Elevated interleukin-6 and interleukin-1	(Tekpli et al., 2013)
	Altered methylation of SORL1 HLA-DRB5, ABCA7, BIN1 and SLC24A4	(Yu et al., 2015)
	UQCRC1 gene hypermethylation (inflammation and oxidative stress)	(Ma et al., 2016)
	Hypo hydroxyl methylation of ANK1 gene	(Ma et al., 2016)
	Hypermethylation of PLD3 gene	(Blanco-Luquin et al., 2018)
Parkinson's disease	Hypermethylation of MOG, KCTD5, HLA-DQA1, VAV2, TRIM10, GFPT2, ARHGEF10, TMEM9, HLA-DRB5, MAPT, MRI1, HLA-DRB6, LASS3, GSTTP2, GSTTP and hypomethylation of DNAJA3, JAKMIP3, FRK, LRRC27, LGALS7, DMBX1, TUBA3E, FOXK1, MAGI2, SLC25A24, ME 886, APBA1, GSTT1, MYOM2, and TMCO3	(Wüllner et al., 2016)
Huntington Disease	Hypermethylation of Sox2, Nes, and Pax6	(Wood, 2013)

7.4 EPIGENETIC MODULATORS FOR POTENTIAL TREATMENT FOR NEURODEGENERATION

Neurodegeneration is a process both disease and gene–environment interactions play important roles. Therefore, potential treatment opportunities lie in early-stage detection as well as intervention targeting epigenetic changes.

The epigenome can provide treatment options for neurodegenerative diseases. DNMTs catalyse DNA methylation, and regulating this process presents a promising therapeutic option in this context. The DNMT inhibitors decitabine and azacitidine were approved by the US FDA for treating malignancies, but they are toxic and quite unstable. In a study on cultured dopaminergic neurons, decitabine aggravated neurotoxicity and upregulated genes related to PD (Wang et al., 2013). In AD, small molecule DNMT inhibitors with substantial potential include those that bind with DNMTs directly and are not integrated into DNA, like the well-known inhibitor 5-AZA in neurons, as DNA methylation modulators is a promising potential effective treatment option even in neurons (Cuadrado-Tejedor et al., 2013). In another study with AD patients, a diet with vitamin B and folates, poor methyl donors, was responsible for PSEN1 gene hypomethylation that produced Aβ, a key factor in AD symptoms. Supplementation of SAM restored PSEN1 methylation, which ameliorated the AD symptoms (Fuso et al., 2012). The HDAC inhibitors vorinostat, panobinostat, and romidepsin increased motor and cognitive functions in animal models with neurodegenerative diseases (Coppedè, 2021; Coppedè, 2014). However, clinical trials in humans are desirable.

7.4.1 Reported Natural Constituents for Epigenetic Mechanisms

Several natural origin constituents were reported to exert good epigenetic mechanisms have been effective in treating or preventing neurodegeneration. Many are available in the diverse foods we consume (Table 7.2).

TABLE 7.2
Natural Constituents and Their Epigenetic Mechanisms

Phytoconstituent	Mechanism	Target Disease	Reference(s)
Curcumin	Inhibits DNMT expression; modulates histone methylation, acetylation, and phosphorylation; regulates miRNAs	AD	(Bahramsoltani et al., 2017; Hassan et al., 2019)
Resveratrol	DNA methylation; histone modifications; miRNA regulation mediated in part by the SIRT1-serine/threonine protein kinase ROCK1 and SIRT1-Akt pathways, which enhance α-secretase activity and reduce Aβ plaque formation	AD	(Chen et al., 2005; Feng et al., 2013; Zhang et al., 2014; Granzotto and Zatta, 2011)
AtreMorine	Increases DNA methylation, enhances catecholamine (dopamine, noradrenaline, adrenaline) levels	AD PD	(Cacabelos, 2017; Cacabelos et al., 2016)
Nosustrophine	Modulates DNA methylation and SIRT and HDAC activity and expression	AD	(Martínez-Iglesias et al., 2022)
Epigallocatechin-3-gallate polyphenolic compounds	Inhibits DNMT to effectively alter the DNA methylation profile, reduces α-synuclein aggregation in PD	AD, PD	(Liu et al., 2005; Lee et al., 2009)
Sulforaphane (obtained from broccoli, sprouts, cabbage and kale)	Reduces DNMT1 and DNMT3a; increases H3 acetylation at K9 and H4, decreases H3 trimethylation at K9 and K27	Apoptosis	(Meeran et al., 2010)
Velvet bean extract	Reduces free radicals, effective antioxidant	PD	(Lampariello et al., 2012)
Maslinic acid	Protects cells from free radicals by upholding VEGF and HIF-1α expression	AD	(Lozano-Mena et al., 2014)
Methyl jasmonate	Inhibits oxidative stress and down regulates α-synuclein expression	PD	(Alabi et al., 2019)
Vitamins			
Vitamin B12	Regulates DNA methylation	AD	(Mohajeri et al., 2015; Athanasopoulos et al., 2016)

7.5 PROSPECTS

With advanced testing facilities and gene expression studies, we are developing more understanding of the role of epigenetics in neurodegeneration, and the already widely available literature on animal and human clinical trials has revealed epigenetic mechanisms and expanded treatment options. Epigenetic markers can improve motor and cognitive function and synaptic plasticity and are potentially favourable novel agents for treatment without toxicity and druggability. Chemical agents and phytochemicals targeted for treating neurodegenerative must be optimized based on their pharmacokinetic profiles and ability to cross the blood–brain barrier, which could open new avenues of drug design and development.

REFERENCES

Alabi, A. O., Ajayi, A. M., Ben-Azu, B., Omorobge, O. & Umukoro, S. 2019. Methyl Jasmonate Ameliorates Rotenone-Induced Motor Deficits in Rats Through Its Neuroprotective Activity and Increased Expression of Tyrosine Hydroxylase Immunopositive Cells. *Metab Brain Dis*, 34, 1723–1736.

Asikainen, S., Rudgalvyte, M., Heikkinen, L., Louhiranta, K., Lakso, M., Wong, G. & Nass, R. 2010. Global MicroRNA Expression Profiling of Caenorhabditis Elegans 'Parkinson's Disease Models. *J Mol Neurosci*, 41, 210–218.

Athanasopoulos, D., Karagiannis, G. & Tsolaki, M. 2016. Recent Findings in Alzheimer Disease and Nutrition Focusing on Epigenetics. *Adv Nutr*, 7, 917–927.

Bahramsoltani, R., Rahimi, R. & Farzaei, M. H. 2017. Pharmacokinetic Interactions of Curcuminoids with Conventional Drugs: A Review. *J Ethnopharmacol*, 209, 1–12.

Ballatore, C., Lee, V. M. & Trojanowski, J. Q. 2007. Tau-Mediated Neurodegeneration in 'Alzheimer's Disease and Related Disorders. *Nat Rev Neurosci*, 8, 663–672.

Biagioli, M., Ferrari, F., Mendenhall, E. M., Zhang, Y., Erdin, S., Vijayvargia, R., Vallabh, S. M., Solomos, N., Manavalan, P., Ragavendran, A., Ozsolak, F., Lee, J. M., Talkowski, M. E., Gusella, J. F., Macdonald, M. E., Park, P. J. & Seong, I. S. 2015. Htt CAG Repeat Expansion Confers Pleiotropic Gains of Mutant Huntingtin Function in Chromatin Regulation. *Hum Mol Genet*, 24, 2442–2457.

Blanco-Luquin, I., Altuna, M., Sánchez-Ruiz de Gordoa, J., Urdánoz-Casado, A., Roldán, M., Cámara, M., Zelaya, V., Erro, M. E., Echavarri, C. & Mendioroz, M. 2018. PLD3 Epigenetic Changes in the Hippocampus of 'Alzheimer's Disease. *Clin Epigenetics*, 10, 116.

Bradley-Whitman, M. A. & Lovell, M. A. 2013. Epigenetic Changes in the Progression of 'Alzheimer's Disease. *Mech Ageing Dev*, 134, 486–495.

Cacabelos, R. 2017. 'Parkinson's Disease: From Pathogenesis to Pharmacogenomics. *Int J Mol Sci*, 18.

Cacabelos, R., Fernández-Novoa, L., Alejo, R., Corzo, L., Rodríguez, S., Alcaraz, M., Nebril, L., Cacabelos, P., Fraile, C., Carrera, I. & Carril, J. C. 2016. E-PodoFavalin-15999 (Atremorine®)-Induced Neurotransmitter and Hormonal Response in Parkinson's Disease. *Journal of Exploratory Research in Pharmacology*, 1, 1–12.

Cai, L., Sutter, B. M., Li, B. & Tu, B. P. 2011. Acetyl-CoA Induces Cell Growth and Proliferation by Promoting the Acetylation of Histones at Growth Genes. *Mol Cell*, 42, 426–437.

Chen, J., Zhou, Y., Mueller-Steiner, S., Chen, L. F., Kwon, H., Yi, S., Mucke, L. & Gan, L. 2005. SIRT1 Protects Against Microglia-Dependent Amyloid-Beta Toxicity Through Inhibiting NF-kappaB Signaling. *J Biol Chem*, 280, 40364–40374.

Coppedè, F. 2012. Genetics and Epigenetics of 'Parkinson's Disease. *Sci World J*, 2012, 489830.

Coppedè, F. 2014. The Potential of Epigenetic Therapies in Neurodegenerative Diseases. *Front Genet*, 5, 220.

Coppedè, F. 2021. Epigenetic Regulation in 'Alzheimer's Disease: Is It a Potential Therapeutic Target? *Expert Opin Ther Targets*, 25, 283–298.

Cuadrado-Tejedor, M., Oyarzabal, J., Lucas, M. P., Franco, R. & García-Osta, A. 2013. Epigenetic Drugs in Alzheimer's Disease. *Biomol Concepts*, 4, 433–445.

Darmopil, S., Martín, A. B., de Diego, I. R., Ares, S. & Moratalla, R. 2009. Genetic Inactivation of Dopamine D1 But Not D2 Receptors Inhibits L-DOPA-Induced Dyskinesia and Histone Activation. *Biol Psychiatry*, 66, 603–613.

Dickson, D. W. 2018. Neuropathology of Parkinson Disease. *Parkinsonism Relat Disord*, 46(Suppl 1), S30–s33.

Dickson, D. W., Braak, H., Duda, J. E., Duyckaerts, C., Gasser, T., Halliday, G. M., Hardy, J., Leverenz, J. B., Del Tredici, K., Wszolek, Z. K. & Litvan, I. 2009. Neuropathological Assessment of 'Parkinson's Disease: Refining the Diagnostic Criteria. *Lancet Neurol*, 8, 1150–1157.

Dorsey, E. R., Constantinescu, R., Thompson, J. P., Biglan, K. M., Holloway, R. G., Kieburtz, K., Marshall, F. J., Ravina, B. M., Schifitto, G., Siderowf, A. & Tanner, C. M. 2007. Projected Number of People with Parkinson Disease in the Most Populous Nations, 2005 Through 2030. *Neurology*, 68, 384–386.

Du, G., Liu, X., Chen, X., Song, M., Yan, Y., Jiao, R. & Wang, C. C. 2010. Drosophila Histone Deacetylase 6 Protects Dopaminergic Neurons Against {Alpha}-Synuclein Toxicity by Promoting Inclusion Formation. *Mol Biol Cell*, 21, 2128–2137.

Eid, A., Bihaqi, S. W., Renehan, W. E. & Zawia, N. H. 2016. Developmental Lead Exposure and Lifespan Alterations in Epigenetic Regulators and Their Correspondence to Biomarkers of 'Alzheimer's Disease. *Alzheimers Dement (Amst)*, 2, 123–131.

Ellison, E. M., Abner, E. L. & Lovell, M. A. 2017. Multiregional Analysis of Global 5-Methylcytosine and 5-Hydroxymethylcytosine Throughout the Progression of 'Alzheimer's Disease. *J Neurochem*, 140, 383–394.

Espay, A. J., Kalia, L. V., Gan-Or, Z., Williams-Gray, C. H., Bedard, P. L., Rowe, S. M., Morgante, F., Fasano, A., Stecher, B., Kauffman, M. A., Farrer, M. J., Coffey, C. S., Schwarzschild, M. A., Sherer, T., Postuma, R. B., Strafella, A. P., Singleton, A. B., Barker, R. A., Kieburtz, K., Olanow, C. W., Lozano, A., Kordower, J. H., Cedarbaum, J. M., Brundin, P., Standaert, D. G. & Lang, A. E. 2020. Disease Modification and Biomarker Development in Parkinson Disease: Revision or Reconstruction? *Neurology*, 94, 481–494.

Feng, X., Liang, N., Zhu, D., Gao, Q., Peng, L., Dong, H., Yue, Q., Liu, H., Bao, L., Zhang, J., Hao, J., Gao, Y., Yu, X. & Sun, J. 2013. Resveratrol Inhibits β-Amyloid-Induced Neuronal Apoptosis Through Regulation of SIRT1-ROCK1 Signaling Pathway. *PLoS ONE*, 8, e59888.

Forlenza, O. V., Diniz, B. S. & Gattaz, W. F. 2010. Diagnosis and Biomarkers of Predementia in 'Alzheimer's Disease. *BMC Med*, 8, 89.

Fournier, A., Sasai, N., Nakao, M. & Defossez, P. A. 2012. The Role of Methyl-Binding Proteins in Chromatin Organization and Epigenome Maintenance. *Brief Funct Genomics*, 11, 251–264.

Francelle, L., Lotz, C., Outeiro, T., Brouillet, E. & Merienne, K. 2017. Contribution of Neuroepigenetics to Huntington's Disease. *Front Hum Neurosci*, 11.

Frieling, H., Gozner, A., Römer, K. D., Lenz, B., Bönsch, D., Wilhelm, J., Hillemacher, T., De Zwaan, M., Kornhuber, J. & Bleich, S. 2007. Global DNA Hypomethylation and DNA Hypermethylation of the Alpha Synuclein Promoter in Females with Anorexia Nervosa. *Mol Psychiatry*, 12, 229–230.

Fuso, A., Nicolia, V., Ricceri, L., Cavallaro, R. A., Isopi, E., Mangia, F., Fiorenza, M. T. & Scarpa, S. 2012. S-Adenosylmethionine Reduces the Progress of the Alzheimer-like Features Induced by B-Vitamin Deficiency in Mice. *Neurobiol Aging*, 33, 1482.e1–e16.

Fuso, A., Seminara, L., Cavallaro, R. A., D'Anselmi, F. & Scarpa, S. 2005. S-Adenosylmethionine/Homocysteine Cycle Alterations Modify DNA Methylation Status with Consequent Deregulation of PS1 and BACE and Beta-Amyloid Production. *Mol Cell Neurosci*, 28, 195–204.

GBD 2019 dementia forecasting collaborators. 2022. Estimation of the Global Prevalence of Dementia in 2019 and Forecasted Prevalence in 2050: An Analysis for the Global Burden of Disease Study 2019. *Lancet Public Health*, 7, e105–e125.

Gräff, J., Rei, D., Guan, J. S., Wang, W. Y., Seo, J., Hennig, K. M., Nieland, T. J., Fass, D. M., Kao, P. F., Kahn, M., Su, S. C., Samiei, A., Joseph, N., Haggarty, S. J., Delalle, I. & Tsai, L. H. 2012. An Epigenetic Blockade of Cognitive Functions in the Neurodegenerating Brain. *Nature*, 483, 222–226.

Granzotto, A. & Zatta, P. 2011. Resveratrol Acts Not Through Anti-Aggregative Pathways But Mainly Via Its Scavenging Properties Against Aβ and Aβ-Metal Complexes Toxicity. *PLoS ONE*, 6, e21565.

Guedes, J. R., Custódia, C. M., Silva, R. J., De Almeida, L. P., Pedroso de Lima, M. C. & Cardoso, A. L. 2014. Early MiR-155 Upregulation Contributes to Neuroinflammation in 'Alzheimer's Disease Triple Transgenic Mouse Model. *Hum Mol Genet*, 23, 6286–6301.

Halfmann, R. & Lindquist, S. 2010. Epigenetics in the Extreme: Prions and the Inheritance of Environmentally Acquired Traits. *Science*, 330, 629–632.

Hardy, J. 2006. 'Alzheimer's Disease: The Amyloid Cascade Hypothesis: An Update and Reappraisal. *J Alzheimers Dis*, 9, 151–153.

Harr, J. C., Gonzalez-Sandoval, A. & Gasser, S. M. 2016. Histones and Histone Modifications in Perinuclear Chromatin Anchoring: From Yeast to Man. *EMBO Rep*, 17, 139–155.

Harrison, I. F. & Dexter, D. T. 2013. Epigenetic Targeting of Histone Deacetylase: Therapeutic Potential in 'Parkinson's Disease? *Pharmacol Ther*, 140, 34–52.

Hassan, F.-U., Rehman, M. S.-U., Khan, M. S., Ali, M. A., Javed, A., Nawaz, A. & Yang, C. 2019. Curcumin as an Alternative Epigenetic Modulator: Mechanism of Action and Potential Effects. *Front Genet*, 10.

Ho, L. & Crabtree, G. R. 2010. Chromatin Remodelling During Development. *Nature*, 463, 474–484.

Hu, Y., Chopra, V., Chopra, R., Locascio, J. J., Liao, Z., Ding, H., Zheng, B., Matson, W. R., Ferrante, R. J., Rosas, H. D., Hersch, S. M. & Scherzer, C. R. 2011. Transcriptional Modulator H2A Histone Family, Member Y (H2AFY) Marks Huntington Disease Activity in Man and Mouse. *Proc Natl Acad Sci U S A*, 108, 17141–17146.

Hwang, J. Y., Aromolaran, K. A. & Zukin, R. S. 2017. The Emerging Field of Epigenetics in Neurodegeneration and Neuroprotection. *Nat Rev Neurosci*, 18, 347–361.

Hyeon, J. W., Kim, A. H. & Yano, H. 2021. Epigenetic Regulation in 'Huntington's Disease. *Neurochem Int*, 148, 105074.

Irier, H. A. & Jin, P. 2012. Dynamics of DNA Methylation in Aging and 'Alzheimer's Disease. *DNA Cell Biol*, 31(Suppl 1), S42–S48.

Jimenez-Sanchez, M., Licitra, F., Underwood, B. R. & Rubinsztein, D. C. 2017. 'Huntington's Disease: Mechanisms of Pathogenesis and Therapeutic Strategies. *Cold Spring Harb Perspect Med*, 7.

Jowaed, A., Schmitt, I., Kaut, O. & Wüllner, U. 2010. Methylation Regulates Alpha-Synuclein Expression and Is Decreased in 'Parkinson's Disease 'Patients' Brains. *J Neurosci*, 30, 6355–6359.

Khan, S. S., Khatik, L. G. & Datusalia, K. A. 2023. Strategies for Treatment of Disease-Associated Dementia Beyond 'Alzheimer's Disease: An Update. *Curr Neuropharmacol*, 21, 309–339.

Kim, D. H., Villeneuve, L. M., Morris, K. V. & Rossi, J. J. 2006. Argonaute-1 Directs siRNA-Mediated Transcriptional Gene Silencing in Human Cells. *Nat Struct Mol Biol*, 13, 793–797.

Kim, J., Inoue, K., Ishii, J., Vanti, W. B., Voronov, S. V., Murchison, E., Hannon, G. & Abeliovich, A. 2007. A MicroRNA Feedback Circuit in Midbrain Dopamine Neurons. *Science*, 317, 1220–1224.

Kim, J., Yoon, H., Horie, T., Burchett, J. M., Restivo, J. L., Rotllan, N., Ramírez, C. M., Verghese, P. B., Ihara, M., Hoe, H. S., Esau, C., Fernández-Hernando, C., Holtzman, D. M., Cirrito, J. R., Ono, K. & Kim, J. 2015. MicroRNA-33 Regulates ApoE Lipidation and Amyloid-β Metabolism in the Brain. *J Neurosci*, 35, 14717–14726.

Kimura, H. 2013. Histone Modifications for Human Epigenome Analysis. *J Hum Genet*, 58, 439–445.

Konki, M., Malonzo, M., Karlsson, I. K., Lindgren, N., Ghimire, B., Smolander, J., Scheinin, N. M., Ollikainen, M., Laiho, A., Elo, L. L., Lönnberg, T., Röyttä, M., Pedersen, N. L., Kaprio, J., Lähdesmäki, H., Rinne, J. O. & Lund, R. J. 2019. Peripheral Blood DNA Methylation Differences in Twin Pairs Discordant for 'Alzheimer's Disease. *Clin Epigenetics*, 11, 130.

Lampariello, L. R., Cortelazzo, A., Guerranti, R., Sticozzi, C. & Valacchi, G. 2012. The Magic Velvet Bean of Mucuna Pruriens. *J Tradit Complement Med*, 2, 331–339.

Landles, C. & Bates, G. P. 2004. Huntingtin and the Molecular Pathogenesis of 'Huntington's Disease. Fourth in Molecular Medicine Review Series. *EMBO Rep*, 5, 958–963.

Lee, J. W., Lee, Y. K., Ban, J. O., Ha, T. Y., Yun, Y. P., Han, S. B., Oh, K. W. & Hong, J. T. 2009. Green Tea (-)-Epigallocatechin-3-Gallate Inhibits Beta-Amyloid-Induced Cognitive Dysfunction Through Modification of Secretase Activity Via Inhibition of ERK and NF-kappaB Pathways in Mice. *J Nutr*, 139, 1987–1993.

Lim, C., Shimizu, J., Kawano, F., Kim, H. J. & Kim, C. K. 2020. Adaptive Responses of Histone Modifications to Resistance Exercise in Human Skeletal Muscle. *PLoS ONE*, 15, e0231321.

Liu, H. L., Chen, Y., Cui, G. H. & Zhou, J. F. 2005. Curcumin, a Potent Anti-Tumor Reagent, Is a Novel Histone Deacetylase Inhibitor Regulating B-NHL Cell Line Raji Proliferation. *Acta Pharmacol Sin*, 26, 603–609.

Livingston, C. & Monroe-Duprey, L. 2023. A Review of Levodopa Formulations for the Treatment of 'Parkinson's Disease Available in the United States. *J Pharm Pract*, 8971900221151194.

Llorens, F., Thüne, K., Martí, E., Kanata, E., Dafou, D., Díaz-Lucena, D., Vivancos, A., Shomroni, O., Zafar, S., Schmitz, M., Michel, U., Fernández-Borges, N., Andréoletti, O., Del Río, J. A., Díez, J., Fischer, A., Bonn, S., Sklaviadis, T., Torres, J. M., Ferrer, I. & Zerr, I. 2018. Regional and Subtype-Dependent MiRNA Signatures in Sporadic Creutzfeldt-Jakob Disease Are Accompanied by Alterations in MiRNA Silencing Machinery and Biogenesis. *PLoS Pathog*, 14, e1006802.

Lozano-Mena, G., Sánchez-González, M., Juan, M. E. & Planas, J. M. 2014. Maslinic Acid, a Natural Phytoalexin-Type Triterpene from Olives — A Promising Nutraceutical? *Molecules*, 19, 11538–11559.

Ma, S. L., Tang, N. L. & Lam, L. C. 2016. Association of Gene Expression and Methylation of UQCRC1 to the Predisposition of 'Alzheimer's Disease in a Chinese Population. *J Psychiatr Res*, 76, 143–147.

MacDonald, M. E., Ambrose, C. M., Duyao, M. P., Myers, R. H., Lin, C., Srinidhi, L., Barnes, G., Taylor, S. A., James, M., Groot, N., Macfarlane, H., Jenkins, B., Anderson, M. A., Wexler, N. S., Gusella, J. F., Bates, G. P., Baxendale, S., Hummerich, H., Kirby, S., North, M., Youngman, S., Mott, R., Zehetner, G., Sedlacek, Z., Poustka, A., Frischauf, A.-M., Lehrach, H., Buckler, A. J., Church, D., Doucette-Stamm, L., O'donovan, M. C., Riba-Ramirez, L., Shah, M., Stanton, V. P., Strobel, S. A., Draths, K. M., Wales, J. L., Dervan, P., Housman, D. E., Altherr, M., Shiang, R., Thompson, L., Fielder, T., Wasmuth, J. J., Tagle, D., Valdes, J., Elmer, L., Allard, M., Castilla, L., Swaroop, M., Blanchard, K., Collins, F. S., Snell, R., Holloway, T., Gillespie, K., Datson, N., Shaw, D. & Harper, P. S. 1993. A Novel Gene Containing a Trinucleotide Repeat That Is Expanded and Unstable on 'Huntington's Disease Chromosomes. *Cell*, 72, 971–983.

Marques, F., Sousa, J. C., Sousa, N. & Palha, J. A. 2013. Blood–Brain-Barriers in Aging and in Alzheimer's Disease. *Mol Neurodegener*, 8, 38.

Martinat, C., Bacci, J. J., Leete, T., Kim, J., Vanti, W. B., Newman, A. H., Cha, J. H., Gether, U., Wang, H. & Abeliovich, A. 2006. Cooperative Transcription Activation by Nurr1 and Pitx3 Induces Embryonic Stem Cell Maturation to the Midbrain Dopamine Neuron Phenotype. *Proc Natl Acad Sci U S A*, 103, 2874–2879.

Martínez-Iglesias, O., Naidoo, V., Carrera, I., Corzo, L. & Cacabelos, R. 2022. Nosustrophine: An Epinutraceutical Bioproduct with Effects on DNA Methylation, Histone Acetylation and Sirtuin Expression in 'Alzheimer's Disease. *Pharmaceutics*, 14.

Mastroeni, D., Grover, A., Delvaux, E., Whiteside, C., Coleman, P. D. & Rogers, J. 2010. Epigenetic Changes in 'Alzheimer's Disease: Decrements in DNA Methylation. *Neurobiol Aging*, 31, 2025–2037.

Meeran, S. M., Patel, S. N. & Tollefsbol, T. O. 2010. Sulforaphane Causes Epigenetic Repression of hTERT Expression in Human Breast Cancer Cell Lines. *PLoS ONE*, 5, e11457.

Mendioroz, M., Celarain, N., Altuna, M., Sánchez-Ruiz de Gordoa, J., Zelaya, M. V., Roldán, M., Rubio, I., Larumbe, R., Erro, M. E., Méndez, I. & Echávarri, C. 2016. CRTC1 Gene Is Differentially Methylated in the Human Hippocampus in 'Alzheimer's Disease. *Alzheimers Res Ther*, 8, 15.

Mladenova, D., Barry, G., Konen, L. M., Pineda, S. S., Guennewig, B., Avesson, L., Zinn, R., Schonrock, N., Bitar, M., Jonkhout, N., Crumlish, L., Kaczorowski, D. C., Gong, A., Pinese, M., Franco, G. R., Walkley, C. R., Vissel, B. & Mattick, J. S. 2018. Adar3 is Involved in Learning and Memory in Mice. *Front Neurosci*, 12.

Mohajeri, M. H., Troesch, B. & Weber, P. 2015. Inadequate Supply of Vitamins and DHA in the Elderly: Implications for Brain Aging and Alzheimer-Type Dementia. *Nutrition*, 31, 261–275.

Monti, N., Cavallaro, R. A., Stoccoro, A., Nicolia, V., Scarpa, S., Kovacs, G. G., Fiorenza, M. T., Lucarelli, M., Aronica, E., Ferrer, I., Coppedè, F., Troen, A. M. & Fuso, A. 2020. CpG and Non-CpG Presenilin1 Methylation Pattern in Course of Neurodevelopment and Neurodegeneration Is Associated with Gene Expression in Human and Murine Brain. *Epigenetics*, 15, 781–799.

Moore, L. D., Le, T. & Fan, G. 2013. DNA Methylation and Its Basic Function. *Neuropsychopharmacology*, 38, 23–38.

Mposhi, A., van der Wijst, M. G., Faber, K. N. & Rots, M. G. 2017. Regulation of Mitochondrial Gene Expression, the Epigenetic Enigma. *Front Biosci (Landmark Ed)*, 22, 1099–1113.

Nagata, K., Mano, T., Murayama, S., Saido, T. C. & Iwata, A. 2018. DNA Methylation Level of the Neprilysin Promoter in 'Alzheimer's Disease Brains. *Neurosci Lett*, 670, 8–13.

Narayan, P. J., Lill, C., Faull, R., Curtis, M. A. & Dragunow, M. 2015. Increased Acetyl and Total Histone Levels in Post-Mortem 'Alzheimer's Disease Brain. *Neurobiol Dis*, 74, 281–294.

Nicholas, A. P., Lubin, F. D., Hallett, P. J., Vattem, P., Ravenscroft, P., Bezard, E., Zhou, S., Fox, S. H., Brotchie, J. M., Sweatt, J. D. & Standaert, D. G. 2008. Striatal Histone Modifications in Models of Levodopa-Induced Dyskinesia. *J Neurochem*, 106, 486–494.

Norsworthy, P. J., Thompson, A. G. B., Mok, T. H., Guntoro, F., Dabin, L. C., Nihat, A., Paterson, R. W., Schott, J. M., Collinge, J., Mead, S. & Viré, E. A. 2020. A Blood MiRNA Signature Associates with Sporadic Creutzfeldt-Jakob Disease Diagnosis. *Nat Commun*, 11, 3960.

Panicker, N., Ge, P., Dawson, V. L. & Dawson, T. M. 2021. The Cell Biology of 'Parkinson's Disease. *J Cell Biol*, 220.

Pi, T., Lang, G., Liu, B. & Shi, J. 2022. Protective Effects of *Dendrobium nobile* Lindl. Alkaloids on 'Alzheimer's Disease-Like Symptoms Induced by High-Methionine Diet. *Curr Neuropharmacol*, 20, 983–997.

Piaceri, I., Raspanti, B., Tedde, A., Bagnoli, S., Sorbi, S. & Nacmias, B. 2015. Epigenetic Modifications in 'Alzheimer's Disease: Cause or Effect? *J Alzheimers Dis*, 43, 1169–1173.

Portela, A. & Esteller, M. 2010. Epigenetic Modifications and Human Disease. *Nat Biotechnol*, 28, 1057–1068.

Pringsheim, T., Day, G. S., Smith, D. B., Rae-Grant, A., Licking, N., Armstrong, M. J., De Bie, R. M. A., Roze, E., Miyasaki, J. M., Hauser, R. A., Espay, A. J., Martello, J. P., Gurwell, J. A., Billinghurst, L., Sullivan, K., Fitts, M. S., Cothros, N., Hall, D. A., Rafferty, M., Hagerbrant, L., Hastings, T., O'Brien, M. D., Silsbee, H., Gronseth, G. & Lang, A. E. 2021. Dopaminergic Therapy for Motor Symptoms in Early Parkinson Disease Practice Guideline Summary: A Report of the AAN Guideline Subcommittee. *Neurology*, 97, 942–957.

Prohovnik, I., Perl, D. P., Davis, K. L., Libow, L., Lesser, G. & Haroutunian, V. 2006. Dissociation of Neuropathology from Severity of Dementia in Late-Onset Alzheimer Disease. *Neurology*, 66, 49–55.

Sadri-Vakili, G., Bouzou, B., Benn, C. L., Kim, M. O., Chawla, P., Overland, R. P., Glajch, K. E., Xia, E., Qiu, Z., Hersch, S. M., Clark, T. W., Yohrling, G. J. & Cha, J. H. 2007. Histones Associated with Downregulated Genes are Hypo-Acetylated in 'Huntington's Disease Models. *Hum Mol Genet*, 16, 1293–1306.

Schipper, H. M., Maes, O. C., Chertkow, H. M. & Wang, E. 2007. MicroRNA Expression in Alzheimer Blood Mononuclear Cells. *Gene Regul Syst Bio*, 1, 263–274.

Sharma, S. & Taliyan, R. 2015. Transcriptional Dysregulation in 'Huntington's Disease: The Role of Histone Deacetylases. *Pharmacol Res*, 100, 157–169.

Södersten, E., Feyder, M., Lerdrup, M., Gomes, A. L., Kryh, H., Spigolon, G., Caboche, J., Fisone, G. & Hansen, K. 2014. Dopamine Signaling Leads to Loss of Polycomb Repression and Aberrant Gene Activation in Experimental Parkinsonism. *PLoS Genet*, 10, e1004574.

Sohrabi, N., Habibi, L. & Akrami, S. M. 2023. Regulatory Role of Insulin on Endogenous L1 ORF1 and NEFM Gene Expression Through PI3K Signaling Pathway Specifically in Neuroblastoma Cell Line. *Iran J Public Health*, 52, 633–645.

Song, C. X., Shi, J., Zhang, P., Zhang, Y., Xu, J., Zhao, L., Zhang, R., Wang, H. & Chen, H. 2022. Immunotherapy for 'Alzheimer's Disease: Targeting β-Amyloid and Beyond. *Transl Neurodegener*, 11, 18.

Song, C. X., Szulwach, K. E., Fu, Y., Dai, Q., Yi, C., Li, X., Li, Y., Chen, C. H., Zhang, W., Jian, X., Wang, J., Zhang, L., Looney, T. J., Zhang, B., Godley, L. A., Hicks, L. M., Lahn, B. T., Jin, P. & He, C. 2011. Selective Chemical Labeling Reveals the Genome-Wide Distribution of 5-Hydroxymethylcytosine. *Nat Biotechnol*, 29, 68–72.

Song, W., Zsindely, N., Faragó, A., Marsh, J. L. & Bodai, L. 2018. Systematic Genetic Interaction Studies Identify Histone Demethylase Utx as Potential Target for Ameliorating 'Huntington's Disease. *Hum Mol Genet*, 27, 649–666.

Spillantini, M. G., Schmidt, M. L., Lee, V. M., Trojanowski, J. Q., Jakes, R. & Goedert, M. 1997. Alpha-Synuclein in Lewy Bodies. *Nature*, 388, 839–840.

Tannorella, P., Stoccoro, A., Tognoni, G., Petrozzi, L., Salluzzo, M. G., Ragalmuto, A., Siciliano, G., Haslberger, A., Bosco, P., Bonuccelli, U., Migliore, L. & Coppedè, F. 2015. Methylation Analysis of Multiple Genes in Blood DNA of 'Alzheimer's Disease and Healthy Individuals. *Neurosci Lett*, 600, 143–147.

Tekpli, X., Landvik, N. E., Anmarkud, K. H., Skaug, V., Haugen, A. & Zienolddiny, S. 2013. DNA Methylation at Promoter Regions of Interleukin 1B, Interleukin 6, and Interleukin 8 in Non-Small Cell Lung Cancer. *Cancer Immunol Immunother*, 62, 337–345.

Toker, L., Tran, G. T., Sundaresan, J., Tysnes, O.-B., Alves, G., Haugarvoll, K., Nido, G. S., Dölle, C. & Tzoulis, C. 2021. Genome-Wide Histone Acetylation Analysis Reveals Altered Transcriptional Regulation in the Parkinson's Disease Brain. *Mol Neurodegener*, 16, 31.

Tsankova, N., Renthal, W., Kumar, A. & Nestler, E. J. 2007. Epigenetic Regulation in Psychiatric Disorders. *Nat Rev Neurosci*, 8, 355–367.

Turner, A. J., Fisk, L. & Nalivaeva, N. N. 2004. Targeting Amyloid-Degrading Enzymes as Therapeutic Strategies in Neurodegeneration. *Ann N Y Acad Sci*, 1035, 1–20.

Valor, L. M., Guiretti, D., Lopez-Atalaya, J. P. & Barco, A. 2013. Genomic Landscape of Transcriptional and Epigenetic Dysregulation in Early Onset Polyglutamine Disease. *J Neurosci*, 33, 10471–10482.

Vashishtha, M., Ng, C. W., Yildirim, F., Gipson, T. A., Kratter, I. H., Bodai, L., Song, W., Lau, A., Labadorf, A., Vogel-Ciernia, A., Troncosco, J., Ross, C. A., Bates, G. P., Krainc, D., Sadri-Vakili, G., Finkbeiner, S., Marsh, J. L., Housman, D. E., Fraenkel, E. & Thompson, L. M. 2013. Targeting h3k4 Trimethylation in Huntington Disease. *Proc Natl Acad Sci U S A*, 110, E3027–E3036.

Viré, E. A. & Mead, S. 2023. Gene Expression and Epigenetic Markers of Prion Diseases. *Cell Tissue Res*, 392, 285–294.

Wang, Y., Wang, X., Li, R., Yang, Z. F., Wang, Y. Z., Gong, X. L. & Wang, X. M. 2013. A DNA Methyltransferase Inhibitor, 5-aza-2'-Deoxycytidine, Exacerbates Neurotoxicity and Upregulates 'Parkinson's Disease-Related Genes in Dopaminergic Neurons. *CNS Neurosci Ther*, 19, 183–190.

Watson, C. T., Roussos, P., Garg, P., Ho, D. J., Azam, N., Katsel, P. L., Haroutunian, V. & Sharp, A. J. 2016. Genome-Wide DNA Methylation Profiling in the Superior Temporal Gyrus Reveals Epigenetic Signatures Associated with Alzheimer's Disease. *Genome Med*, 8, 5.

Wei, J.-W., Huang, K., Yang, C. & Kang, C.-S. 2017. Non-Coding RNAs as Regulators in Epigenetics (Review). *Oncol Rep*, 37, 3–9.

West, R. L., Lee, J. M. & Maroun, L. E. 1995. Hypomethylation of the Amyloid Precursor Protein Gene in the Brain of an 'Alzheimer's Disease Patient. *J Mol Neurosci*, 6, 141–146.

Wood, H. 2013. Neurodegenerative Disease: Altered DNA Methylation and RNA Splicing Could Be Key Mechanisms in Huntington Disease. *Nat Rev Neurol*, 9, 119.

Wu, J., Basha, M. R., Brock, B., Cox, D. P., Cardozo-Pelaez, F., Mcpherson, C. A., Harry, J., Rice, D. C., Maloney, B., Chen, D., Lahiri, D. K & Zawia, N. H. 2008. 'Alzheimer's Disease (AD)-Like Pathology in Aged Monkeys after Infantile Exposure to Environmental Metal Lead (Pb): Evidence for a Developmental Origin and Environmental Link for AD. *J Neurosci*, 28, 3–9.

Wüllner, U., Kaut, O., Deboni, L., Piston, D. & Schmitt, I. 2016. DNA Methylation in 'Parkinson's Disease. *J Neurochem*, 139(Suppl 1), 108–120.

Xie, B., Liu, Z., Liu, W., Jiang, L., Zhang, R., Cui, D., Zhang, Q. & Xu, S. 2017a. DNA Methylation and Tag SNPS of the BDNF Gene in Conversion of Amnestic Mild Cognitive Impairment into 'Alzheimer's Disease: A Cross-Sectional Cohort Study. *J Alzheimers Dis*, 58, 263–274.

Xie, B., Xu, Y., Liu, Z., Liu, W., Jiang, L., Zhang, R., Cui, D., Zhang, Q. & Xu, S. 2017b. Elevation of Peripheral BDNF Promoter Methylation Predicts Conversion from Amnestic Mild Cognitive Impairment to 'Alzheimer's Disease: A 5-Year Longitudinal Study. *J Alzheimers Dis*, 56, 391–401.

Xylaki, M., Atzler, B. & Outeiro, T. F. 2019. Epigenetics of the Synapse in Neurodegeneration. *Curr Neurol Neurosci Rep*, 19, 72.

Yang, X., Liu, M., Li, M., Zhang, S., Hiju, H., Sun, J., Mao, Z., Zheng, M. & Feng, B. 2020. Epigenetic Modulations of Noncoding RNA: A Novel Dimension of Cancer Biology. *Mol Cancer*, 19, 64.

Yu, L., Chibnik, L. B., Srivastava, G. P., Pochet, N., Yang, J., Xu, J., Kozubek, J., Obholzer, N., Leurgans, S. E., Schneider, J. A., Meissner, A., De Jager, P. L. & Bennett, D. A. 2015. Association of Brain DNA Methylation in SORL1, ABCA7, HLA-DRB5, SLC24A4, and BIN1 with Pathological Diagnosis of Alzheimer Disease. *JAMA Neurol*, 72, 15–24.

Zampieri, M., Ciccarone, F., Calabrese, R., Franceschi, C., Bürkle, A. & Caiafa, P. 2015. Reconfiguration of DNA Methylation in Aging. *Mech Ageing Dev*, 151, 60–70.

Zhang, J., Feng, X., Wu, J., Xu, H., Li, G., Zhu, D., Yue, Q., Liu, H., Zhang, Y., Sun, D., Wang, H. & Sun, J. 2014. Neuroprotective Effects of Resveratrol on Damages of Mouse Cortical Neurons Induced by β-Amyloid Through Activation of SIRT1/Akt1 Pathway. *Biofactors*, 40, 258–267.

Zhang, K., Schrag, M., Crofton, A., Trivedi, R., Vinters, H. & Kirsch, W. 2012. Targeted Proteomics for Quantification of Histone Acetylation in 'Alzheimer's Disease. *Proteomics*, 12, 1261–1268.

Zhang, X., Molsberry, S. A., Schwarzschild, M. A., Ascherio, A. & Gao, X. 2022. Association of Diet and Physical Activity with All-Cause Mortality Among Adults with Parkinson Disease. *JAMA Netw Open*, 5, e2227738.

Zsindely, N. & Bodai, L. 2018. Histone Methylation in 'Huntington's Disease: Are Bivalent Promoters the Critical Targets? *Neural Regen Res*, 13, 1191–1192.

8 *In Silico* Tools, Molecular Docking Knowledge, and Epigenetic Drug Discovery

Prabha Thangavelu,[1] Pradeep Kumar MR,[2] Jawahar Natarajan,[3] Veerasamy Ravichandran,[4] and Muthu Kumaradoss Kathiravan[5]
[1]Nandha College of Pharmacy, Erode, Tamil Nadu, India
[2]KLE College of Pharmacy, Hubballi, Karnataka, India
[3]JSS Academy of Higher Education & Research, Ooty, Tamil Nadu, India
[4]Faculty of Pharmacy, AIMST University, Semeling, Malaysia
[5]SRMIST, Kattankulathur, Tamil Nadu, India

8.1 INTRODUCTION

Epigenetics is a term for relationships between environmental factors like nutrition, drugs, air, water, pollution, etc. and gene expression. Epigenetics modulates genomic functions, resulting in heritable modifications to gene expression that occur outside of the DNA sequence. Changes in epigenetic factors dysregulate cell functioning and gene expression, leading to complex diseases. Epigenetics is built on key molecular modifications to nucleosomes—the building blocks of chromatin. These covalent alterations, including methylation, acetylation, phosphorylation, and ubiquitination, precisely regulate downstream patterns of gene expression in a context-specific way. The dynamic regulation of epigenetic variation influence a range of biological processes, including DNA damage, gene transcription, genome reprogramming, and homeostasis management, each contributing to diverse functional effects. The direct link between epigenetic dysfunction and the etiology and progression of diseases such as immune-mediated, metabolic, cardiovascular, and neurodegenerative diseases and cancers emphasizes the significance of these covalent alterations (Ptak and Petronis, 2008; Heerboth et al., 2014; Lu et al., 2018a; Lundstrom, 2017).

Writers, readers, and erasers are the three main categories of proteins that do the alterations. Writers enzymes catalyse modifications including ADP-ribosylation, SUMOylation, phosphorylation, acetylation, ubiquitination, and methylation. Acetylation is exclusively linked to histones, whereas methylation occurs on both DNA and histone proteins. These changes influence not only the recruitment of additional molecules like chromatin remodelers and non-coding RNAs but also the affinity for binding between DNA and histone proteins (Rossetto et al., 2012).

Readers are the enzymes that identify the epigenetic markers inserted by writers to recruit downstream effectors to carry out an extensive range of functions. Methyllysine, methylarginine, phosphoserine, and acetyllysine are a few of the epigenetic readers that have been identified. They possess a specific domain that allows them to recognize and bind to writer-added histones and DNA alterations (Biswas and Rao, 2018; Lu et al., 2018a).

DOI: 10.1201/9781032706511-8

Depending on the demand of the cell to maintain the epigenetic marks in order to change the expression states of the gene locus or alter the transcription condition of the cell, Eraser enzymes remove covalent alterations introduced by writers, adapting to cell demands for maintaining epigenetic marks, altering gene transcription states, and modifying transcription conditoins. Erasers are categorized into many groups based on their respective roles and substrate preferences. These families include histone demethyltransferases (HDMs), DNA demethylases (DNMTs), and histone deacetylases (HDACs), amongst others (Biswas and Rao, 2018; Lu et al., 2018a). Figure 8.1 displays the components of epigenetic protein readers, writers, and erasers that do alterations.

In recent decades, the ability to reverse unwanted epigenetic alterations has prompted curiosity about medications to treat diseases of the epigenome (Francis, 2011). Though the fundamental processes of the epigenome have become better understood, problems with instability *in vivo* and a lack of specificity have made it challenging to design targeted medications (Foulks et al., 2012). Computational techniques as additional aspects of the drug discovery process might help in the creation of epigenetic medicines.

Today, there is much ongoing research on writers, erasers, and readers in epigenetics since they are thought to be potential pharmacological targets for designing compounds. (Ganesan et al., 2019; Lundstrom, 2017). Computers are crucial in drug design research because they enable creative and effective methods that reduce the time and expense associated with drug discovery (Yu and MacKerell, 2017). One computational approach is molecular docking, which enables predicting molecular and energetic patterns that regulate recognition as well as molecular connections (Morris and Lim-Wilby, 2008).

This chapter provides a summary of the main techniques used in computational drug development today. The primary interest is using computational methods to design bioactive compounds that target genes associated with epigenetic control. These targets have been linked to several illnesses, including coronary artery disease, cancer, and mental disorders.

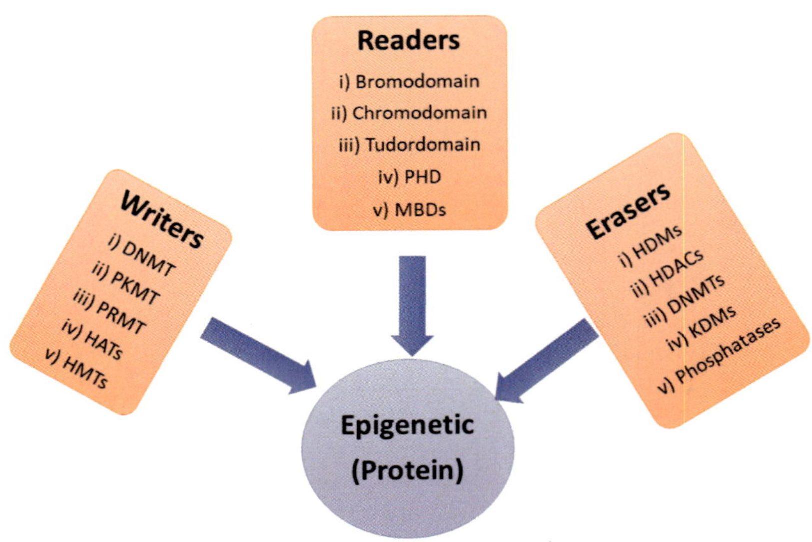

FIGURE 8.1 Components of epigenetic protein readers, writers, and erasers that do alterations.

Notes: DNMT—DNA methyltransferase; HAT—Histone acetyltransferase; HMT—Histone methyltransferase; HDAC—Histone deacetylase; HDM—Histone demethyltransferase; KDM—Histone demethylase; PKMT—Protein lysine methyltransferase; PRMT—Protein arginine methyltransferase.

8.1.1 *In Silico* Tools

In silico tools and molecular docking knowledge play a crucial role in epigenetic drug discovery for several reasons:

- **Efficiency:** Studying the interactions of potential drug compounds with particular epigenetic targets, such as enzymes and proteins, is a crucial step in the development of epigenetic drugs. With the aid of *in silico* tools, researchers can quickly screen huge libraries of compounds to find candidates for further investigation.
- **Cost-effectiveness:** Laboratory experiments can be time-consuming and expensive to run. *In silico* methods are more economical because they eliminate the need for physical compound screening.
- **Target identification:** By examining the structure and operation of epigenetic enzymes and proteins, *in silico* tools can be used to identify and validate potential epigenetic drug targets.
- **Lead optimization:** Using molecular docking, scientists can foresee the molecular interactions of potential drug candidates with their target proteins. This will aid in optimizing lead compounds for greater binding affinity and effectiveness.
- **Virtual screening:** Virtual screening makes it possible to quickly identify compounds that have the potential to bind to particular epigenetic targets.
- **Polypharmacology prediction:** Epigenetic drug development often targets multiple pathways at the same time, guided by polypharmacology predictions. *In silico* tools enable predicting polypharmacological effects, facilitating the design of drugs with broader therapeutic potential.
- **Drug repurposing:** *In silico* techniques support identifying current medications that could potentially have an impact on the epigenome and repurpose them for use in new indications.
- **Reducing side effects:** *In silico* methods can predict how substances will interact with off-target proteins, reducing the possibility of negative side effects.
- **Data integration:** To provide a thorough understanding of epigenetic mechanisms and drug interactions, *in silico* tools can integrate a variety of data sources, including structural biology, genomics, and pharmacology.
- **Cutting down on laboratory work:** By reducing the number of potential compounds and optimizing those using computational techniques, scientists can cut down on the time and materials needed for experimental work.
- **Mechanism of action:** Researchers use molecular docking to understand the molecular processes by which potential drugs interact with epigenetic targets. This makes it easier to comprehend how the substances impact gene expression and epigenetic changes.
- **Inhibition of epigenetic enzymes:** Molecular docking is used to design and evaluate compounds that inhibit the activity of epigenetic enzymes such as HDACs and DNMTs.
- **Prediction of binding sites:** By predicting potential binding sites on epigenetic proteins or enzymes, molecular docking aids in the development of drugs that target specific regions.

8.1.2 Current Computational Approaches in Epigenetic Drug Discovery

Computational techniques are effective tools for drug development and discovery; they offer a way to find possible medications and therapeutic targets while also improving our understanding of epigenetic processes in the context of epigenetic drug development (Mishra, 2011).

In silico molecular docking studies that analyze binding to specific receptors can provide insights into binding shape, pattern, and affinity. The development of epigenetic regulators has been expedited by the use of computational approaches in the lead identification, design, and selection of new molecules. Computational approaches are also helpful for optimization stages and reducing drug

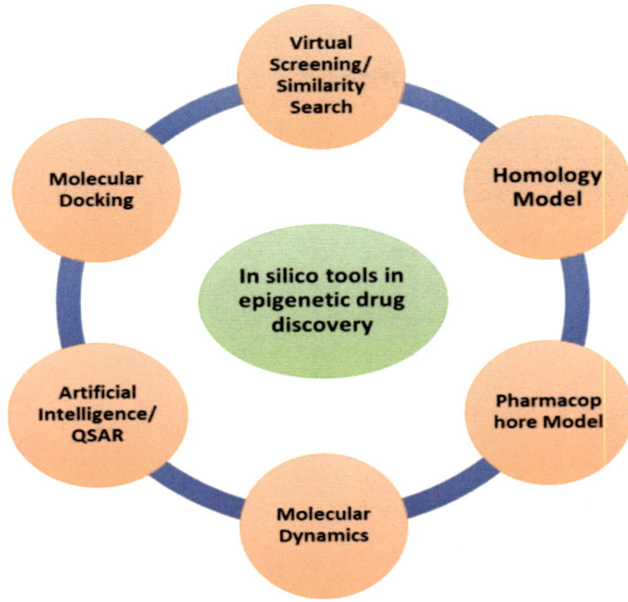

FIGURE 8.2 *In silico* tools used in epigenetic drug discovery.

development costs. *In silico* techniques including homology, virtual screening, docking, pharmacophore generation, and MD simulation have been applied in recent decades to understand the mechanism of action of current medications and to develop novel epigenetic antagonists (Figure 8.2).

Computational biology techniques select previously described molecules to identify compounds with a higher probability of targeting effectiveness in experimental studies. More complex methods—such as molecular docking, structure-activity analysis, and pharmacophore modeling—are applied based on available data to determine binding affinities (Aguayo-Ortiz and Fernández-De Gortari, 2016). Describes the diverse *in silico* tools available in epigenetic drug discovery.

8.2 VIRTUAL SCREENING

Virtual screening has proven to be a successful approach to direct lead optimization and hit discovery. It has also been used to find promising epigenetic medications (Medina-Franco and Caulfield, 2011; Medina-Franco et al., 2015; Yoo and Medina-Franco, 2012). Virtual screening was employed in the early days to find new DNMT inhibitors. By employing molecular modeling, an initial set of 1990 molecules from the National Cancer Institute was screened, and the top two ranked molecules were further confirmed in *in vitro* and *in vivo* studies (Siedlecki et al., 2006). After screening an additional pool of 260,000 molecules from the National Cancer Institute database, 65,000 molecules were chosen. Thirteen of the 24 molecules that were produced by molecular modeling were selected for experimental testing: Seven exhibited measurable DNMT inhibitory action, and six were specific for the DNMT isoform DNMT3B (Kuck et al., 2010; Yoo and Medina-Franco, 2012).

As an integral part of virtual screening, consensual binding has proven to be a useful tool for the systematic search of active natural substances against epigenetic targets. Medina-Franco and Yoo (2013) identified a set of 53 consensual matches using three distinct docking techniques. Others used molecular docking to detect an analogue of methylenedisalicylic acid in the NCI repository (Kuck et al., 2010).

Many drugs have been repurposed after they demonstrated efficacy in treating conditions other than those they were initially approved for. For instance, valproic acid, an anticonvulsant prescribed for seizures, also inhibits HDAC (Vellore and Baron, 2014). Researchers who explored repurposing bioactive substances from food used 32 approved antidepressants to analyze a database off 4600 bioactive food components. They subjected the ten substances that were most closely related to antidepressant medications to experimental screening against an HDAC and found that valproic acid was the most similar. Of the ten substances, two exhibited HDAC inhibition similar to that of valproic acid (Martinez-Mayorga et al., 2013). Méndez-Lucio et al. (2014) conducted a three-dimensional structural similarities analysis of a repository of approved pharmaceuticals to discover a molecule, NSC14778, which had previously been reported as a promising DNMT antagonist. They identified olsalazine as an anti-inflammatory owing to its strong structural resemblance with NSC14778 that likely suppressed DNMT isoforms.

Spannhoff et al. (2007) used homology modelling of human and *Aspergillus nidulans* PRMT1 to conduct the initial target-based strategy for PRMT antagonists. Using the GOLD programme, a total of 1630 molecules that were taken from the NCI have been docked to the binding region of the fungus PRMT framework. A pharmacophore query was employed to choose 36 molecules that were then subjected to an *in vitro* experiment to determine their inhibitory potential against human PRMT1. Among them, AMI-1 and NSC34238 had the minimum *in vitro* IC_{50} readings (1.2 and 1.7 μM, respectively).

Novel NAD+-dependent deacetylase sirtuin (SIRT) antagonists were additionally identified via computational methods including scanning commercially available databases and then docking the results into the binding region. Depending on their strong structural resemblance to NSC112546 (thiobarbiturate substrate) and β-phenylsplitomicin (splitomicin substrate), these investigations produced two new families of antagonists. According to Heinke et al. (2011) and Lawson et al. (2010), the results were used as leads in the optimization and development of novel SIRT antagonists.

Molecule TA1, a thioglycolic amide, was discovered to be a PRMT antagonist through the use of fragment-based virtual screening (Spannhoff et al., 2007). The researchers used GOLD to sort the Hans–Knöll database (which contains around 9000 chemicals) and then docked the resultant 900 molecules to a homology model of the fungus PRMT. Given its lack of chemical stability and druglike properties, the scientists looked for similarly structured molecules that consisted of an α-methylthioglycolic amide. This led to the identification of an internal compound (TA2) as an antagonist of both human and fungus PRMTs, as well as indicating a manner of bisubstrate suppression. This provides fascinating implications for additional structural modifications aimed at developing novel, powerful, and specific PRMT antagonists.

8.3 MOLECULAR DYNAMICS

Simulations can provide information about details that are absent from experiments and determine whether theoretical frameworks can predict the results of experiments. Molecular dynamics (MD) simulations, for instance, provide insights into the atomic-level dynamics of proteins that might enhance the accuracy of both predicted and discovered protein structures. Compared with previous molecular modelling techniques, MD offers more information and is useful for estimating free-energy changes resulting from ligand binding and improving conformational sampling (Lindahl, 2008; Vellore and Baron, 2014).

Hydralazine was discovered in 1988 to inhibit DNA methylation and is normally used as a strong arterial vasodilator. When used together with valproic acid, the drug later showed antitumor effects in clinical studies (Dueñas-González et al., 2008). The binding behavior of a DNMT isoform was modeled using MD to better understand the basic molecular mechanisms of the methylation inhibitory action of hydralazine. Based on this modeling, hydralazine showed similar binding behavior to that of nucleoside analogues, which play a significant role in DNA methylation processes (Medina-Franco and Caulfield, 2011).

The catalytic regions of DNMTs upon binding to S-adenosyl methionine (SAM) were modeled using MD simulations to better understand DNMTs. The diverse DNMT isoforms were represented by crystal structures along with various molecular modeling methods. However, no noticeable conformational changes were seen following the binding of SAM on a nanosecond timescale. However, the study revealed details about the protein movements of the DNMTs and how they bind to this cofactor (Evans and Bronowska, 2010).

8.4 HOMOLOGY MODELLING

The extremely stable catalytic region of DNMTs is responsible for the success of several homology models. Numerous structure-based design investigations, including docking, were predicated on homology models that employed the crystal structures of bacterial DNMTs prior to the availability of DNMT crystal structures. This kind of homology modeling proved crucial in the discovery of new inhibitors of DNMT. Employing virtual screening together with a homology model of a human DNMT, RG108 became the first DNMT inhibitor (Kuck et al., 2010; Yoo et al., 2012). The same homology model was then utilized to find two more DNMT inhibitors that were then utilized to optimize the development of further DNMT inhibitors. The crystal structure of a similar human DNMT isoform was used to model the catalytic domain of a human DNMT isoform. N4-fluoroacetyl-5-azacytidine was developed using this methodology, and it effectively suppressed DNA methylation in human cancer cells (Siedlecki et al., 2003).

To create a hydralazine binding system, this homology modelling was paired with MD and docking (Singh et al., 2009). Following that, two homology models of the catalytic region of a human DNMT were built employing the crystal structures of bacterial DNMTs. Despite having distinct homologous templates, the two homology models have crucial interactions at the catalytic site in common. The models were later verified by overlaying them onto the recently released crystal structure, and the homology model was found to be compatible with the proposed mechanism of DNA methylation.

8.5 PHARMACOPHORE MODELLING

Yoo and Medina-Franco (2011b) presented a computational approach for pharmacophore development that could be utilized in conjunction with empirical data and MD. Employing MD and crystal structure, an active conformation of a DNMT inhibitor's catalytic binding region was predicted. The next step was employing recognized active substances at the catalytic site to perform molecular docking. The model was validated by comparing the molecular docking results with earlier studies. Researchers suggested using the data from this model to figure out the binding pattern of DNMTs, which could lead to the development of potent non-nucleoside inhibitors.

Ensemble techniques allow scientists to investigate the ligand-binding interactions of DNMTs and identify potential inhibitors (Medina-Franco and Yoo, 2013). Pharmacophore models for 16 reported DNMT inhibitors were generated by Yoo et al. (2012). According to research employing the highest-scoring docking poses and a homology model of the catalytic binding site, the researchers found that most of the inhibitors resembled the pharmacophore properties of the model and discovered aurin tricarboxylic acid to be a new DNMT inhibitor.

Utilising the HypoGen software plugin of Discovery Studio, Kandakatla and Ramakrishnan (2014) generated a pharmacophore query model that utilised an entire set of 48 HDAC2 antagonists. The most effective model included four pharmacophore characteristics that were used in a search for virtual screening using the NCI/NIH and Maybridge records. Eight lead molecules were identified as newly discovered candidate HDAC2 antagonists by docking experiments. Molecules NSC108392 and MFCD01935795 exhibited the highest estimated activity levels and docking results.

Wang et al. (2012) merged radioactive methylation tests with pharmacophore modelling to discover small molecules as PRMT1 antagonists. Of the 90,000 molecules screened from the SPECS repositories, approximately 102 molecules showed inhibitory action against PRMT1. Top-ranked molecules were picked, and eventually, six hits were found to be PRMT1 antagonists with significant suppressive action; the two substances with the lowest *in vitro* IC_{50} values were AE-848/32000014 and AE-848/34542007.

Employing SIRT1 activators, Sakkiah et al. (2009) conducted pharmacophore modelling research with the goal of treating age-related illnesses such as overweight, type 2 diabetes, and Parkinson's disease. While screening 607 sirtuin activator compounds, the most suitable model produced the pharmacophoric properties of two hydrogen bond acceptors and one aromatic center. Moreover, virtual screening in the Maybridge repositories identified seven compounds that had every pharmacophoric characteristic. Of these, two novel molecules—CD02704 and HTS01501—exhibited similarities as training molecules and can be utilized as new frameworks to find novel SIRT1 activators.

8.6 DISCOVERING DRUGS WITH DOCKING FOR EPIGENETIC MODULATORS

8.6.1 Molecular Docking

A prominent computational chemistry method for simulating protein–ligand complexes and employing it to develop drugs is called molecular docking. When determining the potential energy of the complex, the ultimate objective is to predict the conformations of the ligand within the target. Considering the intricate nature of these systems and the limitations of our present computing power, it is difficult to accurately accomplish these goals. Nevertheless, molecular docking's affordable price makes it possible to sort through huge databases of chemical substances to find both active and inactive compounds.

It is crucial to remember that inaccurate scoring will lead to inaccurate assessment of energy and structure. Furthermore, the best possible candidates will be overlooked due to substandard conformational analysis (Novikov and Chilov, 2009). Docking has demonstrated its use in several applications, such as determining if a ligand would better function as an enzyme inhibitor than a competitive mechanism by identifying the most significant protein–ligand interactions and ligand molecular poses. Docking is also a component of virtual screening.

For successful docking, the target must be three-dimensional in structure, requiring nuclear magnetic resonance or crystallography. In the absence of the required data, we can leverage homology modelling to generate it, as suggested by Ewing and Kuntz (1997). It's noteworthy that the structure derived from crystallography doesn't always align with the structure found in a living organism. One way to address this discrepancy is by analysing and minimizing the target's structure using MD or any existing minimization algorithms. Once the target and ligand structures have been minimized, it's crucial to remember that docking only minimally considers the system's conformational adaptability; while it often allows for ligand movement, it rarely accounts for changes in the target structure. Furthermore, because docking ignores the solvent, the system may be poorly characterized if the solvent molecules are integral to the protein structure or play a crucial role in the process of coupling between the ligand and its binding site (Novikov and Chilov, 2009). An approximate posture inside the binding location of interest can be predicted if all these docking factors are considered. AutoDock, AutoDockVina, and Dock are the most often used docking software (Morris et al., 2009; Trott and Olson, 2010; Ewing et al., 2001).

Docking is also useful in drug discovery for clarifying the molecular reactions connected with potential drugs and can also be used to better understand currently available epigenetic drugs. It was utilized for investigating the protein–ligand relationships of known DNMT inhibitors (Mishra, 2011). Fourteen molecules, including both nucleoside and non-nucleoside inhibitors with varying structural classes were employed for docking in order to better understand the various binding positions of DNMTs. Docking reveals similar binding interactions between the various chemicals, and the binding site is essential to DNA methylation (Yoo and Medina-Franco, 2012).

Numerous docking investigations pertaining to epigenetic targets have been reported (Zhao et al., 2014; Ragno et al., 2007). The chemicals HZ1 and HZ2 were discovered by Zhao et al. (2014), employing fragment-based virtual screening as two new substances that connect to the initial bromodomain of BRD4 via favourable binding energy levels and low μM affinity (IC_{50} = 7.0 and 7.5, correspondingly). Additionally, the molecules demonstrate selectivity in different kinds of bromodomains. From this valuable knowledge obtained, one can develop new drugs for the management of solid tumours and haematological cancers, in which BRD4 is a target for therapy.

According to Aldawsari et al. (2016), the two isoforms of DNMT, 3A and 3B, are specifically inhibited by a family of resveratrol molecules. This work demonstrated that the enzymatic and catalytic region of DNMT and related variants had favourable interactions with one other, compared to the DNMT1 region. These findings showed that the DNMT3A inhibitors FSA9 and FSA10 performed better than resveratrol.

Chen et al. (2014) performed virtual screening-based docking using a database consisting of 111,121 molecules (SPECS) via Glide that had been selected according to their physicochemical characteristics for the DNMT1 enzyme (PDB ID: 4DA4). From the docking result, among the 51 compounds that were produced, the molecule DC_05 discovered had an IC_{50} of 10.3 μM and was effective against DNMT1. The authors conducted structural homology screening research and selectivity experiments on different isoforms employing DC_05.

Fang et al. (2003) used docking to validate experimental data. The authors specifically stated that one of the main ingredients in green tea, (–)-epigallocatechin-3-gallate (EGCG), is an inhibitor of DNMT and may stimulate genes that have been repressed in cancer cells. They also suggested that the energy obtained from the MD computation and the postures that develop between different sites in the DNMT active region are what cause this influence. The research of Medina-Franco and Yoo (2013) offers more evidence for the importance of docking, showing that SID49645275, among the most potent antagonists of DNMT, binds to the position of this enzyme cofactor.

Four molecules that inhibited DNMT3A were identified as a result of synthesised analogues of the virtual screening hit NSC319745 (Kabro et al., 2013; Kuck et al., 2010). Among them, AK1 had an EC_{50} of 36 μM towards DNMT3A. This molecule exhibits an intense attraction to the SAM pocket and could be utilised as a lead for the design and optimisation of novel selective antagonists targeting DNMT3A. Using the DNMT1 catalytic site, Yoo and Medina-Franco (2012) conducted a flexible docking investigation of the antagonist trimethyl aurintricarboxylic acid (NSC97317). According to the docking investigation, NSC97317 can attach to the catalytic region by binding to the amino acids involved in DNA methylation.

For instance, Chang et al. (2009) used docking to identify ligand crystallographic structures. They published the SET region of a G9a crystal structure in combination with SAH (S-adenosyl-L-homocysteine) and BIX-01294. Then they assessed the corresponding energy levels of the potential chair–boat, boat–boat, and chair–chair configurations in relation to BIX-01294's piperidine and diazepane rings. Extra-precision flexible Glide docking was performed on the hydrogenated forms of the smallest energy chair–chair conformer with the binding region of G9a protein. The mechanism behind this binding was demonstrated by the strong hydrogen bond between the hydrogenated piperidine moiety and the ASP1131 amino acid (PDB ID: 3FPD).

To further describe the pharmacophoric properties related to the heterocycle, Srimongkolpithak et al. (2014) prepared a number of G9a protein antagonists based on the crystallographic structures of G9a protein–BIX-01294 and G9a protein-UNC0224. Through those attempts, two strong and specific G9a/GLP inhibitors that utilize a quinazoline scaffold were identified: NS41 and NS42. The binding region of G9a–UNC0224 (PDB ID: 3K5K) showed how crucial it was to maintain the hydrogenated basic nitrogen at the first position and the di-(OCH_3) group on the aromatic ring in order to maintain the hydrogen bond relationship with G9a (Liu et al., 2009). Docking is helpful for sirtuin cofactor NAD simulation, as reported by Butler et al. (2010) in the early phases of their research on the development of HDAC antagonists.

Subha and Kumar (2008) studied docking with category-I histone deacetylases and 12 identified antagonists that had already been under phase I/II studies. Only the transition state analogue displayed the ideal binding mechanism: The hydroxyl group firmly fit into the binding region and satisfied almost all of the requirements (i.e., drug similarity value, drug score, and binding affinity score). Based on their findings, the researchers proposed developing improved hydroxamate molecules as possible class-I HDAC inhibitors. The result of this method can be used as a starting point to ascertain how a known active chemical would respond mechanically against a particular target.

Singh et al. (2009) identified the important DNA methyl transferase amino acid sites interacting with hydralazine and the moiety responsible for the beneficial interaction.

In subsequent work, known DNMT inhibitors were docked using the crystal structure of a DNMT coupled to unmethylated cytosine-phosphate-guanine (CpG) sites in DNA. First, the crystal structure's catalytic binding site was modelled using MD to change it from an inactive to an active conformation, and the inhibitors' binding positions interacted with the DNMT's catalytic region that related to the mechanism of DNA methylation. Yoo et al. (2013) performed docking of large-framework molecules, namely SGI-1027 and CBC12, incorporating the cofactor and substrate binding regions of the MTase region of human DNMT1 and DNMT3A. The docking result showed that both molecules resided in the cofactor and substrate binding locations based on the availability of extra regions. Furthermore, in the binding paradigm of DNMT3A, SGI-1027 resided in the cofactor interaction site, while CBC12 was docked in both the substrate-interacting site and the cofactor domain. Additionally, the docking analysis demonstrates that SGI-1027 acts as a non-selective antagonist, and its binding score is better than that of SAH. These outcomes showed a high degree of concurrence with the empirical results (Yoo et al., 2013).

The fields of cheminformatics and molecular simulation have significantly advanced the development of drugs (Martinez-Mayorgaet al., 2020) Numerous important medications, including oseltamivir (a neuraminidase antagonist employed to treat influenza), imatinib (a kinase antagonist employed to treat certain kinds of malignancies), dorzolamide (a carbonic anhydrase II antagonist employed to manage excessive eye pressure, including glaucoma), and many more, were developed using computational methods. An innovative antiviral drug called enfuvirtide is used in combination treatment to treat HIV-1 infection (Talele et al., 2010; Leelananda and Lindert, 2016).

Many groups have applied various computational methods to epigenetics research in the last few decades. Researchers recently utilized examples from epigenetics outlined the benefits and drawbacks of several structure-based techniques in multi-target drug design (Sivakumar et al., 2020). For instance, Kuang et al. (2015) published in-depth modelling research on many BRD isoforms using MD models. The scientists used these proteins' non-negligible ligand-binding dynamics properties to improve our understanding of the BRD family's binding site.

For epigenetic targets, computational methods can be employed to identify hotspots for interactions between proteins and hidden allosteric binding sites (Lu et al., 2018b; Yang and Wang, 2010). Computational methods can enhance the pharmacodynamics and pharmacokinetics at the hit-to-lead stage. Using quantum and molecular mechanics, Letfus et al. (2020) recently produced toxoflavin-based antagonists that inhibited human histone lysine demethylase enzymes (a cellular domain relevant to prostate and breast cancer). These studies yielded compounds with low nM action in biological tests and low mM action in cell-based assays, as well as enhanced *in vitro* pharmacokinetic properties.

In epigenetic drug development, computational biology primarily aids in discovering and validating hits for experimental testing. Using molecular docking and pharmacophore generation, Tao et al. (2019) recently identified a novel small chemical with anticancer activity that showed submicromolar *in vitro* enzymatic suppression of the protein arginine N-methyltransferase 5. In a different study, Song et al. (2019) used pharmacophore queries, molecular docking, and MD modelling to find drugs with specific efficacy in developing cancerous cells. See Table 8.1 for a sample of the current epigenetic targets and drugs discovered. Table 8.2 gives the chemical structures of certain epigenetic inhibitors.

8.7 OTHER *IN SILICO* METHODS

Recently, finding epigenetic targets has drawn special attention from the pharmaceutical industry. However, a lack of complete knowledge of the biology behind several epigenetic targets limited the study (Comley, 2015). To better understand genome functioning, researchers have come up with an approach that uses database analysis to identify epigenetic targets. The National Centre for Biotechnology Information (NCBI), the University of California Santa Cruz (UCSC) genome browser, and the Encyclopedia of DNA Elements Consortium (ENCODE) were among the databases used in a 2015 study to gather data that might inform *in vivo* research in the future. The ENCODE project seeks to identify and describe functional components of the human genome. The UCSC genome browser was chosen for discovering epigenetic alterations based on this data, and functional genomic research from NCBI was layered on top of that. By applying this procedure, scholars better understood relevant mechanisms and employ this data to direct *in vivo* epigenetic target investigations (Hay et al., 2017).

Researchers believe that QSAR modelling will remain a crucial tool for expediting the identification of epidrugs. Current machine learning and artificial intelligence algorithms have significantly improved this method in recent years. Researcher have documented the creation of QSAR models and the detection of virtual hits for epigenetic targets (Alves et al., 2021; Capuzzi et al., 2018; Hajjo et al., 2010; Gozalbes et al., 2005; Lima et al., 2018; Ballante et al., 2017).

A set of 29 substituted hydroxamic acid-based HDAC antagonists were subjected to docking with the human HDAC1 binding site (Guo et al., 2005). A 3D-QSAR model that had standard R^2 levels of 0.982 for CoMFA and 0.954 for COMSIA was developed via binding. Comparable prediction abilities enable both methods to discover new compounds with enhanced activity.

TABLE 8.1
Diverse *In Silico* Models for Epigenetic Drug Discovery

In Silico Tools	Compounds	Epigenetic Target	Reference(s)
Similarity search	Olsalazine	DNMT	Méndez-Lucio et al. (2014)
Virtual screening	RG108		Kuck et al. (2010)
Virtual screening	N4-fluoroacetyl-5-azacytidine		Siedlecki et al. (2003)
Virtual screening	NSC14778		Méndez-Lucio et al. (2014)
Pharmacophore modelling	Aurin tricarboxylic acid		Yoo et al. (2012)
Docking	FSA9 and FSA10		Aldawsari et al. (2016)
Docking	DC_05		Chen et al. (2014)
Docking	(−)-epigallocatechin-3-gallate		Fang et al. (2003)
Docking	SID49645275		Medina-Franco and Yoo (2013)
Virtual Screening	NSC319745		Kabro et al. (2013)
Docking	AK1		Kabro et al. (2013)
Docking	Trimethylaurintricarboxylic acid (NSC97317)		Yoo and Medina-Franco (2012)
Induced fit docking	SW155246		Medina-Franco and Yoo (2013)
Docking	SGI-1027 and CBC12		Yoo et al. (2013)
Similarity search	NSC112546 and β-phenylsplitomicin	SIRT	Heinke et al. (2011); Lawson et al. (2010)
Pharmacophore model	CD02704 and HTS01501		Sakkiah et al. (2009)
Pharmacophore modelling/ Docking	NSC108392 and MFCD01935795	HDAC	Kandakatla and Ramakrishnan (2014)
Docking	NS41 and NS42	PKMT G9a	Srimongkolpithak et al. (2014)
Docking	BIX-01294	G9a	Chang et al. (2009)
Docking-based screening	HZ1 and HZ2	BRD	Zhao et al. (2014)
Pharmacophore modelling/ Docking	AE-848/32000014 and AE-848/34542007	PRMT	Wang et al. (2012)
Docking-based screening	AMI-1		Spannhoff et al. (2007)
Pharmacophore modelling/ Docking	NSC34238		Spannhoff et al. (2007)
Fragment-based screening	TA1 and TA2		Spannhoff et al. (2007)

TABLE 8.2
Chemical Structure of Diverse Epigenetic Inhibitors

Code	Compound	Compound IUPAC Structure
DNMTs		
NCS 14778	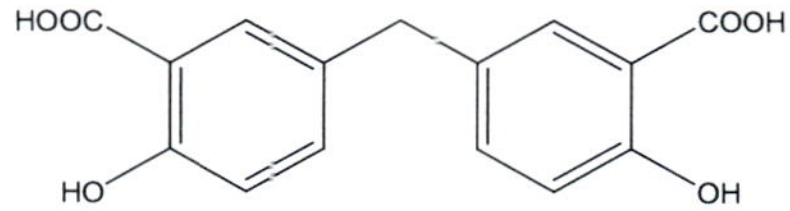	-
Oslalazine	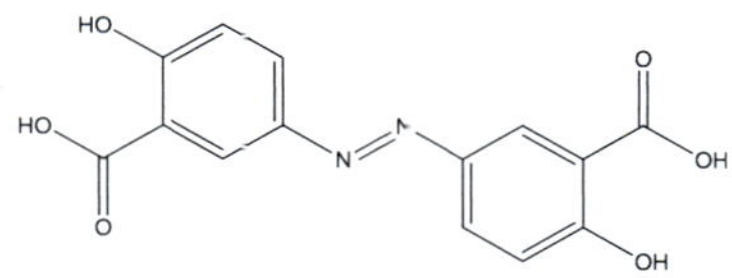	5-[(3-carboxy-4-hydroxyphenyl) diazenyl]-2-hydroxybenzoic acid
SID49645275	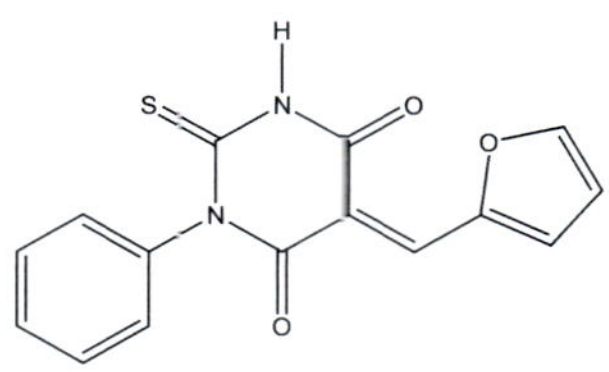	(E)-5-(furan-2-ylmethylene)-3-phenyl-2-thioxo-dihydropyrimidine-4,6(1H,5H)-dione
NSC 97317	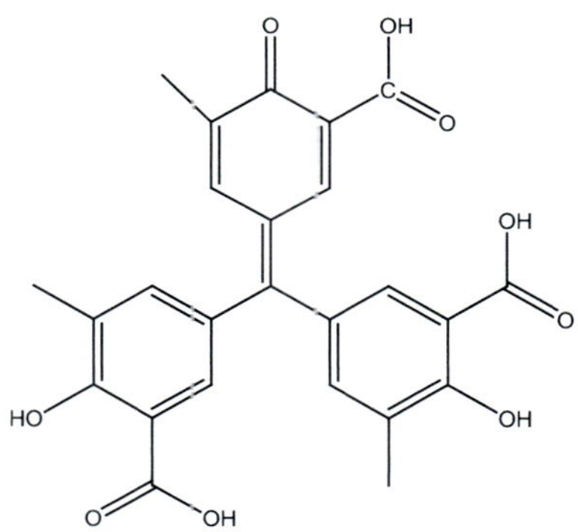	Trimehtylaurintricarboxilic acid
RG1O8	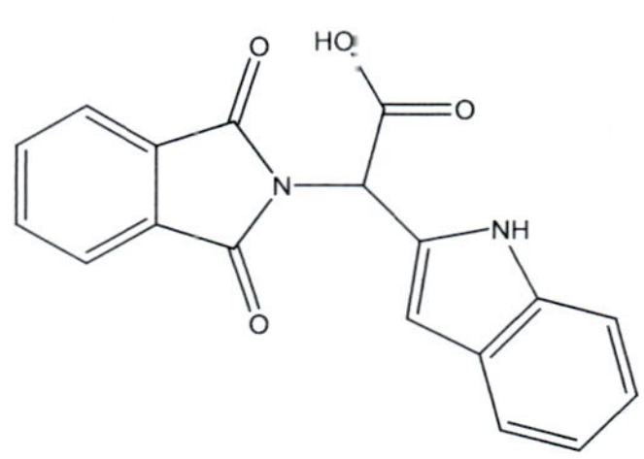	2-(1,3-dioxoisoindolin-2-yl)-2-(1H-indol-2-yl)acetic acid
SW155246	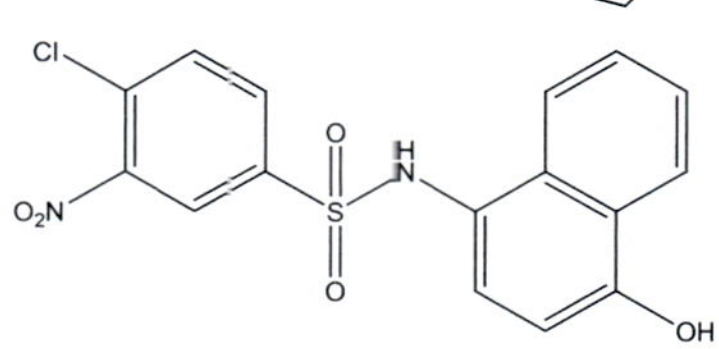	4-chloro-N-(4-hydroxynaphthalen-1-yl)-3-nitrobenzenesulfonamide
CBC12	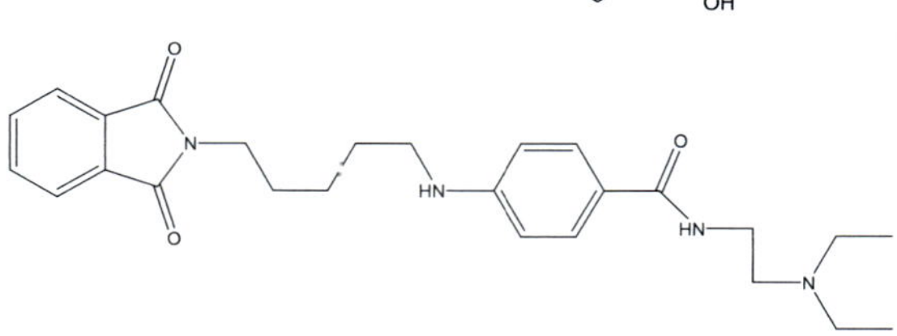	N-(2-(diethylamino)ethyl)-4-(5-(1,3-dioxoisoindolin-2-yl) pentylamino)benzamide

(*Continued*)

TABLE 8.2 *(Continued)*
Chemical Structure of Diverse Epigenetic Inhibitors

Code	Compound	Compound IUPAC Structure
EGCG	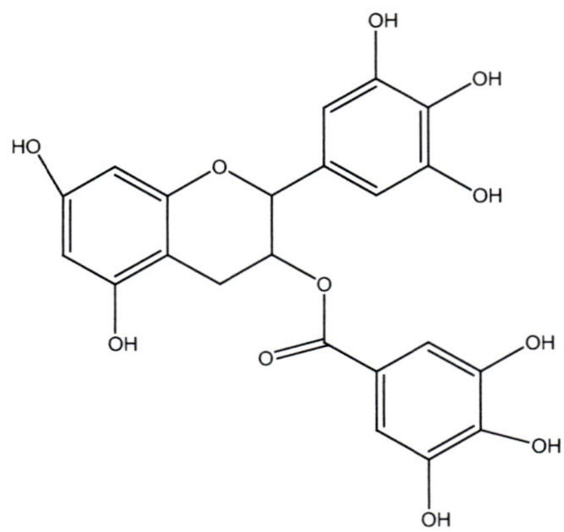	5,7-dihydroxy-2-(3,4,5-trihydroxyphenyl)-3,4-dihydro-2H-chromen-3-yl 3,4,5-trihydroxybenzoate
DC_05	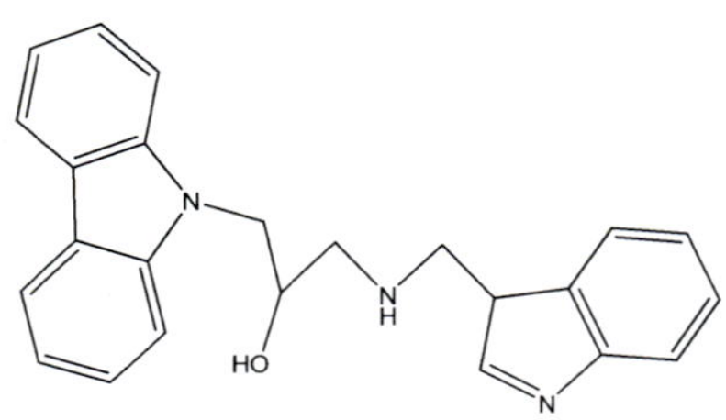	1-((3H-indol-3-yl)methylamino)-3-(9H-carbazol-9-yl)propan-2-ol
SGI-102	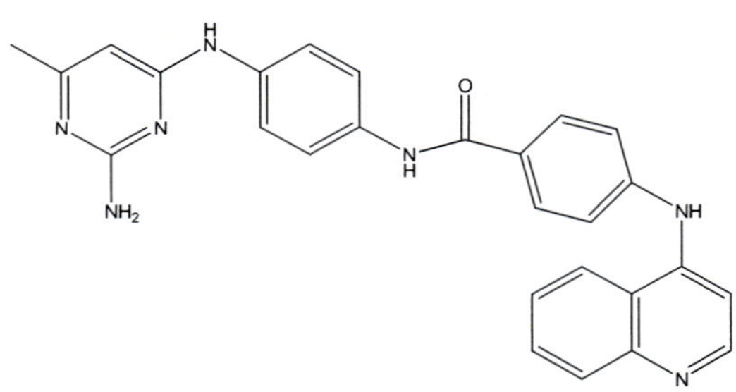	N-(4-(2-amino-6-methylpyrimidin-4-ylamino)phenyl)-4-(quinolin-4-ylamino)benzamide
SIRT Inhibitors		
NSC 112546	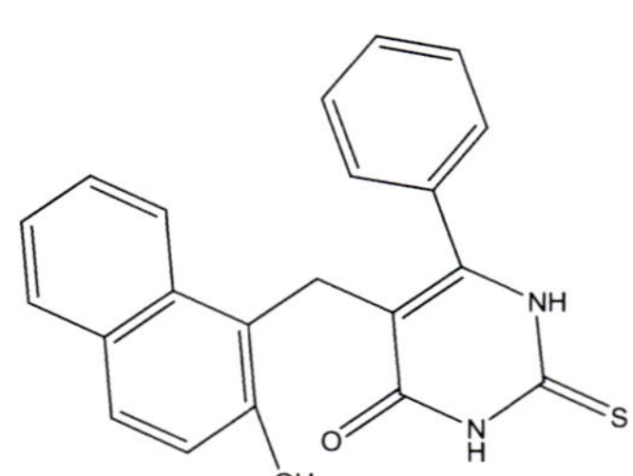	5-((2-hydroxynaphthalen-1-yl)methyl)-6-phenyl-2-thioxo-2,3-dihydropyrimidin-4(1H)-one

(Continued)

TABLE 8.2 *(Continued)*
Chemical Structure of Diverse Epigenetic Inhibitors

Code	Compound	Compound IUPAC Structure
NSC684571	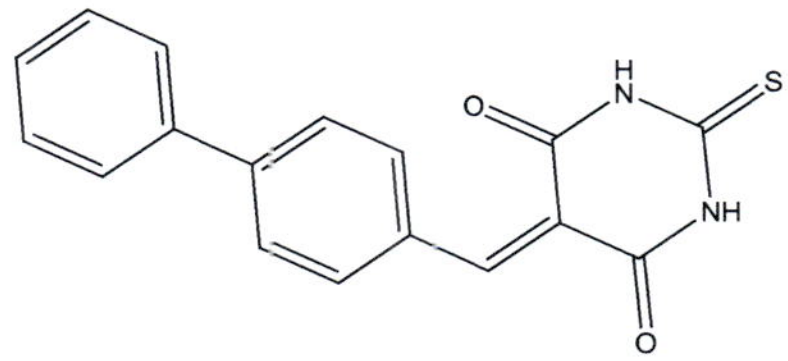	
B-Phenylsplitomicin	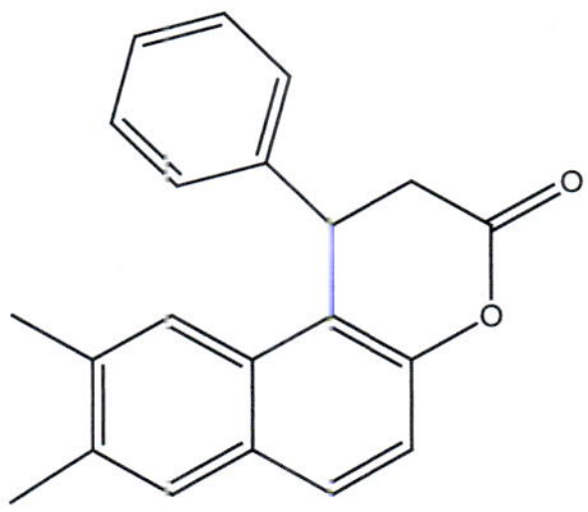	8,9-dimethyl-1-phenyl-1,2-dihydrobenzo[f]chromen-3-one
HDAC Inhibitors		
Suberoylanilide hydroxamic acid	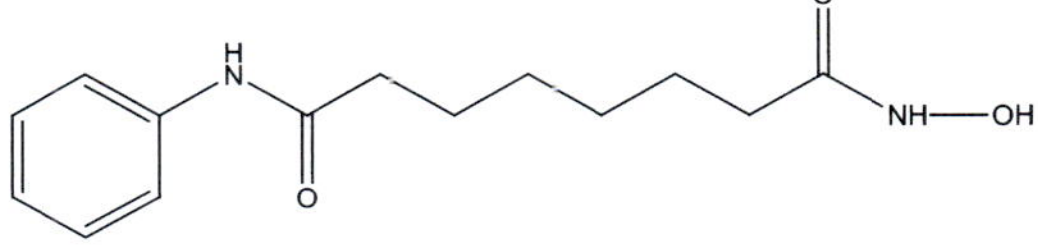	N1-hydroxy-N8-phenyloctanediamide
Trichostatin A	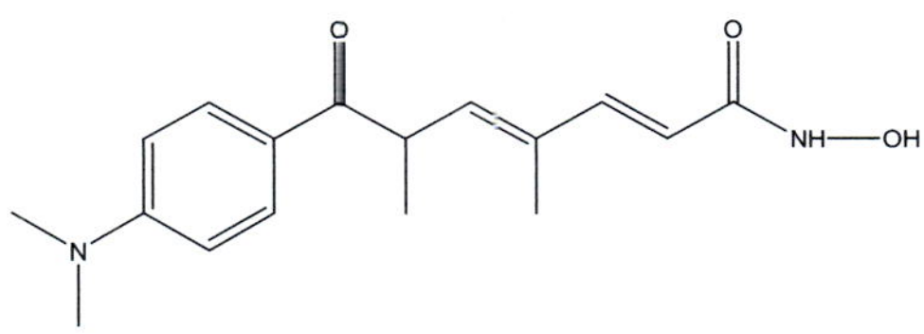	(2E,4E)-7-(4-(dimethylamino) phenyl)-N-hydroxy-4,6-dimethyl-7-oxohepta-2,4-dienamide
BRD4 Inhibitors		
HZ1	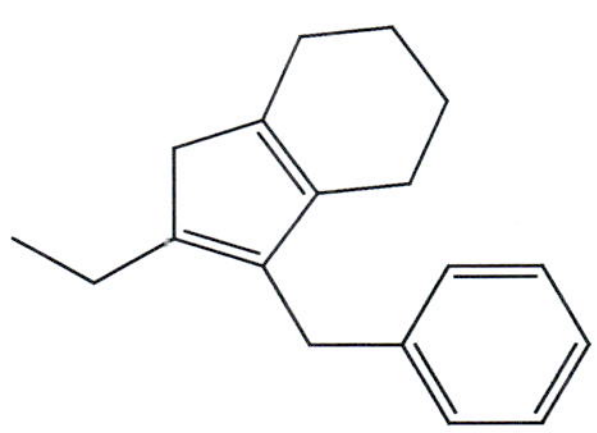	3-benzyl-2-ethyl-4,5,6,7-tetrahydro-1H-indene
HZ2	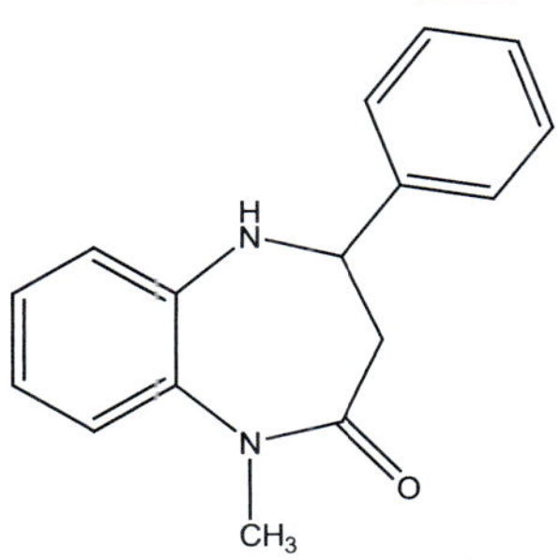	1-methyl-4-phenyl-4,5-dihydro-1H-benzo[b][1,4]diazepin-2(3H)-one
PRMT1 Inhibitors		

(Continued)

TABLE 8.2 *(Continued)*
Chemical Structure of Diverse Epigenetic Inhibitors

Code	Compound	Compound IUPAC Structure
AMI-I	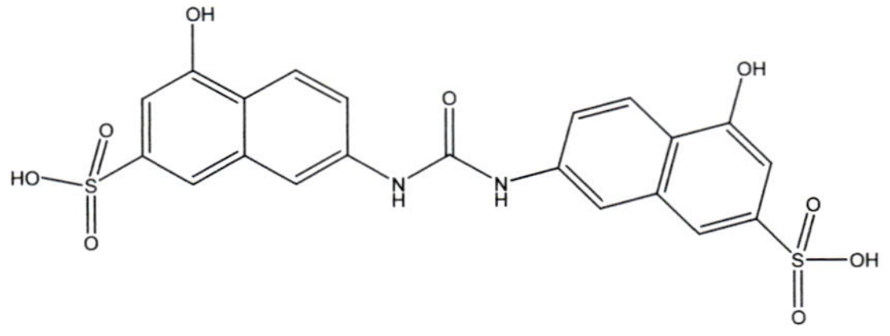	-
NSC34238	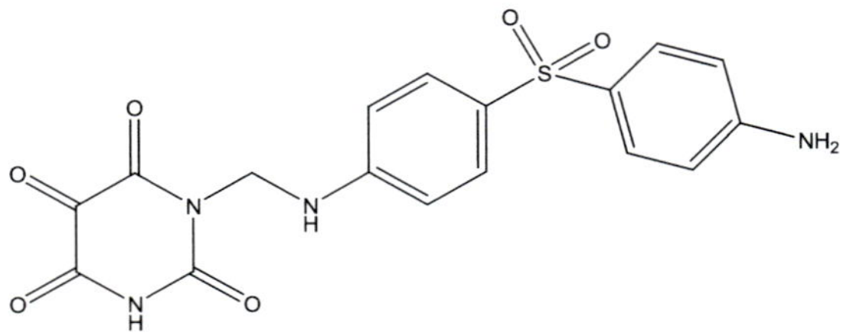	1-((4-(4-aminophenylsulfonyl) phenylamino)methyl) pyrimidine-2,4,5,6(1H,3H)-tetraone
TA1	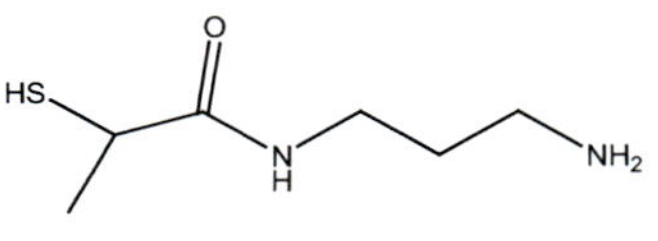	N-(3-aminopropyl)-2-mercaptopropanamide
TA2	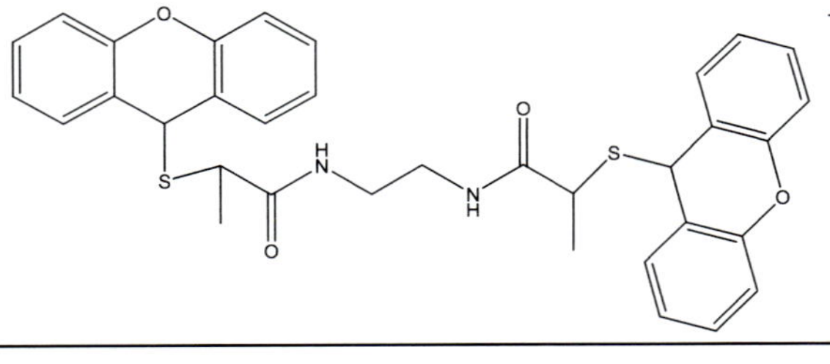	-

Four sets of HDAC antagonists were subjected to docking followed by a 3D-QSAR investigation (Ragno et al., 2006). The investigators used two external test sets, consisting of 7 and 12 freshly synthesised substances, to assess the predictive accuracy of the model; the strongest accurate model had SDEPs of 0.96 and 0.99, correspondingly, and the model was interpreted to link the docking with the development of improved HDAC antagonists. For the purpose of depicting the training sets to construct 3D-QSAR designs, the same researchers employed three distinct interaction techniques with a targeted library of small molecules within the catalytic region of human PRMT1.

Ortore et al. (2009) performed docking with the HDAC1 and HDAC8 binding regions of around 40 compounds that were previously evaluated for their suppressive efficacy towards the two isoenzymes. The 3D-QSAR models demonstrated a reasonable prediction coefficient of correlation (q^2 ranged from 0.690 to 0.793) and excellent R^2 values between 0.921 and 0.946). This data proved helpful in obtaining quantitative details on the relationships between ligands and receptors.

Computing power has increased due to the advancement of graphics processing units, and today, research has concentrated on AI methods along with molecular docking and MD modelling. For instance, molecular docking on 170,000 chemicals was recently carried out by Lyu et al. (2019) and led to the empirical discovery of a new framework of AmpC phenolate antagonists. After optimization, one of the compounds exhibited 77 nM inhibitory activity. Researchers have also discovered

30 compounds that inhibit the dopamine receptor D4. For instance, Gentile et al., 2020 developed the hybrid platform Deep Docking, which enables the rapid and reliable estimation of docking results for thousands of molecules. It is predicated on QSAR models trained on docking results using deep learning.

Researchers have explored a number of recent uses of generative neural network models and deep learning. For example, ReLesSE (Reinforcement Learning for Structural Evolution) is a deep reinforcement learning system that can assist in designing chemical databases with an orientation towards desirable inhibitory action (Popova et al., 2018). Employing a similar methodology, Zhavoronkov et al. (2019) revealed the identification of a strong candidate for the discoidin domain receptor tyrosine kinase 1 target that is linked to fibrosis and other diseases.

One crucial factor that directly relates to the discovery of epigenetic drugs is the possibility of polypharmacological frameworks enabled by epigenetic targets. Drugs with measurable toxicity often have unintended side effects (Bantscheff et al., 2009); indeed, the main factors contributing to drug turnover are toxicity and ineffectiveness (Van Norman, 2019). Recent research, however, has emphasized the benefits of multi-target design (MTD), utilizing the concept of selective synergism (Peters, 2013). In MTDs, structure-based and machine-learning techniques were effective (Zhang et al., 2017), and because multiple targets are engaged in the same pathway, epigenetic modification is appropriate for MTD and can result in synergistic effects (Bechter and Schöffski, 2020). Promising MTD approaches for epigenetic drug development should concentrate on producing molecules that block the reading, writing, and erasing of the same gene modification (Atkinson et al., 2014; Yuan et al., 2019; San José-Enériz et al., 2017). In brief, it is clear that a viable path for the discovery of new epigenetic medications lies in epigenetic polypharmacology.

8.8 CONCLUSIONS AND PROSPECTS

There are still many unanswered questions in the complicated field of epigenetics and its connection to disease. The discovery of new therapies and the availability of epigenetic treatments have been constrained by the cost, although awareness of epigenetic processes is increasing. It has been demonstrated that the combination of epigenetics and computational techniques offers great promise for drug development.

This chapter covered current developments in cutting-edge computer-aided drug design methods that can be utilized to facilitate innovative findings in the area, as well as the application of cheminformatics in epigenetic drug discovery. The chapter authors demonstrated throughout this chapter the important role of molecular docking in the search for epigenetic medicines. New epigenetic proteins as possible therapeutic targets are found as research advances, although the already available epigenetic medications need to be developed in order to boost their potency and specificity, lessen their adverse effects, and enhance other drug-like qualities. By aiding in the awareness of the binding characteristics (mode and energy) at the molecular level, docking studies contribute to the discovery and development of improved drugs. The application of docking in epigenetic drug development (hit or lead molecule) and hit molecule optimization to provide drug candidates with better attributes has been demonstrated in a number of cases. Docking studies are useful for virtual screening, pharmacophore generation, and drug optimization. They are used independently or in combination with other *in silico* techniques, including molecular dynamic modelling and QSAR.

Currently, the collaborative work using these computational techniques to create epigenetic drugs for combating different illnesses is significantly increasing. Few attempts at drug development include both genetic and epigenetic data, despite the fact that this research has greatly advanced epigenetic drug discovery. Even though the availability of crystal structures for numerous proteins implicated in epigenetic processes has grown, many molecular simulation techniques still depend on their obtainability. Many of these methods can be computationally costly, and because computer science is such a broad area, there are still many other computational methods beyond molecular modelling that have not yet been well investigated. However, the field of epigenetic drug development appears to have a bright future with the advent of new modelling techniques, technology, and studies on genomics.

REFERENCES

Aguayo-Ortiz, Rodrigo, and Eli Fernández-de Gortari. "Overview of computer-aided drug design for epigenetic targets." In *Epi-informatics*, pp. 21–52. Academic Press, 2016.

Aldawsari, Fahad S., Rodrigo Aguayo-Ortiz, Kanishk Kapilashrami, Jakyung Yoo, Minkui Luo, José L. Medina-Franco, and Carlos A. Velázquez-Martínez. "Resveratrol-salicylate derivatives as selective DNMT3 inhibitors and anticancer agents." *Journal of Enzyme Inhibition and Medicinal Chemistry* 31, no. 5 (2016): 695–703.

Alves, Vinicius M., Tesia Bobrowski, Cleber C. Melo-Filho, Daniel Korn, Scott Auerbach, Charles Schmitt, Eugene N. Muratov, and Alexander Tropsha. "QSAR modeling of SARS-CoV Mpro inhibitors identifies sufugolix, cenicriviroc, proglumetacin, and other drugs as candidates for repurposing against SARS-CoV-2." *Molecular Informatics* 40, no. 1 (2021): 2000113.

Atkinson, Stephen J., Peter E. Soden, Davina C. Angell, Marcus Bantscheff, Chun-wa Chung, Kathryn A. Giblin, Nicholas Smithers et al. "The structure based design of dual HDAC/BET inhibitors as novel epigenetic probes." *Medicinal Chemistry Communications* 5, no. 3 (2014): 342–351.

Ballante, Flavio, D. Rajasekhar Reddy, Nancy J. Zhou, and Garland R. Marshall. "Structural insights of SmKDAC8 inhibitors: Targeting Schistosoma epigenetics through a combined structure-based 3D QSAR, *in vitro* and synthesis strategy." *Bioorganic & Medicinal Chemistry* 25, no. 7 (2017): 2105–2132.

Bantscheff, Marcus, Arjen Scholten, and Albert J.R. Heck. "Revealing promiscuous drug–target interactions by chemical proteomics." *Drug Discovery Today* 14, no. 21–22 (2009): 1021–1029.

Bechter, Oliver, and Patrick Schöffski. "Make your best BET: The emerging role of BET inhibitor treatment in malignant tumors." *Pharmacology & Therapeutics* 208 (2020): 107479.

Biswas, Subhankar, and C. Mallikarjuna Rao. "Epigenetic tools (the writers, the readers and the erasers) and their implications in cancer therapy." *European Journal of Pharmacology* 837 (2018): 8–24.

Butler, Kyle V., Jay Kalin, Camille Brochier, Guilio Vistoli, Brett Langley, and Alan P. Kozikowski. "Rational design and simple chemistry yield a superior, neuroprotective HDAC6 inhibitor, tubastatin A." *Journal of the American Chemical Society* 132, no. 31 (2010): 10842–10846.

Capuzzi, Stephen J., Wei Sun, Eugene N. Muratov, Carles Martínez-Romero, Shihua He, Wenjun Zhu, Hao Li et al. "Computer-aided discovery and characterization of novel Ebola virus inhibitors." *Journal of Medicinal Chemistry* 61, no. 8 (2018): 3582–3594.

Chang, Yanqi, Xing Zhang, John R. Horton, Anup K. Upadhyay, Astrid Spannhoff, Jin Liu, James P. Snyder, Mark T. Bedford, and Xiaodong Cheng. "Structural basis for G9a-like protein lysine methyltransferase inhibition by BIX-01294." *Nature Structural & Molecular Biology* 16, no. 3 (2009): 312–317.

Chen, Shijie, Yulan Wang, Wen Zhou, Shanshan Li, Jianlong Peng, Zhe Shi, Junchi Hu et al. "Identifying novel selective non-nucleoside DNA methyltransferase 1 inhibitors through docking-based virtual screening." *Journal of Medicinal Chemistry* 57, no. 21 (2014): 9028–9041.

Comley, J. "Epigenetic targets: On the verge of becoming a major new category for successful drug research." *Drug Discovery World*. https://www.ddw-online.com/epigenetic-targets-on-the-verge-of-becoming-a-major-new-category-for-successful-drug-research-1805-201508/ Accessed 15 (2015).

Dueñas-González, Alfonso, Patricia García-López, Luis Alonso Herrera, Jose Luis Medina-Franco, Aurora González-Fierro, and Myrna Candelaria. "The prince and the pauper. A tale of anticancer targeted agents." *Molecular cancer* 7 (2008): 1–33.

Evans, David A., and Agnieszka Katarzyna Bronowska. "Implications of fast-time scale dynamics of human DNA/RNA cytosine methyltransferases (DNMTs) for protein function." *Theoretical Chemistry Accounts* 125 (2010): 407–418.

Ewing, Todd J.A., and Irwin D. Kuntz. "Critical evaluation of search algorithms for automated molecular docking and database screening." *Journal of Computational Chemistry* 18, no. 9 (1997): 1175–1189.

Ewing, Todd J.A., Shingo Makino, A. Geoffrey Skillman, and Irwin D. Kuntz. "DOCK 4.0: Search strategies for automated molecular docking of flexible molecule databases." *Journal of Computer-Aided Molecular Design* 15 (2001): 411–428.

Fang, Ming Zhu, Yimin Wang, Ni Ai, Zhe Hou, Yi Sun, Hong Lu, William Welsh, and Chung S. Yang. "Tea polyphenol (−)-epigallocatechin-3-gallate inhibits DNA methyltransferase and reactivates methylation-silenced genes in cancer cell lines." *Cancer Research* 63, no. 22 (2003): 7563–7570.

Foulks, Jason M., K. Mark Parnell, Rebecca N. Nix, Suzanna Chau, Krzysztof Swierczek, Michael Saunders, Kevin Wright et al. "Epigenetic drug discovery: Targeting DNA methyltransferases." *Journal of Biomolecular Screening* 17, no. 1 (2012): 2–17.

Francis, Richard C. *Epigenetics: How Environment Shapes Our Genes*. WW Norton & Company, 2011.

Ganesan, A., Paola B. Arimondo, Marianne G. Rots, Carmen Jeronimo, and María Berdasco. "The timeline of epigenetic drug discovery: From reality to dreams." *Clinical Epigenetics* 11 (2019): 1–17.

Gentile, Francesco, Vibudh Agrawal, Michael Hsing, Anh-Tien Ton, Fuqiang Ban, Ulf Norinder, Martin E. Gleave, and Artem Cherkasov. "Deep docking: A deep learning platform for augmentation of structure based drug discovery." *ACS Central Science* 6, no. 6 (2020): 939–949.

Gozalbes, Rafael, Catherine Rolland, Eric Nicolai, Marie-France Paugam, Laurent Coussy, Dragos Horvath, Frédérique Barbosa, Boryeu Mao, Frédéric Revah, and Nicolas Froloff. "QSAR strategy and experimental validation for the development of a GPCR focused library." *QSAR & Combinatorial Science* 24, no. 4 (2005): 508–516.

Guo, Yanshen, Jingfa Xiao, Zongru Guo, Fengming Chu, Yonghao Cheng, and Song Wu. "Exploration of a binding mode of indole amide analogues as potent histone deacetylase inhibitors and 3D-QSAR analyses." *Bioorganic & Medicinal Chemistry* 13, no. 18 (2005): 5424–5434.

Hajjo, Rima, Christopher M. Grulke, Alexander Golbraikh, Vincent Setola, Xi-Ping Huang, Bryan L. Roth, and Alexander Tropsha. "Development, validation, and use of quantitative structure– activity relationship models of 5-Hydroxytryptamine (2B) receptor ligands to identify novel receptor binders and putative valvulopathic compounds among common drugs." *Journal of Medicinal Chemistry* 53, no. 21 (2010): 7573–7586.

Hay, Elizabeth A., Philip Cowie, and Alasdair MacKenzie. "Determining epigenetic targets: A beginner's guide to identifying genome functionality through database analysis." *Population Epigenetics: Methods and Protocols* (2017): 29–45.

Heerboth, Sarah, Karolina Lapinska, Nicole Snyder, Meghan Leary, Sarah Rollinson, and Sibaji Sarkar. "Use of epigenetic drugs in disease: An overview." *Genetics & Epigenetics* 6 (2014): GEG-S12270.

Heinke, Ralf, Luca Carlino, Srinivasaraghavan Kannan, Manfred Jung, and Wolfgang Sippl. "Computer-and structure-based lead design for epigenetic targets." *Bioorganic & Medicinal Chemistry* 19, no. 12 (2011): 3605–3615.

Kabro, Anzhelika, Hugo Lachance, Iris Marcoux-Archambault, Valérie Perrier, Vicky Doré, Christina Gros, Véronique Masson et al. "Preparation of phenylethylbenzamide derivatives as modulators of DNMT3 activity." *Medicinal Chemistry Communications* 4, no. 12 (2013): 1562–1570.

Kandakatla, Naresh, and Geetha Ramakrishnan. "Ligand based pharmacophore modeling and virtual screening studies to design novel HDAC2 inhibitors." *Advances in Bioinformatics* 2014 (2014).

Kuang, Ming, Jingwei Zhou, Laiyou Wang, Zhihong Liu, Jiao Guo, and Ruibo Wu. "Binding kinetics versus affinities in BRD4 inhibition." *Journal of Chemical Information and Modeling* 55, no. 9 (2015): 1926–1935.

Kuck, Dirk, Narender Singh, Frank Lyko, and Jose L. Medina-Franco. "Novel and selective DNA methyltransferase inhibitors: Docking-based virtual screening and experimental evaluation." *Bioorganic & Medicinal Chemistry* 18, no. 2 (2010): 822–829.

Lawson, Michael, Urszula Uciechowska, Jörg Schemies, Tobias Rumpf, Manfred Jung, and Wolfgang Sippl. "Inhibitors to understand molecular mechanisms of NAD+-dependent deacetylases (sirtuins)." *Biochimica et Biophysica Acta (BBA)-Gene Regulatory Mechanisms* 1799, no. 10–12 (2010): 726–739.

Leelananda, Sumudu P., and Steffen Lindert. "Computational methods in drug discovery." *Beilstein Journal of Organic Chemistry* 12, no. 1 (2016): 2694–2718.

Letfus, Vatroslav, Dubravko Jelić, Ana Bokulić, Adriana Petrinić Grba, and Sanja Koštrun. "Rational design, synthesis and biological profiling of new KDM4C inhibitors." *Bioorganic & Medicinal Chemistry* 28, no. 1 (2020): 115128.

Lima, Marilia N.N., Cleber C. Melo-Filho, Gustavo C. Cassiano, Bruno J. Neves, Vinicius M. Alves, Rodolpho C. Braga, Pedro V.L. Cravo et al. "QSAR-driven design and discovery of novel compounds with antiplasmodial and transmission blocking activities." *Frontiers in Pharmacology* 9 (2018): 146.

Lindahl, Erik R. "Molecular dynamics simulations." *Molecular Modeling of Proteins* 443 (2008): 3–23.

Liu, Feng, Xin Chen, Abdellah Allali-Hassani, Amy M. Quinn, Gregory A. Wasney, Aiping Dong, Dalia Barsyte et al. "Discovery of a 2, 4-diamino-7-aminoalkoxyquinazoline as a potent and selective inhibitor of histone lysine methyltransferase G9a." *Journal of Medicinal Chemistry* 52, no. 24 (2009): 7950–7953.

Lu, Shaoyong, Mingfei Ji, Duan Ni, and Jian Zhang. "Discovery of hidden allosteric sites as novel targets for allosteric drug design." *Drug Discovery Today* 23, no. 2 (2018b): 359–365.

Lu, Wenchao, Rukang Zhang, Hao Jiang, Huimin Zhang, and Cheng Luo. "Computer-aided drug design in epigenetics." *Frontiers in Chemistry* 6 (2018a): 57.

Lundstrom, Kenneth. "Epigenetics: New possibilities for drug discovery." *Future Medicinal Chemistry* 9, no. 5 (2017): 437–441.

Lyu, Jiankun, Sheng Wang, Trent E. Balius, Isha Singh, Anat Levit, Yurii S. Moroz, Matthew J. O'Meara et al. "Ultra-large library docking for discovering new chemotypes." *Nature* 566, no. 7743 (2019): 224–229.

Martinez-Mayorga, Karina, Abraham Madariaga-Mazon, José L. Medina-Franco, and Gerald Maggiora. "The impact of chemoinformatics on drug discovery in the pharmaceutical industry." *Expert Opinion on Drug Discovery* 15, no. 3 (2020): 293–306.

Martinez-Mayorga, Karina, Terry L. Peppard, Fabian López-Vallejo, Austin B. Yongye, and Jose L. Medina-Franco. "Systematic mining of generally recognized as safe (GRAS) flavor chemicals for bioactive compounds." *Journal of Agricultural and Food Chemistry* 61, no. 31 (2013): 7507–7514.

Medina-Franco, José L., and Thomas Caulfield. "Advances in the computational development of DNA methyltransferase inhibitors." *Drug Discovery Today* 16, no. 9–10 (2011): 418–425.

Medina-Franco, José L., Oscar Méndez-Lucio, Alfonso Dueñas-González, and Jakyung Yoo. "Discovery and development of DNA methyltransferase inhibitors using *in silico* approaches." *Drug Discovery Today* 20, no. 5 (2015): 569–577.

Medina-Franco, José L., and Jakyung Yoo. "Docking of a novel DNA methyltransferase inhibitor identified from high-throughput screening: Insights to unveil inhibitors in chemical databases." *Molecular Diversity* 17 (2013): 337–344.

Méndez-Lucio, Oscar, Jeremy Tran, José L. Medina-Franco, Nathalie Meurice, and Mark Muller. "Toward drug repurposing in epigenetics: Olsalazine as a hypomethylating compound active in a cellular context." *Chem Med Chem* 9, no. 3 (2014): 560–565.

Mishra, Nitish Kumar. "Computational modeling of P450s for toxicity prediction." *Expert Opinion on Drug Metabolism & Toxicology* 7, no. 10 (2011): 1211–1231.

Morris, Garrett M., Ruth Huey, William Lindstrom, Michel F. Sanner, Richard K. Belew, David S. Goodsell, and Arthur J. Olson. "AutoDock4 and AutoDockTools4: Automated docking with selective receptor flexibility." *Journal of Computational Chemistry* 30, no. 16 (2009): 2785–2791.

Morris, Garrett M., and Marguerita Lim-Wilby. "Molecular docking." *Molecular Modeling of Proteins* (2008): 365–382.

Novikov, Fedor N., and Ghermes G. Chilov. "Molecular docking: Theoretical background, practical applications and perspectives." *Mendeleev Communications* 5, no. 19 (2009): 237–242.

Ortore, Gabriella, Francesco Di Colo, and Adriano Martinelli. "Docking of hydroxamic acids into HDAC1 and HDAC8: A rationalization of activity trends and selectivities." *Journal of Chemical Information and Modeling* 49, no. 12 (2009): 2774–2785.

Peters, Jens-Uwe. "Polypharmacology–foe or friend?" *Journal of Medicinal Chemistry* 56, no. 22 (2013): 8955–8971.

Popova, Mariya, Olexandr Isayev, and Alexander Tropsha. "Deep reinforcement learning for de novo drug design." *Science Advances* 4, no. 7 (2018): eaap7885.

Ptak, Carolyn, and Arturas Petronis. "Epigenetics and complex disease: From etiology to new therapeutics." *Annual Review of Pharmacology and Toxicology* 48 (2008): 257–276.

Ragno, Rino, Silvia Simeoni, Sabrina Castellano, Caterina Vicidomini, Antonello Mai, Antonella Caroli, Anna Tramontano et al. "Small molecule inhibitors of histone arginine methyltransferases: Homology modeling, molecular docking, binding mode analysis, and biological evaluations." *Journal of Medicinal Chemistry* 50, no. 6 (2007): 1241–1253.

Ragno, Rino, Silvia Simeoni, Sergio Valente, Silvio Massa, and Antonello Mai. "3-D QSAR studies on histone deacetylase inhibitors. A GOLPE/GRID approach on different series of compounds." *Journal of Chemical Information and Modeling* 46, no. 3 (2006): 1420–1430.

Rossetto, Dorine, Nikita Avvakumov, and Jacques Côté. "Histone phosphorylation: A chromatin modification involved in diverse nuclear events." *Epigenetics* 7, no. 10 (2012): 1098–1108.

Sakkiah, Sugunadevi, Navaneethakrishnan Krishnamoorthy, Poornima Gajendrarao, Sundarapandian Thangapandian, Yun-O. Lee, Song-Mi Kim, Jung-Keun Suh, Hyong-Ha Kim, and Keun-Woo Lee. "Pharmacophore mapping and virtual screening for SIRT1 activators." *Bulletin of the Korean Chemical Society* 30, no. 5 (2009): 1152–1156.

San José-Enériz, Edurne, Xabier Agirre, Obdulia Rabal, Amaia Vilas-Zornoza, Juan A. Sanchez-Arias, Estibaliz Miranda, Ana Ugarte et al. "Discovery of first-in-class reversible dual small molecule inhibitors against G9a and DNMTs in hematological malignancies." *Nature Communications* 8, no. 1 (2017): 15424.

Siedlecki, Pawel, Regine Garcia Boy, Slobodan Comagic, Ralf Schirrmacher, Manfred Wiessler, Piotr Zielenkiewicz, Sandor Suhai, and Frank Lyko. "Establishment and functional validation of a structural homology model for human DNA methyltransferase 1." *Biochemical and Biophysical Research Communications* 306, no. 2 (2003): 558–563.

Siedlecki, Pawel, Regine Garcia Boy, Tanja Musch, Bodo Brueckner, Sandor Suhai, Frank Lyko, and Piotr Zielenkiewicz. "Discovery of two novel, small-molecule inhibitors of DNA methylation." *Journal of Medicinal Chemistry* 49, no. 2 (2006): 678–683.

Singh, Narender, Alfonso Dueñas-González, Frank Lyko, and Jose L. Medina-Franco. "Molecular modeling and molecular dynamics studies of hydralazine with human DNA methyltransferase 1." *Chem Med Chem: Chemistry Enabling Drug Discovery* 4, no. 5 (2009): 792–799.

Sivakumar, Krishnankutty Chandrika, Jin Haixiao, C. Benjamin Naman, and T. P. Sajeevan. "Prospects of multitarget drug designing strategies by linking molecular docking and molecular dynamics to explore the protein–ligand recognition process." *Drug Development Research* 81, no. 6 (2020): 685–699.

Song, Qiuhang, Tingting Liu, Yucui Liu, Shuyue Wang, Cong Fan, Lihua Zheng, Yongli Bao et al. "An improved protocol for the virtual screening discovery of novel histone deacetylase inhibitors." *Chemical and Pharmaceutical Bulletin* 67, no. 10 (2019): 1076–1081.

Spannhoff, Astrid, Ralf Heinke, Ingo Bauer, Patrick Trojer, Eric Metzger, Ronald Gust, Roland Schüle, Gerald Brosch, Wolfgang Sippl, and Manfred Jung. "Target-based approach to inhibitors of histone arginine methyltransferases." *Journal of Medicinal Chemistry* 50, no. 10 (2007): 2319–2325.

Srimongkolpithak, Nitipol, Sandeep Sundriyal, Fengling Li, Masoud Vedadi, and Matthew J. Fuchter. "Identification of 2, 4-diamino-6, 7-dimethoxyquinoline derivatives as G9a inhibitors." *Medicinal Chemistry Communications* 5, no. 12 (2014): 1821–1828.

Subha, Kalyanamoorthy, and Gopal Ramesh Kumar. "Assessment for the identification of better HDAC inhibitor class through binding energy calculations and descriptor analysis." *Bioinformation* 3, no. 5 (2008): 218.

Talele, Tanaji T., Santosh A. Khedkar, and Alan C. Rigby. "Successful applications of computer aided drug discovery: Moving drugs from concept to the clinic." *Current Topics in Medicinal Chemistry* 10, no. 1 (2010): 127–141.

Tao, Hongrui, Xue Yan, Kongkai Zhu, and Hua Zhang. "Discovery of novel PRMT5 inhibitors by virtual screening and biological evaluations." *Chemical and Pharmaceutical Bulletin* 67, no. 4 (2019): 382–388.

Trott, Oleg, and Arthur J. Olson. "AutoDock Vina: Improving the speed and accuracy of docking with a new scoring function, efficient optimization, and multithreading." *Journal of Computational Chemistry* 31, no. 2 (2010): 455–461.

Van Norman, Gail A. "Phase II trials in drug development and adaptive trial design." *JACC: Basic to Translational Science* 4, no. 3 (2019): 428–437.

Vellore, Nadeem A., and Riccardo Baron. "Epigenetic molecular recognition: A biomolecular modeling perspective." *Chem Med Chem* 9, no. 3 (2014): 484–494.

Wang, Juxian, Limin Chen, Sarmistha Halder Sinha, Zhongjie Liang, Huifang Chai, Sakthivel Muniyan, Yu-Wei Chou et al. "Pharmacophore-based virtual screening and biological evaluation of small molecule inhibitors for protein arginine methylation." *Journal of Medicinal Chemistry* 55, no. 18 (2012): 7978–7987.

Yang, Chao-Yie, and Shaomeng Wang. "Computational analysis of protein hotspots." *ACS Medicinal Chemistry Letters* 1, no. 3 (2010): 125–129.

Yoo, Jakyung, Sun Choi, and Jose L. Medina-Franco. "Molecular modeling studies of the novel inhibitors of DNA methyltransferases SGI-1027 and CBC12: Implications for the mechanism of inhibition of DNMTs." *PLoS ONE* 8, no. 4 (2013): e62152.

Yoo, Jakyung, Joo Hee Kim, Keith D. Robertson, and Jose L. Medina-Franco. "Molecular modeling of inhibitors of human DNA methyltransferase with a crystal structure: Discovery of a novel DNMT1 inhibitor." *Advances in Protein Chemistry and Structural Biology* 87 (2012): 219–247.

Yoo, Jakyung, and José L. Medina-Franco. "Discovery and optimization of inhibitors of DNA methyltransferase as novel drugs for cancer therapy." *Drug Development—A Case Study Based Insight into Modern Strategies* (2011a): 3–22.

Yoo, Jakyung, and José L. Medina-Franco. "Homology modeling, docking and structure-based pharmacophore of inhibitors of DNA methyltransferase." *Journal of Computer-Aided Molecular Design* 25 (2011b): 555–567.

Yoo, J., and J. L Medina-Franco. "Inhibitors of DNA methyltransferases: Insights from computational studies." *Current Medicinal Chemistry* 19, no. 21 (2012): 3475–3487.

Yu, Wenbo, and Alexander D. MacKerell. "Computer-aided drug design methods." *Antibiotics: Methods and Protocols* (2017): 85–106.

Yuan, Zigao, Shaopeng Chen, Chunmei Gao, Qiuzi Dai, Cunlong Zhang, Qinsheng Sun, Jin-Shun Lin, Chun Guo, Yuzong Chen, and Yuyang Jiang. "Development of a versatile DNMT and HDAC inhibitor C02S modulating multiple cancer hallmarks for breast cancer therapy." *Bioorganic Chemistry* 87 (2019): 200–208.

Zhang, Weilin, Jianfeng Pei, and Luhua Lai. "Computational multitarget drug design." *Journal of Chemical Information and Modeling* 57, no. 3 (2017): 403–412.

Zhao, Hongtao, Lisa Gartenmann, Jing Dong, Dimitrios Spiliotopoulos, and Amedeo Caflisch. "Discovery of BRD4 bromodomain inhibitors by fragment-based high-throughput docking." *Bioorganic & Medicinal Chemistry Letters* 24, no. 11 (2014): 2493–2496.

Zhavoronkov, Alex, Yan A. Ivanenkov, Alex Aliper, Mark S. Veselov, Vladimir A. Aladinskiy, Anastasiya V. Aladinskaya, Victor A. Terentiev et al. "Deep learning enables rapid identification of potent DDR1 kinase inhibitors." *Nature Biotechnology* 37, no. 9 (2019): 1038–1040.

9 The Role of Indian Traditional Knowledge in Epigenetic Drug Discovery

Rajan Logesh,[1] *Maruveettill Arjun,*[2] *Saladula Hima,*[3] *Prabitha Prabhakaran,*[3] *and Kumarappan Chidambaram*[4]
[1]Department of Pharmacognosy, JSS College of Pharmacy, JSS Academy of Higher Education and Research, Mysuru, Karnataka, India
[2]Department of Pharmaceutics, Centre of Excellence in Regulatory Sciences, JSS College of Pharmacy, JSS Academy of Higher Education and Research, Mysuru, Karnataka, India
[3]Department of Pharmaceutical Chemistry, JSS College of Pharmacy, JSS Academy of Higher Education and Research, Mysuru, Karnataka, India
[4]Department of Pharmacology, College of Pharmacy, King Khalid University, Al-Qara, Asir Province, Saudi Arabia

9.1 INTRODUCTION

The term epigenetics was coined 75 years ago by a British biologist named Conrad Waddington described how a phenotype adapts to the environment (Waddington, 1942). Later, epigenetics was defined as phenotypic variations that do not initiate from changes in the genotype. Multicellular eukaryotic organisms display clear variation at the genome level. For example, cells in humans carry specific genomes that produce distinct tissues and organs via variations in genetic expression (Deans and Maggert, 2015).

Chronic diseases have become a serious burden to healthcare systems in both developing and developed countries (Choi et al., 2008), including cancer, diabetes, and other diseases that are difficult to cure (Власик et al., 2017; Ndagi et al., 2017; Foretz et al., 2014). Therefore, understanding the mechanism is necessary for identifying therapeutic leads.

The pathogenesis of epigenetics plays a pivotal role in the growth of diseases (Traboulsi et al., 2017). The main epigenetic alterations are DNA methylation, noncoding RNA, and histone modifications (Xiao et al., 2019). Noncoding RNA regulates genetic transcription by altering DNA methylation and histone levels (Xiao et al., 2018; Xu et al., 2013). Thus, altering these alters human diseases (Guo et al., 2021; Yan et al., 2021). Epigenetic alterations increase cancer risk (Zheng et al., 2016) and can lead to obesity and type 2 diabetes mellitus (Ling and Rönn, 2019). Therefore, epigenetics plays an important role in combating various diseases, creating opportunities for developing clinical therapies (Hamm and Costa, 2015).

In India, Ayurveda is one of the most important, and oldest, traditional systems of medicine that is still in use (Jaiswal et al., 2017). In India, herbal medicine plays a key role in drug discovery and development, where identifying new pharmacological leads enhances therapeutic potential. Traditional medicinal plants, guided by historical knowledge, are studied using modern methods such as ethnobotany or pharmacophylogeny (Souza, Williamson & Hawkins, 2018). Many researchers have reported that medicinal plants can influence disease progression by alters epigenetic mechanisms, with a focus on cancer (Dammann et al., 2017; Sreenivasmurthy et al., 2017; Zhou et al.,

DOI: 10.1201/9781032706511-9

2019). Therefore, understanding the role of epigenetics using medicinal plants could help us develop or discover novel epigenetic drugs (Hsieh et al., 2013). In this chapter, we review existing literature on the role of Indian traditional medicine in modulating epigenetic mechanisms to combat chronic diseases. Additionally, we highlight key research gaps relevant to epigenetic drug discovery.

9.2 INDIAN TRADITIONAL MEDICINE IN DRUG DISCOVERY

Plants provide benefits to healthcare for up to 80% of the world's population (Saslis-Lagoudakis et al., 2014). Among the types, Ayurveda, India's traditional medicine system, is regarded as one of the most important. Medicinal plants have been evolving for thousands of years, with their development closely linked to plant phylogeny—the evolutionary relationships among species. Species that exhibit morphological similarities also share genetic links. Similarly, their physiological and biochemical characteristics often align, with phytochemical compositions remaining consistent among both closely related and genetically distant species. The active constituents in medicinal plants that give them their curative effects are secondary metabolites (Hao and Xiao, 2020).

Ayurveda is based on the principle that individuals achieve physical, mental, and emotional health in coexistence with nature (Liu et al. 2005); it translates from the Sanskrit as "science of life" and dates to the ancient Vedic period in India (Sharma and Keith Wallace, 2020). With the side effects from synthetic drugs, researchers are increasingly investigating traditional medicine systems including medicinal plants. The WHO reported that about 80% of the world's population uses herbal medicine as their primary treatment (Adekson, 2016). India is one of the richest plant sources in the world, being home to over 3000 medicinal plants—about 75% of plant species—and containing more than 1000 herbal industries (Liang et al., 2005). Scientific support for Indian medicine is increasing, but marketing these herbal medicines still faces many challenges. For instance, proper identification and quality control of plant materials are crucial for maintaining the integrity of traditional medicine systems. Additionally, herbal medicines are highly susceptible to adulteration and deterioration, and details are lacking on their side effects (Dubey et al., 2004).

9.3 EPIGENETIC MECHANISMS OF HERBAL MEDICINE

In the realm of human malignancies, the involvement of DNA methyltransferases (DNMTs) and histone deacetylases (HDACs) is well established, with ongoing exploration of DNMT and HDAC inhibitors as potential anticancer drugs in clinical trials (Giaginis et al., 2014; Yuan et al., 2019). DNMTs play a crucial role in orchestrating DNA methyl transfers, which leads to the epigenetic silencing of multiple genes (Mirza et al., 2013; Verdin et al., 2003).

The primary strategy for utilizing medicinal plants and their phytochemicals against diseases involves modulating DNMT and HDAC expression to achieve effective epigenetic regulation. Additionally, as depicted in Figure 9.1, medicinal plants influence gene expression by impacting histone methylation, acetylation, phosphorylation, and demethylation of modified DNA.

Medicinal plants influence cell factors directly, impacting histone methylation, acetylation, and DNA methylation, as depicted in the grey, purple and green sections of the figure. Additionally, herbal medicine influences histone ubiquitination and phosphorylation.

9.4 EPIGENETICS IN MEDICINAL PLANTS AND BIOACTIVE COMPOUNDS

The memory encoded in epigenetics is crucial for sustaining life, dictating the predetermined functional traits of cells. Recent findings indicate that epigenetic modifications are a potential bridge to alterations in gene expression, playing a role in the onset of numerous chronic illnesses. Targeting the epigenome has emerged as a viable approach for disease treatment. Traditional herbal medicine is gaining attention among researchers due to its reduced toxicity and efficacy in addressing

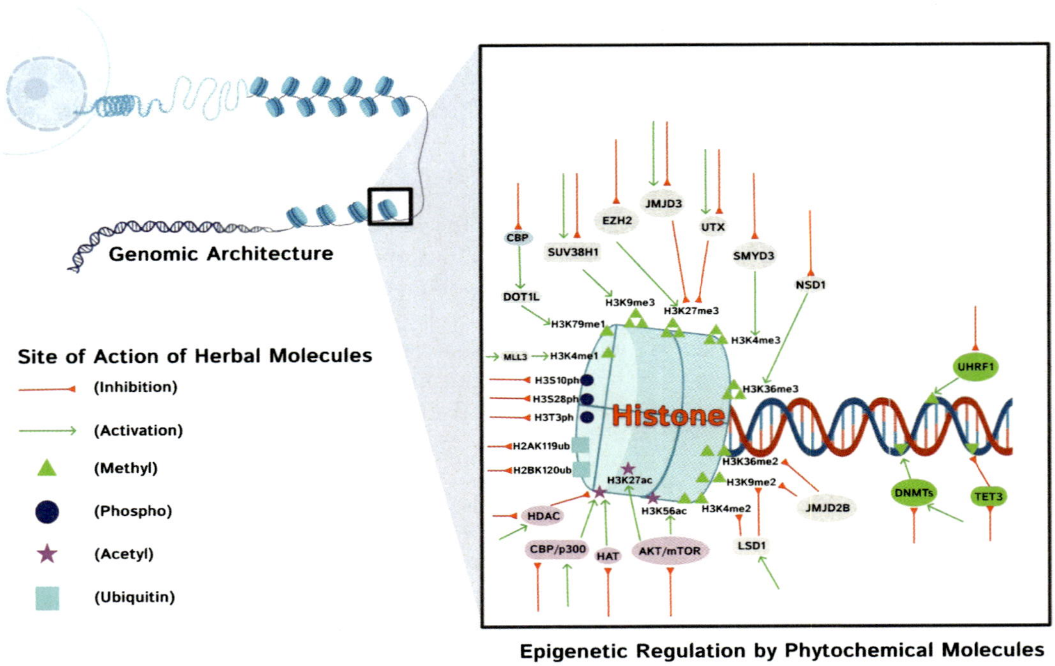

FIGURE 9.1 Epigenetic regulation of phytochemical molecules.

ailments. Notably, researchers have discovered inherent in epigenetic modifications in medicinal plants that can impede various diseases (Giaginis et al., 2014; Yuan et al., 2019).

Herbal medicines contain thousands of molecules, but only a few produce effects (Zhang et al., 2019). The medicinal plants and phytochemicals with epigenetic modification effects are explained in detail in Tables 9.1 and 9.2. Given the efficacy of plant-derived phytochemicals such as paclitaxel and artemisinin, medicinal plants are an increasingly important source in drug discovery (Qiao et al., 2020).

9.5 RECENT UPDATES IN MEDICINAL PLANT GENOMES

In of 21st century, we witnessed a profound transformation in genomic technologies, and these advancements influence every aspect of plant biology, particularly plant metabolism. This is primarily attributed to the accessibility of genome sequences and the substantial datasets linked to them, encompassing data like expression patterns and metabolite profiles. These resources greatly expedite the discovery of the pathways responsible for producing specific metabolites and entire categories of metabolites (Alami et al., 2022).

A crucial aspect within genomics is comparative, wherein biological investigations compare the structural and genomic characteristics of different genomes to comprehend genetic relationships across different species (Bornstein et al., 2023). The progress in genome sequencing, assembly technologies, and associated tools has added renewed dynamism to the extensive advancement of genomics and molecular biology in medicinal plants, facilitating the attainment of high-quality genomes for these plants (Alami et al., 2022). For instance, genetic markers associated with high eugenol and ursolic acid content were identified and used in breeding programs to develop tulsi varieties with higher concentrations of these compounds. Tulsi, also known as holy basil (*Ocimum sanctum Linn*), is a widely used medicinal plant for various diseases (Upadhyay et al., 2015).

TABLE 9.1
Bioactives from Medicinal Plants that Alter Epigenetic Effects

Phytoconstituent	Structure	Source	References
Allicin	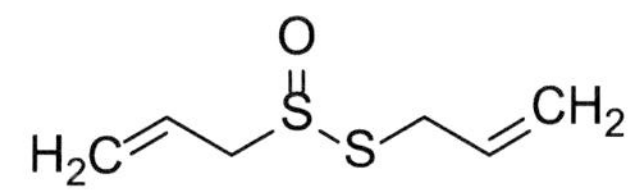	*Allium sativum*	(Chhabria et al., 2015)
Baicalin	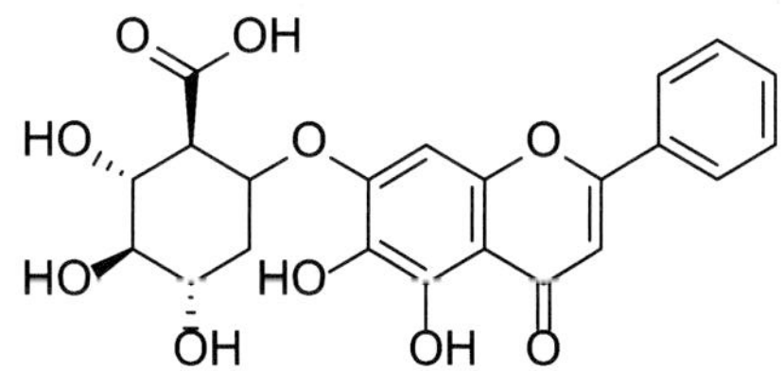	*Scutellaria baicalensis*	(Jiang et al., 2022)
Calebin-A	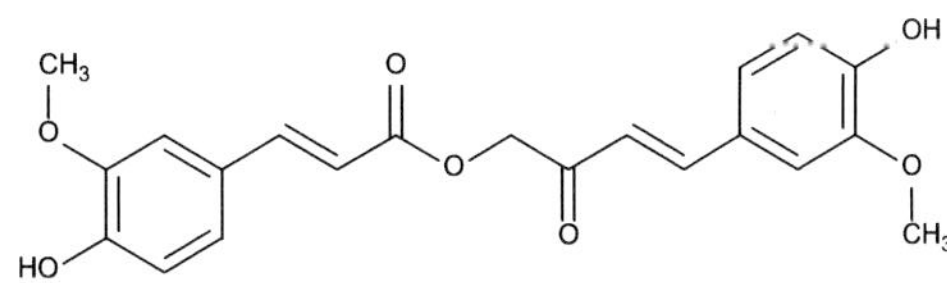	*Curcuma longa*	(Lee et al., 2019)
Cannabidiol	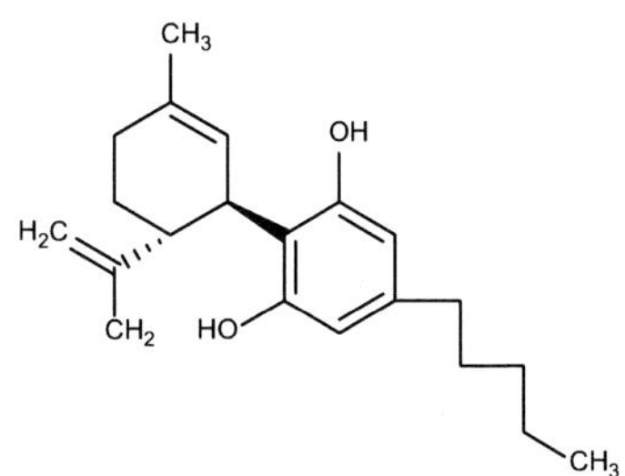	*Cannabis sativa*	(Pucci et al., 2013)
Cannabigerol	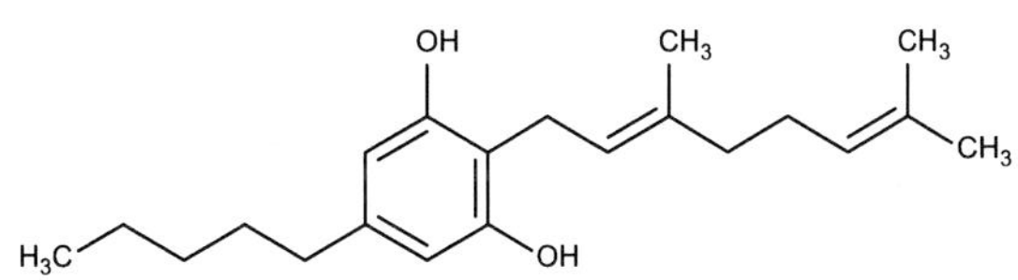	*Cannabis sativa*	(Pucci et al., 2013)

(*Continued*)

TABLE 9.1 *(Continued)*
Bioactives from Medicinal Plants that Alter Epigenetic Effects

Phytoconstituent	Structure	Source	References
Cucurbitacin B	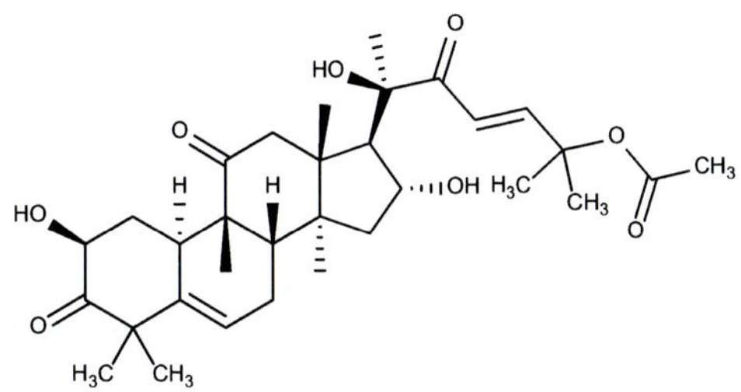	*Trichosanthes cucumerina*	(Dittharot et al., 2019)
Tanshinone I	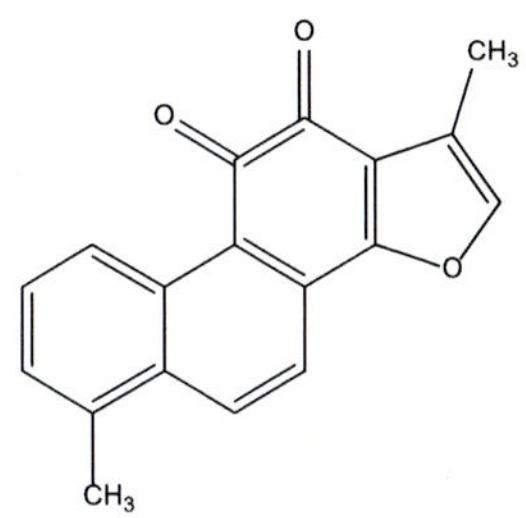	*Salvia miltirrhiza*	(Yong et al., 2011)
Tetramethylpyrazine	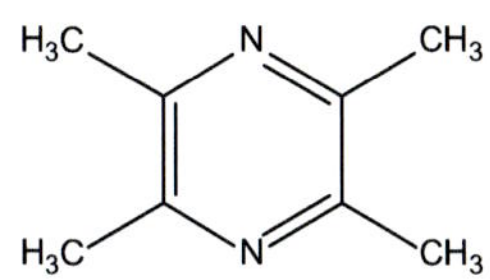	*Chuanxiong rhizoma*	(Yan Y et al., 2014)
Triptolide	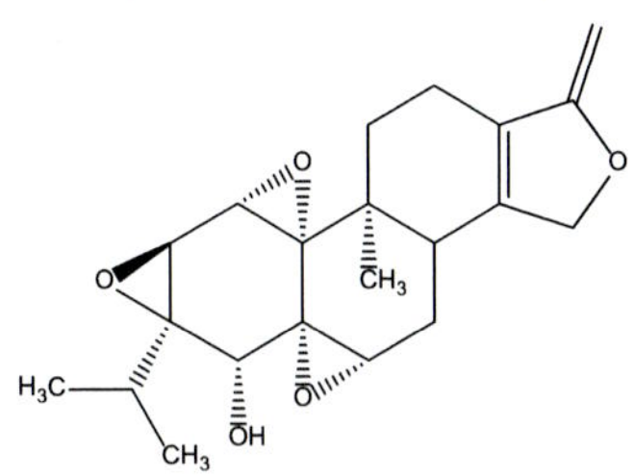	*Tripterygium wilfordia*	(Zhao F et al., 2010)

Valeric Acid	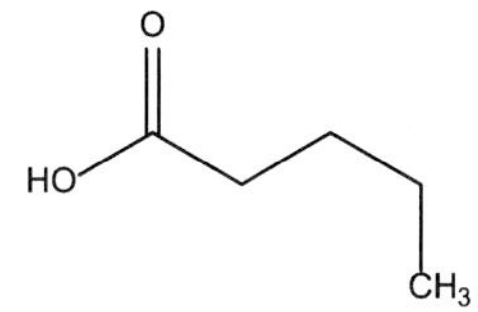	*Valeriana officinalis*	(Shi et al., 2021)
Zaluzanin D	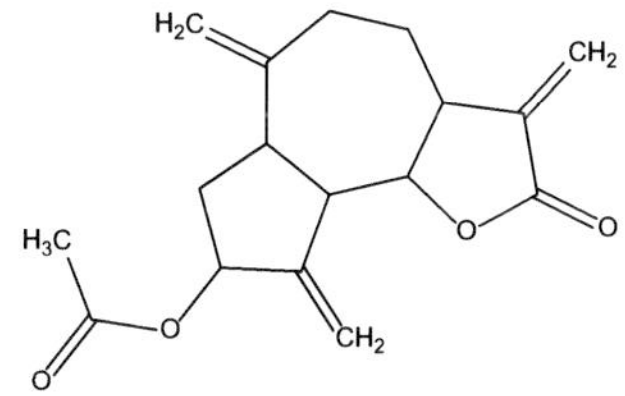	*Gymnanthemum cass*	(Cheeran & Munuswamy-Ramanujam, 2020)
Z-ligustilide	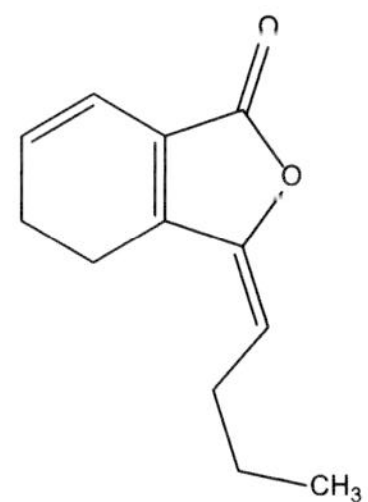	*Chuanxiong rhizoma*	(Wang et al., 2021a)
Curcumin	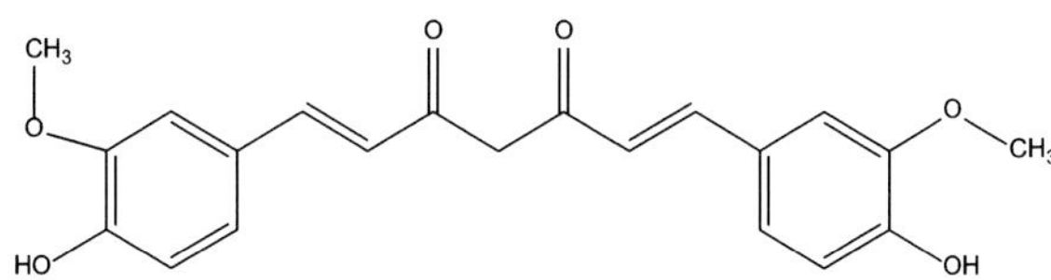	*Curcuma longa*	(Wu et al., 2013)

(*Continued*)

TABLE 9.1 *(Continued)*
Bioactives from Medicinal Plants that Alter Epigenetic Effects

Phytoconstituent	Structure	Source	References
Curcumol	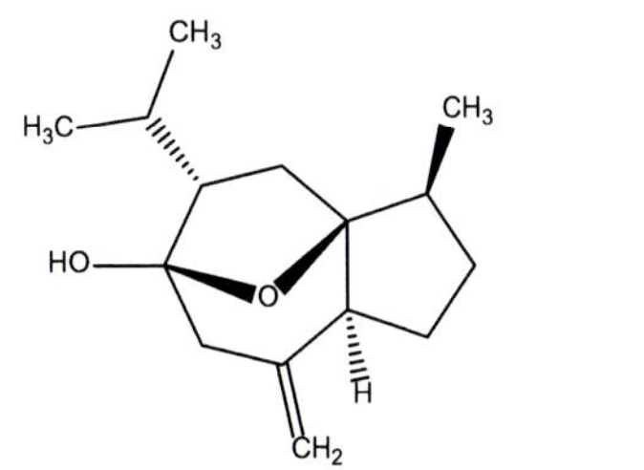	*Curcuma kwangsiensis*	(Tian et al., 2021)
Esculetin	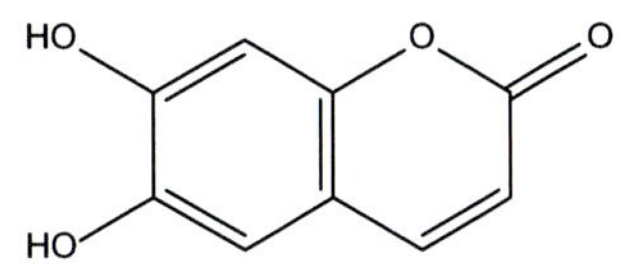	*Citrus limonia*	(Kadakol et al., 2017)
Hesperetin	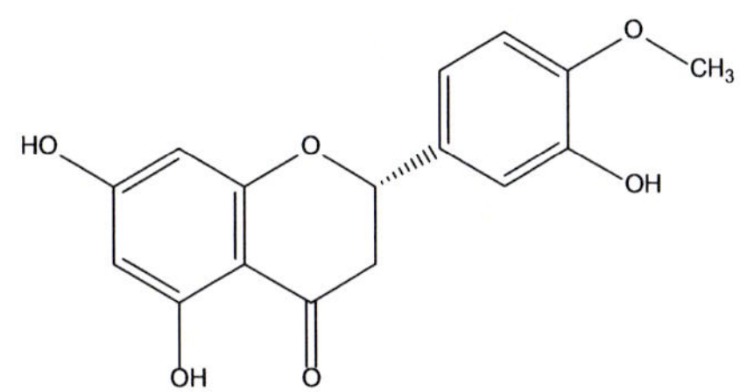	*Citrus species*	(Muhammad et al., 2019)
Hydroxysafor yellow A	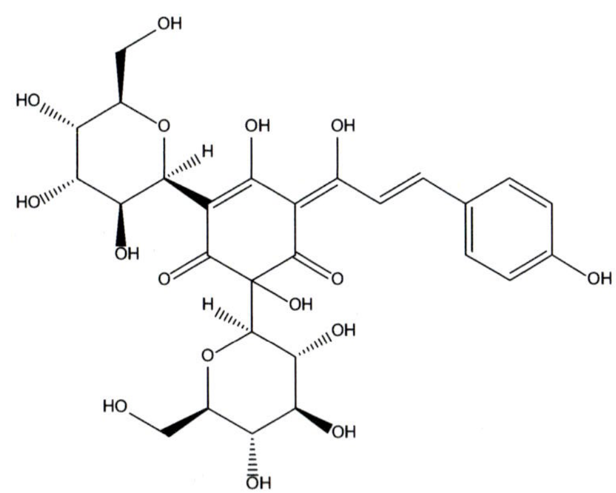	*Flos carthami*	(Wang et al., 2021b)

Juglanin	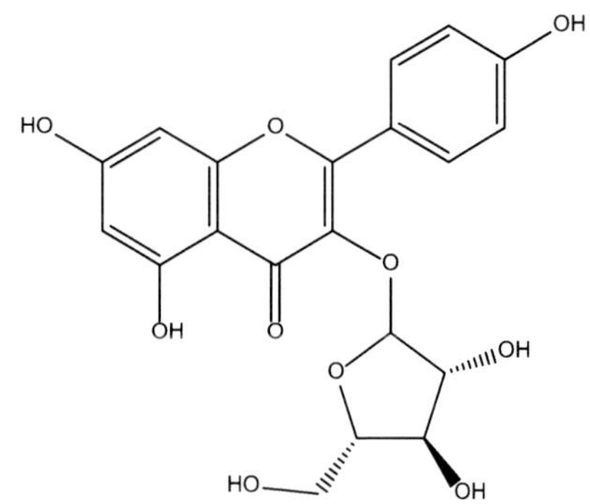	*Polygonum aviculare*	(Li Q et al., 2021)
Kazinol Q	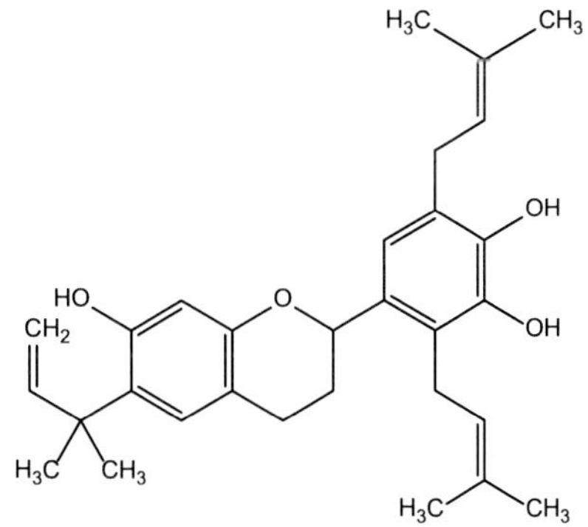	-	(Weng et al., 2014)

(*Continued*)

TABLE 9.1 *(Continued)*
Bioactives from Medicinal Plants that Alter Epigenetic Effects

Phytoconstituent	Structure	Source	References
Kaempferol	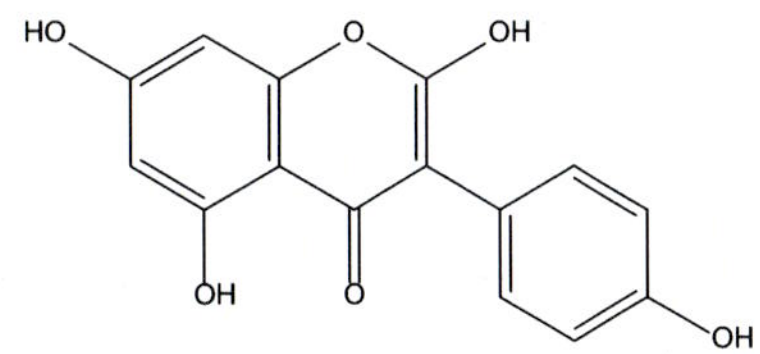	*Gingko biloba*	(Park et al., 2022)
Luteolin	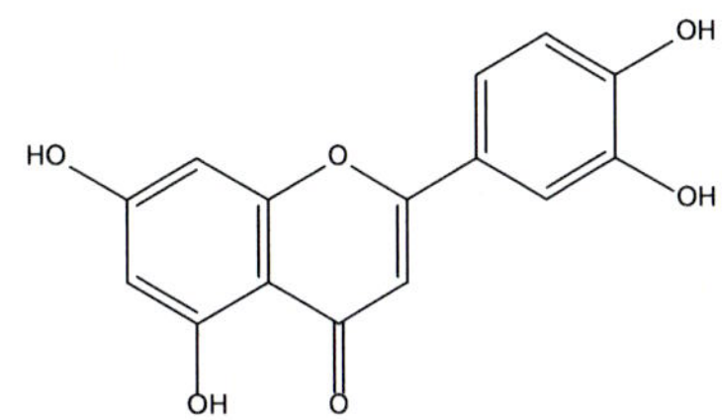	*Apium graveolens*	(Kim et al., 2014)
Moringa isothiocyanate	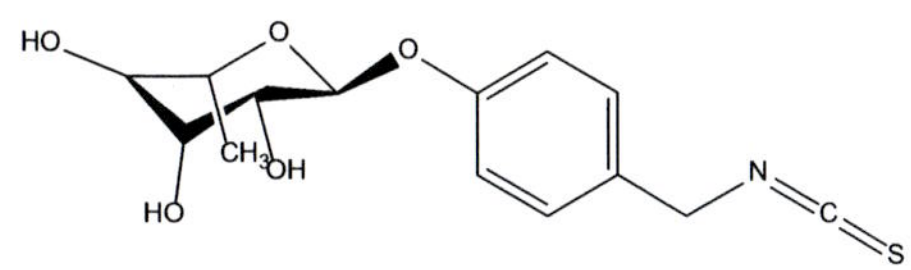	*Moringa oleifera*	(Wang et al., 2019)
Naringenin	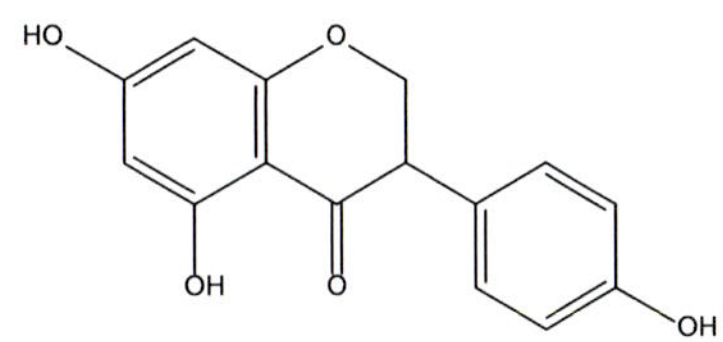	*Aurantii fructus*	(Wang et al., 2021c)
N-butylidenephthalide	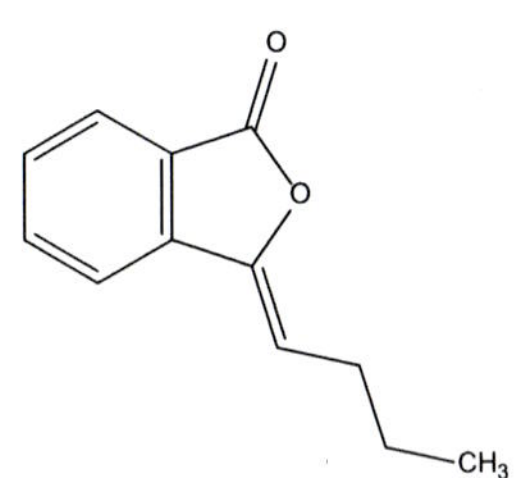	*Angelica sinensis*	(Yu et al., 2010)

Nimbolide	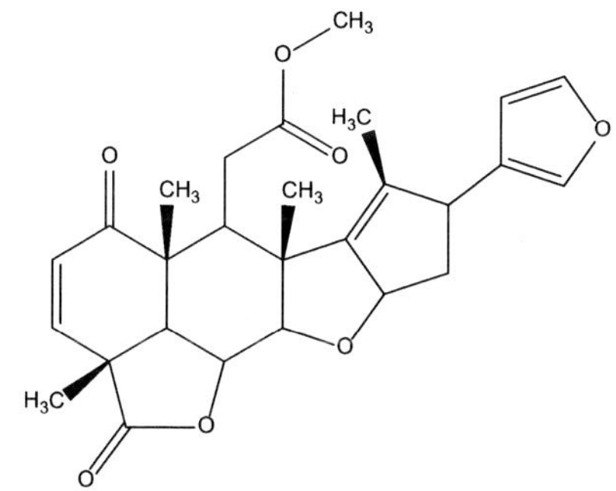	*Azadirachta indica*	(Pooladanda et al., 2018)
Nordihydroguaiaretic acid	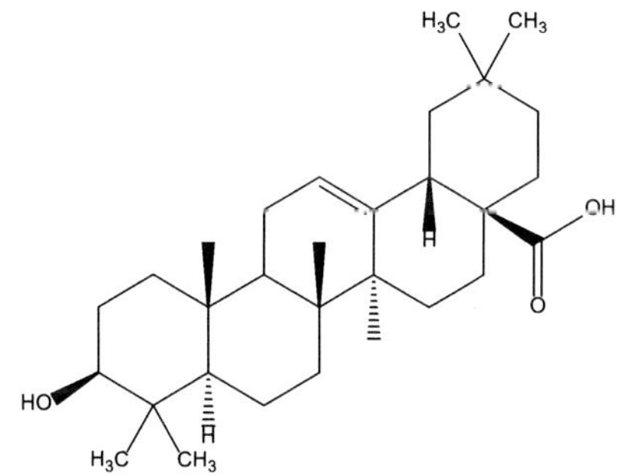	*Larrea tridentata*	(Cu et al., 2008)
Oleanolic acid	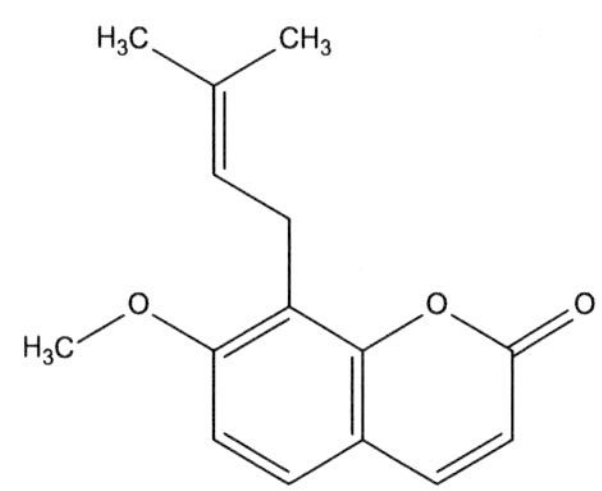	*Olea europaea*	(Lu et al., 2021)

(*Continued*)

TABLE 9.1 *(Continued)*

Bioactives from Medicinal Plants that Alter Epigenetic Effects

Phytoconstituent	Structure	Source	References
Parthenolide	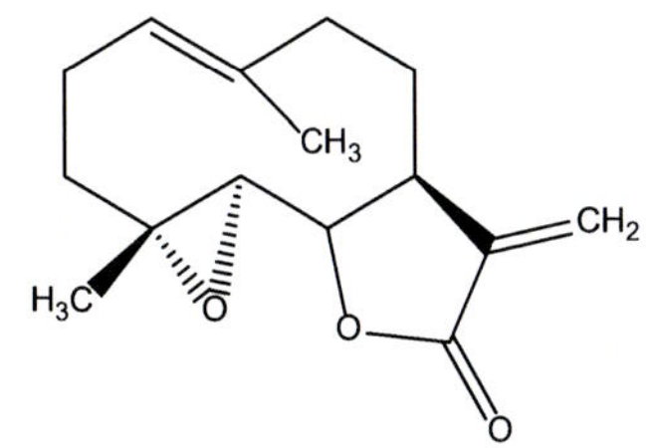	*Tanacetum parthenium*	(Sun et al., 2017)
Piperine	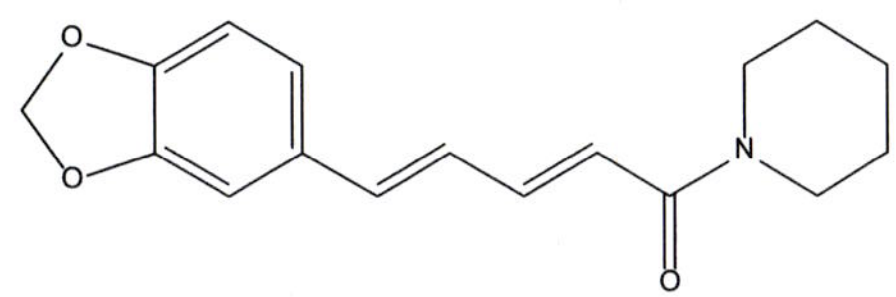	*Piper nigrum*	(Ghantous et al., 2012)
Sennoside A	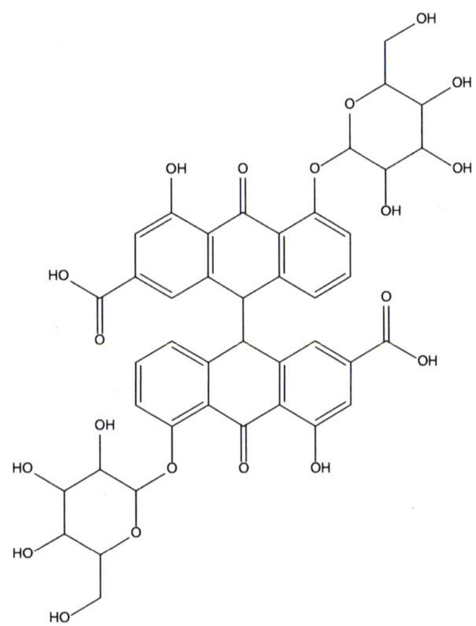	*Rheum rhabarbarum*	(Park et al., 2019)

Silibinin	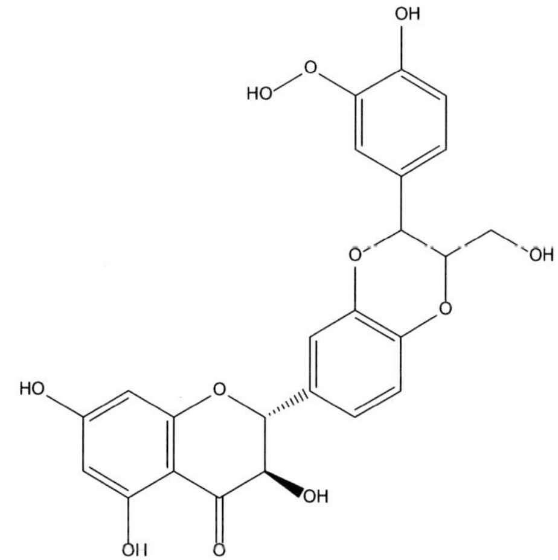	*Silybum marianum*	(Zhu et al., 2021)

TABLE 9.2
Herbal Medicine-Induced Epigenetic Modifications and Chronic Diseases

Extract/Component	Epigenetic Target	Epigenetic Alteration	Reference
Brain tumor			
Curcumin	DNA	Inhibits DNMT1 activity	(Wu et al., 2013)
Tetramethylpyrazine	Histone	Enhances ac-H3 and H4 to the promoter gene	(Yan Y et al., 2014)
Calebin-A	Histone	Reduces H3 histone acetylation	(Lee et al., 2019)
Leonurus sibiricus transgenic root extract	DNA, histone	Reduces the expression of UHRF1 and DNMT1	(Sitarek et al., 2018)
Rhaponticum carthamoides transformed root extract	DNA, histone	Reduces the expression of UHRF1 and DNMT1	(Skała et al., 2018)
Thoracic tumor			
Nordihydroguaiaretic acid	DNA	Reverses p161NK4a hypermethylation	(Cui et al., 2008a)
Thymus serpyllum extract	DNA; histone	Inhibits DNMT and HDAC activity	(Bozkurt et al., 2012)
Tanshinone	Histone	Reduces ac-H3	(Li et al., 2015)
Kazinol Q	DNA	Inhibits DNMT1 activity	(Weng et al., 2014)
Tien-Hsien liquid	DNA	Reduces DNMT1 and 3a	(Yao et al., 2016)
Jinfukang	Histone	Downregulates H3K4me3	(Lu et al., 2016)
Z-ligustilide	Histone	Decreases HDAC1, 2, 4, 5, 7	(Ma et al., 2017)
Nimbolide	Histone	Reduces HDAC2 and increases H3K27ac	(Pooladanda et al., 2018)
Luteolin	Histone	Decreases H3K27ac and H3K56ac	(Wu et al., 2021)
Cucurbitacin B	DNA	Increases DNMT1 methylation	(Dittharot et al., 2019)
Valeric acid	DNA; histone	Reduces HDAC activity and DNA hypomethylation	(Shi et al., 2021)
Digestive system tumor			
N-butylidenephthalide	DNA	Downregulates mRNA and protein levels	(Yu et al., 2010)
Cotinus coggygria Scop. extract	DNA	Inhibits DNMT1 and 3a expression	(Gospodinova et al., 2021)
Ginkgo biloba extract	DNA	Affects methylation of the hepatocellular carcinoma gene	(Kovi et al., 2019)
Aged citrus peel extract	DNA, histone	Decreases the protein expression of HDAC8, DNMT1, and 3a and inhibits CpG (cytosine-phosphate-guanine) methyltransferase activity	(Lin et al., 2019)
Azadirachta indica extract; Nimbolide	DNA; Histone	Inhibits HDAC and DNMT activity	(Qiu et al., 2019)
Silibinin	DNA	Induces DNA hypomethylation	(Zhu et al., 2021)
Valeric acid	Histone	Inhibits HDAC1, 2, and 3 activities	(Han et al., 2020)
Oleanolic acid	DNA	Decreases of TET3 expression	(Lu et al., 2021)

The genomes of many therapeutic plants have been used in sequencing. For example, the *Panax ginseng* genome was sequenced and provided insights into the biosynthesis of ginsenosides, a plant's bioactive compounds with potential health benefits (Xu et al., 2017). Genomic studies have been crucial in identifying genes involved in the biosynthesis of bioactive compounds of *Catharanthus roseus*, which produces vinblastine and vincristine, important anticancer alkaloids (She et al., 2019), (Stander et al, 2020). *Artemisia annua*, the source of artemisinin for malaria treatment, has been studied using combined metabolomics and transcriptomics data approaches to elucidate the regulatory mechanisms governing artemisinin biosynthesis. *Glycyrrhiza uralensis* has been selectively bred using markers associated with glycyrrhizin content to develop high-yielding varieties (Li et al., 2018).

CRISPR-Cas9 offers the potential to alter specific genes in medicinal plants to investigate the production of effective compounds, choose traits to enhance yield, and further explore biosynthetic pathways and regulatory mechanisms. This becomes particularly feasible with the complete sequencing of genomes in certain medicinal plants (Tavakoli et al., 2021).

Recent advancements in sequencing technology, including second-generation, high-throughput chromosome conformation capture sequencing, have facilitated the publication of many newly published medicinal plant genomes such as *Salvia miltiorrhiza* Bunge (Song et al., 2020), *Ophiorrhiza pumila* (Rai et al., 2021), *Coptis chinensis* Franch (Liu et al., 2021), tea (Chen et al., 2020), *Sarcophaga peregrina* (Ren et al., 2021), *Aquilegia oxysepala* var. kansuensis (Xie et al., 2020), *Tripterygium wilfordii* (Tu et al., 2020), and more (Alami et al., 2022).

9.6 CONCLUSION

Recent advances in medicinal plant genomics have transformed our understanding of plant metabolism, and genome sequences and datasets facilitate the exploration of metabolite pathways. By integrating traditional knowledge with modern scientific methods, this chapter shows the symbiotic relationship between Indian traditional knowledge and epigenetic drug discovery, which not only emphasizes the value of traditional systems but also advances novel ways to solve complex modern-day health issues. A new era of personalized medicine may be ushered in by combining conventional insight with advanced genomics and novel therapeutic interventions.

REFERENCES

Adekson, Mary Olufunmilayo. "Similarities and differences between Yoruba traditional healers (YTH) and Native American and Canadian Healers (NACH)." *Journal of Religion and Health* 55 (2016): 1717–1728.

Alami, Mohammad Murtaza, Zhen Ouyang, Yipeng Zhang, Shaohua Shu, Guozheng Yang, Zhinan Mei, and Xuekui Wang. "The current developments in medicinal plant genomics enabled the diversification of secondary metabolites' biosynthesis." *International Journal of Molecular Sciences* 23, no. 24 (2022): 15932.

Власик, Л. Й., Ганна Я. Ступницька, and Анастасія Л. Сухолотюк. "Peculiarities of primary health care organization for patients suffering from chronic obstructive pulmonary disease jointly with coronary heart disease." *Wiadomosci Lekarskie* (Warsaw, Poland: 1960) 70, no. 3, pt 1 (2017): 466–469.

Bornstein, K., G. Gryan, E. S. Chang, A. Marchler-Bauer, and V. A. Schneider. "The NIH comparative genomics resource: Addressing the promises and challenges of comparative genomics on human health." *BMC Genomics* 24, no. (1) (2023): 575.

Bozkurt, Emir, Harika Atmaca, Asli Kisim, Selim Uzunoglu, Ruchan Uslu, and Burcak Karaca. "Effects of Thymus serpyllum extract on cell proliferation, apoptosis and epigenetic events in human breast cancer cells." *Nutrition and Cancer* 64, no. 8 (2012): 1245–1250.

Cheeran, Vinnie, and Ganesh Munuswamy-Ramanujam. "Sesquiterpene lactone Zaluzanin D alters MMP-9 promoter methylation in differentiated THP-1 monocytic cells and down regulates inflammatory cytokines IL-1β and TNF-α." *International Immunopharmacology* 87 (2020): 106803.

Chen, Jie-Dan, Chao Zheng, Jian-Qiang Ma, Chen-Kai Jiang, Sezai Ercisli, Ming-Zhe Yao, and Liang Chen. "The chromosome-scale genome reveals the evolution and diversification after the recent tetraploidization event in tea plant." *Horticulture Research* 7 (2020).

Chhabria, Sagar V., Mohammad A. Akbarsha, Albert P. Li, Prashant S. Kharkar, and Krutika B. Desai. "In situ allicin generation using targeted alliinase delivery for inhibition of MIA PaCa-2 cells via epigenetic changes, oxidative stress and cyclin-dependent kinase inhibitor (CDKI) expression." *Apoptosis* 20 (2015): 1388–1409.

Choi, B. C. K., D. V. McQueen, P. Puska, K. A. Douglas, M. Ackland, S. Campostrini, A. Barceló et al. "Enhancing global capacity in the surveillance, prevention, and control of chronic diseases: Seven themes to consider and build upon." *Journal of Epidemiology & Community Health* 62, no. 5 (2008): 391–397.

Cui, Yi, Cailing Lu, Aijun Kang, Liyong Liu, Sainan Tan, Daguang Sun, Jianguo Hu, and Xu Ma. "Nordihydroguaiaretic acid restores expression of silenced E-cadherin gene in human breast cancer cell lines and xenografts." *Anti-Cancer Drugs* 19, no. 5 (2008a): 487–494.

Cui, Yi, Cailing Lu, Liyong Liu, Daguang Sun, Nan Yao, Sainan Tan, Shanshan Bai, and Xu Ma. "Reactivation of methylation-silenced tumor suppressor gene p16INK4a by nordihydroguaiaretic acid and its implication in G1 cell cycle arrest." *Life Sciences* 82, no. 5–6 (2008b): 247–255.

Dammann, Reinhard H., Antje M. Richter, Adriana P. Jiménez, Michelle Woods, Miriam Küster, and Chamindri Witharana. "Impact of natural compounds on DNA methylation levels of the tumor suppressor gene RASSF1A in cancer." *International Journal of Molecular Sciences* 18, no. 10 (2017): 2160.

Deans, Carrie, and Keith A. Maggert. "What do you mean, 'epigenetic'?" *Genetics* 199, no. 4 (2015): 887–896.

Dittharot, Kanthanadon, Sumana Dakeng, Parichat Suebsakwong, Apichart Suksamrarn, Pimpicha Patmasiriwat, and Moltira Promkan. "Cucurbitacin B induces hypermethylation of oncogenes in breast cancer cells." *Planta Medica* 85, no. 05 (2019): 370–378.

Dubey, N. K., Rajesh Kumar, and Pramila Tripathi. "Global promotion of herbal medicine: 'India's opportunity". *Current Science* 86, no. 1 (2004): 37–41.

Foretz, Marc, Bruno Guigas, Luc Bertrand, Michael Pollak, and Benoit Viollet. "Metformin: From mechanisms of action to therapies." *Cell Metabolism* 20, no. 6 (2014): 953–966.

Ghantous, Akram, Melody Saikali, Tilman Rau, Hala Gali-Muhtasib, Regine Schneider-Stock, and Nadine Darwiche. "Inhibition of tumor promotion by parthenolide: Epigenetic modulation of p21." *Cancer Prevention Research* 5, no. 11 (2012): 1298–1309.

Giaginis, Constantinos, Paraskevi Alexandrou, Ioanna Delladetsima, Ioanna Giannopoulou, Efstratios Patsouris, and Stamatios Theocharis. "Clinical significance of histone deacetylase (HDAC)-1, HDAC-2, HDAC-4, and HDAC-6 expression in human malignant and benign thyroid lesions." *Tumor Biology* 35 (2014): 61–71.

Gospodinova, Zlatina I., Istvan Zupkó, Noémi Bózsity, Vasilissa I. Manova, Mariyana S. Georgieva, Svetla J. Todinova, Stefka G. Taneva, Imre Ocsovszki, and Maria E. Krasteva. "Cotinus coggygria Scop. Induces cell cycle arrest, apoptosis, genotoxic effects, thermodynamic and epigenetic events in MCF7 breast cancer cells." *Zeitschrift für Naturforschung* C 76, no. 3–4 (2021): 129–140.

Guo, Wenqian, Han Ma, Chong-Zhi Wang, Jin-Yi Wan, Haiqiang Yao, and Chun-Su Yuan. "Epigenetic studies of Chinese herbal medicine: Pleiotropic role of DNA methylation." *Frontiers in Pharmacology* 12 (2021): 790321.

Hamm, Christopher A., and Fabricio F. Costa. "Epigenomes as therapeutic targets." *Pharmacology & Therapeutics* 151 (2015): 72–86.

Han, Rui, Olivia Nusbaum, Xinyi Chen, and Yong Zhu. "Valeric acid suppresses liver cancer development by acting as a novel HDAC inhibitor." *Molecular Therapy-Oncolytics* 19 (2020): 8–18.

Hao, Da-cheng, and Pei-gen Xiao. "Pharmaceutical resource discovery from traditional medicinal plants: Pharmacophylogeny and pharmacophylogenomics." *Chinese Herbal Medicines* 12, no. 2 (2020): 104–117.

Hsieh, Hsin-Ying, Pei-Hsun Chiu, and Sun-Chong Wang. "Histone modifications and traditional Chinese medicinals." *BMC Complementary and Alternative Medicine* 13 (2013): 1–11.

Jaiswal, Yogini S., and Leonard L. Williams. "A glimpse of Ayurveda–The forgotten history and principles of Indian traditional medicine." *Journal of Traditional and Complementary Medicine* 7, no. 1 (2017): 50–53.

Jiang, Hongpeng, Qianqian Yao, Yongbo An, Linlin Fan, Jing Wang, and Huiying Li. "Baicalin suppresses the progression of type 2 diabetes-induced liver tumor through regulating METTL3/m6A/HKDC1 axis and downstream p-JAK2/STAT1/clevaged Capase3 pathway." *Phytomedicine* 94 (2022): 153823.

Kadakol, Almesh, Santosh Kumar Goru, Vajir Malek, and Anil Bhanudas Gaikwad. "Esculetin ameliorates vascular perturbation by intervening in the occupancy of H2BK120Ub at At1, At2, Tgfβ1 and Mcp1 promoter gene in thoracic aorta of IR and T2D rats." *Biomedicine & Pharmacotherapy* 95 (2017): 1461–1468.

Kim, Hye Joo, Wooje Lee, and Jung-Mi Yun. "Luteolin inhibits hyperglycemia-induced proinflammatory cytokine production and its epigenetic mechanism in human monocytes." *Phytotherapy Research* 28, no. 9 (2014): 1383–1391.

Kovi, Ramesh C., Sachin Bhusari, Deepak Mav, Ruchir R. Shah, Thai Vu Ton, Mark J. Hoenerhoff, Robert C. Sills, and Arun R. Pandiri. "Genome-wide promoter DNA methylation profiling of hepatocellular carcinomas arising either spontaneously or due to chronic exposure to Ginkgo biloba extract (GBE) in B6C3F1/N mice." *Archives of Toxicology* 93 (2019): 2219–2235.

Lee, Ming-Jen, Yi-Jane Tsai, May-Yao Lin, Huey-Ling You, Nagabhushanam Kalyanam, Chi-Tang Ho, and Min-Hsiung Pan. "Calebin-A induced death of malignant peripheral nerve sheath tumor cells by activation of histone acetyltransferase." *Phytomedicine* 57 (2019): 377–384.

Li, Guobing, Changyu Shan, Lei Liu, Ting Zhou, Jing Zhou, Xiaoye Hu, Yibiao Chen, Hongjuan Cui, and Ning Gao. "Tanshinone IIA inhibits HIF-1α and VEGF expression in breast cancer cells via mTOR/p70S6K/RPS6/4E-BP1 signaling pathway." *PLoS ONE* 10, no. 2 (2015): e0117440.

Li, Qiang, Chenxu Ge, Jun Tan, Yan Sun, Qin Kuang, Xianling Dai, Shaoyu Zhong et al. "Juglanin protects against high fat diet-induced renal injury by suppressing inflammation and dyslipidemia via regulating NF-κB/HDAC3 signaling." *International Immunopharmacology* 95 (2021): 107340.

Li, Yanpeng, Lu Geng, Jing Chen, Ying Huang, Jing Qiao, Guangxi Ren, and Chunsheng Liu. "Molecular marker of high glycyrrhizic acid content based on the SNPs of liquorice IPT gene." 3 *Biotech* 8 (2018): 1–7.

Liang, Y., D. M. Kong, S. T. Chen, S. Y. Wang, and Y. Lv. "Overview of traditional medicine in asia-pacific region." *Asia Pacific Tradit Med.* (2005): 14e79.

Lin, Zi-Han, Yen-Fan Chan, Min-Hsiung Pan, Yen-Chen Tung, and Zheng-Yuan Su. "Aged citrus peel (chenpi) prevents acetaminophen-induced hepatotoxicity by epigenetically regulating Nrf2 pathway." *The American Journal of Chinese Medicine* 47, no. 08 (2019): 1833–1851.

Ling, C. "Rö nn T. epigenetics in human obesity and type 2 diabetes." *Cell Metab* 29, no. 5 (2019): 1028–1044.

Liu, X. M., J. Q. Zou, Z. X. Shen, and G. Q. Su. "Overview of traditional Indian medicine." *Modernization of Traditional Chinese Medicine and Materia Medica——World Science and Technology* 7, no. 6 (2005): 86e88.

Liu, Yifei, Bo Wang, Shaohua Shu, Zheng Li, Chi Song, Di Liu, Yan Niu et al. "Analysis of the Coptis chinensis genome reveals the diversification of protoberberine-type alkaloids." *Nature Communications* 12, no. 1 (2021): 3276.

Lu, Jun, Xiaoli Zhang, Tingting Shen, Chao Ma, Jun Wu, Hualei Kong, Jing Tian, Zhifeng Shao, Xiaodong Zhao, and Ling Xu. "Epigenetic profiling of H3K4Me3 reveals herbal medicine Jinfukang-induced epigenetic alteration is involved in anti-lung cancer activity." *Evidence-Based Complementary and Alternative Medicine* 2016 (2016).

Lu, Xirong, Yuyi Li, Wei Yang, Minghao Tao, Yanmiao Dai, Jinkang Xu, and Qianfei Xu. "Inhibition of NF-κB is required for oleanolic acid to downregulate PD-L1 by promoting DNA demethylation in gastric cancer cells." *Journal of Biochemical and Molecular Toxicology* 35, no. 1 (2021): e22621.

Ma, Hui, Li Li, Guojun Dou, Chengqiang Wang, Juan Li, Hui He, Mingxia Wu, and Hongyi Qi. "Z-ligustilide restores tamoxifen sensitivity of ERα negative breast cancer cells by reversing MTA1/IFI16/HDACs complex mediated epigenetic repression of ERα." *Oncotarget* 8, no. 17 (2017): 29328.

Mirza, Sameer, Gayatri Sharma, Rajinder Parshad, Sidhartha Datta Gupta, Pranav Pandya, and Ranju Ralhan. "Expression of DNA methyltransferases in breast cancer patients and to analyze the effect of natural compounds on DNA methyltransferases and associated proteins." *Journal of Breast Cancer* 16, no. 1 (2013): 23–31.

Muhammad, Tahir, Muhammad Ikram, Rahat Ullah, Shafiq Ur Rehman, and Myeong Ok Kim. "Hesperetin, a citrus flavonoid, attenuates LPS-induced neuroinflammation, apoptosis and memory impairments by modulating TLR4/NF-κB signaling." *Nutrients* 11, no. 3 (2019): 648.

Ndagi, Umar, Ndumiso Mhlongo, and Mahmoud E. Soliman. "Metal complexes in cancer therapy–an update from drug design perspective." *Drug Design, Development and Therapy* (2017): 599–616.

Park, Ui-Hyun, Jin-Taek Hwang, HyeSook Youn, Eun-Joo Kim, and Soo-Jong Um. "Piperine inhibits adipocyte differentiation via dynamic regulation of histone modifications." *Phytotherapy Research* 33, no. 9 (2019): 2429–2439.

Park, Ui-Hyun, Jin-Taek Hwang, HyeSook Youn, Eun-Joo Kim, and Soo-Jong Um. "Kaempferol antagonizes adipogenesis by repressing histone H3K4 methylation at PPARγ target genes." *Biochemical and Biophysical Research Communications* 617 (2022): 48–54.

Pooladanda, Venkatesh, Soumya Bandi, Sandhya Rani Mondi, Krishna Mohan Gottumukkala, and Chandraiah Godugu. "Nimbolide epigenetically regulates autophagy and apoptosis in breast cancer." *Toxicology in Vitro* 51 (2018): 114–128.

Pucci, Mariangela, Cinzia Rapino, Andrea Di Francesco, Enrico Dainese, Claudio D'Addario, and Mauro Maccarrone. "Epigenetic control of skin differentiation genes by phytocannabinoids." *British Journal of Pharmacology* 170, no. 3 (2013): 581–591.

Qiao, Li, Maosen Han, Shijie Gao, Xinxin Shao, Xiaoming Wang, Linlin Sun, Xianjun Fu, and Qingcong Wei. "Research progress on nanotechnology for delivery of active ingredients from traditional Chinese medicines." *Journal of Materials Chemistry B* 8, no. 30 (2020): 6333–6351.

Qiu, Zhifang, Kristina Andrijauskaite, Jay Morris, and Michael J. Wargovich. "Disruption of epigenetic silencing in human colon cancer cells lines utilizing a novel supercritical CO2 extract of neem leaf (Azadirachta indica)." *Anticancer Research* 39, no. 10 (2019): 5473–5481.

Rai, Amit, Hideki Hirakawa, Ryo Nakabayashi, Shinji Kikuchi, Koki Hayashi, Megha Rai, Hiroshi Tsugawa et al. "Chromosome-level genome assembly of Ophiorrhiza pumila reveals the evolution of camptothecin biosynthesis." *Nature Communications* 12, no. 1 (2021): 405.

Ren, Lipin, Yanjie Shang, Li Yang, Shiwen Wang, Xiang Wang, Shan Chen, Zhigui Bao et al. "Chromosome-level de novo genome assembly of Sarcophaga peregrina provides insights into the evolutionary adaptation of flesh flies." *Molecular Ecology Resources* 21, no. 1 (2021): 251–262.

Saslis-Lagoudakis, C. Haris, Julie A. Hawkins, Simon J. Greenhill, Colin A. Pendry, Mark F. Watson, Will Tuladhar-Douglas, Sushim R. Baral, and Vincent Savolainen. "The evolution of traditional knowledge: Environment shapes medicinal plant use in Nepal." *Proceedings of the Royal Society B: Biological Sciences* 281, no. 1780 (2014): 20132768.

Sharma, Hari, and Robert Keith Wallace. "Ayurveda and epigenetics." *Medicina* 56, no. 12 (2020): 687.

She, Jiajie, Hengyu Yan, Jiaotong Yang, Wenying Xu, and Zhen Su. "croFGD: Catharanthus roseus functional genomics database." *Frontiers in Genetics* 10 (2019): 238.

Shi, Fengqin, Ya Li, Rui Han, Alan Fu, Ronghua Wang, Olivia Nusbaum, Qin Qin, Xinyi Chen, Li Hou, and Yong Zhu. "Valerian and valeric acid inhibit growth of breast cancer cells possibly by mediating epigenetic modifications." *Scientific Reports* 11, no. 1 (2021): 2519.

Sitarek, Przemysław, Tomasz Kowalczyk, Simona Santangelo, Adam J. Białas, Monika Toma, Joanna Wieczfinska, Tomasz Śliwiński, and Ewa Skała. "The extract of Leonurus sibiricus transgenic roots with AtPAP1 transcriptional factor induces apoptosis via DNA damage and down regulation of selected epigenetic factors in human cancer cells." *Neurochemical Research* 43 (2018): 1363–1370.

Skała, Ewa, Monika Toma, Tomasz Kowalczyk, Tomasz Śliwiński, and Przemysław Sitarek. "Rhaponticum carthamoides transformed root extract inhibits human glioma cells viability, induces double strand DNA damage, H2A. X phosphorylation, and PARP1 cleavage." *Cytotechnology* 70 (2018): 1585–1594.

Song, Zhenqiao, Caicai Lin, Piyi Xing, Yuanyuan Fen, Hua Jin, Changhao Zhou, Yong Q. Gu, Jianhua Wang, and Xingfeng Li. "A high-quality reference genome sequence of Salvia miltiorrhiza provides insights into tanshinone synthesis in its red rhizomes." *The Plant Genome* 13, no. 3 (2020): e20041.

Souza, Estevão N.F., Elizabeth M. Williamson, and Julie A. Hawkins. "Which plants used in ethnomedicine are characterized? Phylogenetic patterns in traditional use related to research effort." *Frontiers in Plant Science* 9 (2018): 834.

Sreenivasmurthy, Sravan Gopalkrishnashetty, Jing-Yi Liu, Ju-Xian Song, Chuan-Bin Yang, Sandeep Malampati, Zi-Ying Wang, Ying-Yu Huang, and Min Li. "Neurogenic traditional Chinese medicine as a promising strategy for the treatment of Alzheimer's disease." *International Journal of Molecular Sciences* 18, no. 2 (2017): 272.

Stander, Emily Amor, Liuda Johana Sepúlveda, Thomas Duge de Bernonville, Inês Carqueijeiro, Konstantinos Koudounas, Pamela Lemos Cruz, Sébastien Besseau et al. "Identifying genes involved in alkaloid biosynthesis in vinca minor through transcriptomics and gene co-expression analysis." *Biomolecules* 10, no. 12 (2020): 1595.

Sun, Jin, Zhiwei Dong, Yang Zhang, Xiaoning He, Dongdong Fei, Fang Jin, Lin Yuan, Bei Li, and Yan Jin. "Osthole improves function of periodontitis periodontal ligament stem cells via epigenetic modification in cell sheets engineering." *Scientific Reports* 7, no. 1 (2017): 5254.

Tavakoli, Kamand, Alireza Pour-Aboughadareh, Farzad Kianersi, Peter Poczai, Alireza Etminan, and Lia Shooshtari. "Applications of CRISPR-Cas9 as an advanced genome editing system in life sciences." *BioTech* 10, no. 3 (2021): 14.

Tian, Nan-Nan, Yan-Biao Zheng, Zhi-Peng Li, Feng-wei Zhang, and Jin-fang Zhang. "Histone methylatic modification mediates the tumor-suppressive activity of curcumol in hepatocellular carcinoma via an Hotair/EZH2 regulatory axis." *Journal of Ethnopharmacology* 280 (2021): 114413.

Traboulsi, Hussein, Necola Guerrina, Matthew Iu, Dusica Maysinger, Parisa Ariya, and Carolyn J. Baglole. "Inhaled pollutants: The molecular scene behind respiratory and systemic diseases associated with ultrafine particulate matter." *International Journal of Molecular Sciences* 18, no. 2 (2017): 243.

Tu, Lichan, Ping Su, Zhongren Zhang, Linhui Gao, Jiadian Wang, Tianyuan Hu, Jiawei Zhou et al. "Genome of Tripterygium wilfordii and identification of cytochrome P450 involved in triptolide biosynthesis." *Nature Communications* 11, no. 1 (2020): 971.

Upadhyay, Atul K., Anita R. Chacko, A. Gandhimathi, Pritha Ghosh, K. Harini, Agnel P. Joseph, Adwait G. Joshi et al. "Genome sequencing of herb Tulsi (Ocimum tenuiflorum) unravels key genes behind its strong medicinal properties." *BMC Plant Biology* 15, no. 1 (2015): 1–20.

Verdin, Eric, Franck Dequiedt, and Herbert G. Kasler. "Class II histone deacetylases: Versatile regulators." *TRENDS in Genetics* 19, no. 5 (2003): 286–293.

Waddington, Conrad H. "The epigenotype." *Endeavour* 1 (1942): 18–20.

Wang, Chao, Renyi Wu, Davit Sargsyan, Meinizi Zheng, Shanyi Li, Ran Yin, Shan Su, Ilya Raskin, and Ah-Ng Kong. "CpG methyl-seq and RNA-seq epigenomic and transcriptomic studies on the preventive effects of Moringa isothiocyanate in mouse epidermal JB6 cells induced by the tumor promoter TPA." *The Journal of Nutritional Biochemistry* 68 (2019): 69–78.

Wang, Chengqiang, Gen Liu, Guojun Dou, Yi Yang, Lu Chen, Hui Ma, Zhuyun Jiang et al. "Z-Ligustilide selectively targets AML by restoring nuclear receptors Nur77 and NOR-1-mediated apoptosis and differentiation." *Phytomedicine* 82 (2021a): 153448.

Wang, Peng, Min Wang, Tingling Zhuo, Ying Li, Weiping Lin, Lingli Ding, Meng Zhang et al. "Hydroxysafflor yellow a promotes osteogenesis and bone development via epigenetically regulating β-catenin and prevents ovariectomy-induced bone loss." *The International Journal of Biochemistry & Cell Biology* 137 (2021b): 106033.

Wang, Si-wei, Hao Sheng, Yong-feng Bai, Yuan-yuan Weng, Xue-yu Fan, Fang Zheng, Jing-qi Fu, and Feng Zhang. "Inhibition of histone acetyltransferase by naringenin and hesperetin suppresses Txnip expression and protects pancreatic β cells in diabetic mice." *Phytomedicine* 88 (2021c): 153454.

Weng, Jing-Ru, I-Lu Lai, Hsiao-Ching Yang, Chun-Nan Lin, and Li-Yuan Bai. "Identification of kazinol Q, a natural product from Formosan plants, as an inhibitor of DNA methyltransferase." *Phytotherapy Research* 28, no. 1 (2014): 49–54.

Wu, Bingshan, Xueqin Yao, Xiaohu Nie, and Ruxiang Xu. "Epigenetic reactivation of RANK in glioblastoma cells by curcumin: Involvement of STAT3 inhibition." 2013, 32(6):292–297. doi: 10.1089/dna.2013.2042.

Wu, Han-Tsang, Joseph Lin, Yi-En Liu, Hsiao-Fan Chen, Kai-Wen Hsu, Shu-Hsuan Lin, Kai-Yen Peng, Kuo-Juei Lin, Chang-Chi Hsieh, and Dar-Ren Chen. "Luteolin suppresses androgen receptor-positive triple-negative breast cancer cell proliferation and metastasis by epigenetic regulation of MMP9 expression via the AKT/mTOR signaling pathway." *Phytomedicine* 81 (2021): 153437.

Xiao, Yuhang, Min Su, Wei Ou, Hui Wang, Bo Tian, Junliang Ma, Jinming Tang et al. "Involvement of non-coding RNAs in epigenetic modifications of esophageal cancer." *Biomedicine & Pharmacotherapy* 117 (2019): 109192.

Xiao, Yun, Jinyuan Xu, and Wenkang Yin. "Aberrant epigenetic modifications of non-coding RNAs in human disease." *Non-coding RNAs in Complex Diseases: A Bioinformatics Perspective* (2018): 65–75.

Xie, Jinghe, Haifeng Zhao, Kunpeng Li, Rui Zhang, Yongchao Jiang, Meimei Wang, Xuelian Guo et al. "A chromosome-scale reference genome of Aquilegia oxysepala var. kansuensis." *Horticulture Research* 7 (2020).

Xu, Chi, Jing Tian, and Beixin Mo. "siRNA-mediated DNA methylation and H3K9 dimethylation in plants". *Protein & Cell* 4, no. 9 (2013): 656–663.

Xu, Jiang, Yang Chu, Baosheng Liao, Shuiming Xiao, Qinggang Yin, Rui Bai, He Su et al. "Panax ginseng genome examination for ginsenoside biosynthesis." *Gigascience* 6, no. 11 (2017): gix093.

Yan, Rui, Xu-Li Chen, Yan-Ming Xu, and Andy TY Lau. "Epimutational effects of electronic cigarettes." *Environmental Science and Pollution Research* 28, no. 14 (2021): 17044–17067.

Yan, Y., J. Zhao, C. Cao, Z. Jia, N. Zhou, S. Han, Y. Wang, Y. Xu, and H. Cui. "Tetramethylpyrazine promotes SH-SY5Y cell differentiation into neurons through epigenetic regulation of Topoisomerase IIβ." *Neuroscience* 278 (2014): 179–193.

Yao, Chih-Jung, Jyh-Ming Chow, Chia-Ming Yang, Hui-Ching Kuo, Chia-Lun Chang, Hsin-Lun Lee, I. Lai, Shuang-En Chuang, and Gi-Ming Lai. "Chinese herbal mixture, Tien-Hsien liquid, induces G2/M cycle arrest and radiosensitivity in MCF-7 human breast cancer cells through mechanisms involving DNMT1 and Rad51 downregulation". *Evidence-Based Complementary and Alternative Medicine* 2016 (2016).

Yu, Yung-Luen, Sung-Liang Yu, Kuo-Jung Su, Chyou-Wei Wei, Min-Hui Jian, Po-Cheng Lin, I-Hsuan Tseng et al. "Extended O6-methylguanine methyltransferase promoter hypermethylation following n-butylidenephthalide combined with 1, 3-bis (2-chloroethyl)-1-nitrosourea (BCNU) on inhibition of human hepatocellular carcinoma cell growth." *Journal of Agricultural and Food Chemistry* 58, no. 3 (2010): 1630–1638.

Yuan, Zigao, Shaopeng Chen, Chunmei Gao, Qiuzi Dai, Cunlong Zhang, Qinsheng Sun, Jin-Shun Lin, Chun Guo, Yuzong Chen, and Yuyang Jiang. "Development of a versatile DNMT and HDAC hibitor C02S modulating multiple cancer hallmarks for breast cancer therapy." *Bioorganic Chemistry* 87 (2019): 200–208.

Zhang, Hui-Jie, Ya-Ting An, Ying Liang, Xin-Jian Yang, Yu-Qiang Fan, and Xiao-Liang Ren. "Research progress on biopharmaceutics of active ingredients in Chinese materia medica under multi-component system." *Chinese Journal of Experimental Traditional Medical Formulae* (2019): 196–202.

Zhao, Fei, Yan Chen, Linglan Zeng, Rui Li, Rong Zeng, Lu Wen, Yuan Liu, and Chun Zhang. "Role of triptolide in cell proliferation, cell cycle arrest, apoptosis and histone methylation in multiple myeloma U266 cells." *European Journal of Pharmacology* 646, no. 1–3 (2010): 1–11.

Zheng, Shijie C., Martin Widschwendter, and Andrew E. Teschendorff. "Epigenetic drift, epigenetic clocks and cancer risk." *Epigenomics* 8, no. 5 (2016): 705–719.

Zhou, Shao Hu, Yu Fei Deng, Zhi Wei Weng, Hao Wei Weng, and Zhi Dan Liu. "Traditional Chinese medicine as a remedy for male infertility: A review." *The World Journal of 'Men's Health* 37, no. 2 (2019): 175–185.

Zhu, Hong, Huizi Zhao, Songbing Xu, Yuan Zhang, Yuhao Ding, Jun Li, Cheng Huang, and Taotao Ma. "Sennoside A alleviates inflammatory responses by inhibiting the hypermethylation of SOCS1 in CCl4-induced liver fibrosis." *Pharmacological Research* 174 (2021): 105926.

10 The Importance of Omics in Epigenetic Drug Discovery

Kella Alekhya, Anuciya R, Jawahar Rohith R, and Kaviyarasan D
SRM Institute of Science and Technology, Kattankulathur, Chengalpattu District, Tamil Nadu, India

10.1 INTRODUCTION

Omics, which is defined as examining and evaluating vast amounts of data that show the composition, function and structure of a system of life at a specific level, has significantly transformed the methods for analyzing biological systems. Since the invention of the first high-throughput instrument, the DNA microarray, technologies for studying omics have developed quickly and have been employed to record static genetic modifications, post-translational modifications, spatiotemporal proteome dynamics and temporal transcriptome perturbations and alternative splicing [1]. The main nontargeted and non-biased functions of omics include analyzing genome and protein sequencing, mRNAs (transcriptomics) and metabolomics (Figures 10.1 and 10.2). Systems biology is the study of interactions among networks of biological processes. Omics technology is primarily used to accurately comprehend normal physiological processes and to learn about disease processes, including diagnosis, prognosis and screening, which all advance our knowledge of the causes of illnesses [2].

10.2 GENOMICS

Since Watson and Crick's discovery of DNA structure in 1953, our understanding of the human genome has advanced significantly [3]. The study and reference of the genome used to be highly specialized, but the introduction of messenger RNA (mRNA) vaccines made genetics more accessible to the general public. The use of recombinant DNA technologies to modify DNA expanded possibilities in the 1970s. The inheritance patterns of single-gene illnesses provided us with the first insights into medical genetics. McKusick started the Mendelian Inheritance in Man database in the 1960s[4]; as of January 5, 2021, 4368 genes were linked to mutations that cause specific phenotypes. However, the number of diseases with a monogenic etiology is quite modest.

Genomics is the study of the entire genome and the vast amounts of data that come from it. It can lead to better understanding of how our genetic composition affects our health, particularly in relation to diseases and conditions with complicated etiologies. An enormous amount of both individual and population data including genetic data can be used to give the "right treatment to the right patient at the right time" because of the growth of big data and omics technology over the past ten years.

10.3 TRANSCRIPTOMICS

A tissue's or cell's entire collection of RNA transcripts is referred to as its transcriptome. Alternative splice variants are present in the transcriptome [5]. One of the main reasons for transcriptome investigation is that most human genes undergo a process called alternative splicing. Exons and introns

DOI: 10.1201/9781032706511-10

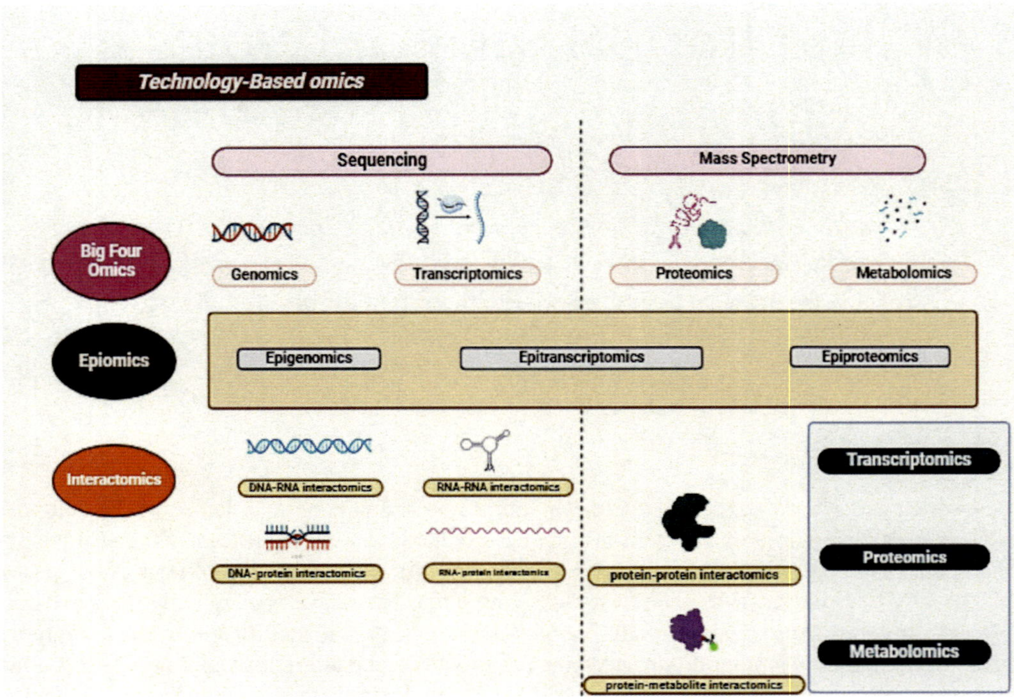

FIGURE 10.1 Technology-based omics.

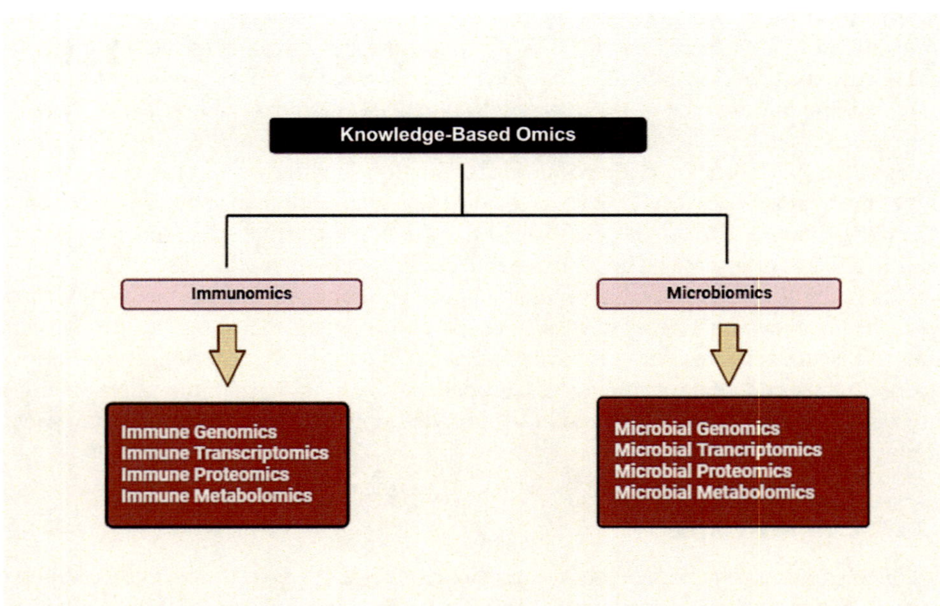

FIGURE 10.2 Knowledge-based omics.

make up the majority of eukaryotic genes, as is well known; after transcription, introns are eliminated, and only exons are translated into protein products.

An organism's information is expressed via RNA (the transcriptome) and stored in DNA (the genome). Finding out how environmental factors (drugs, hormones, etc.) or disease affect a tissue, cell or living organism's transcripts is the primary goal of transcriptomics. A further critical component of transcriptomics is noncoding RNA. These functional components are crucial in the development of many diseases as well as how well in they respond to therapy [6].

10.4 PROTEOMICS

Biological markers, commonly referred to as biomarkers, are indicators that can be used to identify different conditions and determine the presence or absence of biological processes or activities [7]. Proteins are thought of as a class of biomarkers because of their characteristics. Biomarkers are used in medicine to identify diseases, find treatments and track patient care; they are divided into four categories: prognostic, predictive, diagnostic and dispositional [8]. Molecular biomarkers are categorized as lipids, proteins, peptides, nucleic acid, lipids, metabolites and smaller molecules according to their physicochemical properties [9].

Proteins are the most sensitive of these biomarkers, able to detect a specific disease type in its early stages from a very small sample size. Proteomics is the most auspicious approach for identifying proteins as disease biomarkers [10]. Therefore, new applications, tactics, and techniques were needed to identify and create reliable biomarkers for a variety of diseases employing proteins. Recognizing a proteome is essential to understanding proteomics [11]. Scientists and researchers studying biology and medicine used proteomics to find biomarkers that would assist with diagnosis, prognosis and therapy of many diseases [12].

10.5 METABOLOMICS

The thorough assessment of low-molecular-weight compounds and metabolites in a biological material is the definition of the newly arising field of metabolomics. By enabling the profiling of a broader range of metabolites than current clinical laboratory methods, metabolomics serves as a vital lens in the molecular microscope of precision medicine. This leads to thorough explanation of metabolic pathways and biological activities [13].

Unlike genomics and proteomics, which detect molecules with different physical properties, metabolomics poses a considerable analytical barrier [14]. Metabolomics offers a wide viewpoint for measuring physiological reactions to disturbances such as illnesses and for observing the complex interplay between physiology and outside events and circumstances. Metabolites serve as proximal reporters of disease because of their strong association with biological specimens and underlying pathogenic mechanisms [15]. Advantages and disadvantages of omics technologies are shown in Table 10.1.

10.6 APPLICATIONS OF OMICS

10.6.1 Toxicology

Omics has been found useful in toxicology investigations to examine whether substances alter the entire genome, which could result in harmful effects and to determine the potential harmful chemicals according to pathways known to cause adverse outcomes [16].

10.6.2 Pharmacogenomics

Components-based genetics is widely used in precision medicine; pharmacogenomics-informed pharmacotherapy is an emerging application of omics technology that tailors drug selection and

TABLE 10.1
Advantages and Disadvantages of Omics

Advantages	Disadvantages
• Low cost	• Some sequences could not be captured precisely
• Comparatively advanced technology	• Requires a large sample
• High resolution	• Long stages of preparation can result in errors
• High reliability and throughput	• Time consuming
• Direct detection of epigenetic variations	• Periodic modifications to kits, flow cells and software versions
• Offers long read lengths and high accuracy	• Less uniformity

dosing according to a patient's genetic profile. Variability in drug effects can be attributed to diverse pharmacological factors including differences in drug targets, gene transcription and pharmacokinetics. Leveraging advanced statistical and bioinformatic techniques enables more precise therapy [17].

10.6.3 Breast Cancer Research

Multi-omics techniques were used to help decode breast cancer molecular complexity. Additionally, it can aid in the classification of various tumor types and breast cancer subtypes as well as the prediction of survival outcomes and breast cancer heterogeneity [18].

10.6.4 Nanotechnology

The integration of omics and nanotechnology holds promise for unraveling the complicated mechanisms of disease development. The combination allows for examining the molecular processes behind disease development. Through meticulous examination of the minuscule interplay among biomolecules, we can construct all-encompassing disease pathway models that culminate in a deeper awareness of disease mechanisms [19].

10.6.5 Prenatal Diagnosis

Unknown genetic abnormalities passed down from parents can pose a major risk to a baby's health. Omics technologies are being increasingly integrated into the prenatal environment to address these defects [20]. A team of researchers discovered in 1997 fetal DNA without cells in a mother's blood [21]. This discovery created new opportunities for noninvasive prenatal diagnostics. The combination of next-generation sequencing and DNA amplification enabled identifying the DNA [22].

10.6.6 Quantification of Protein

Stable isotope labelling, iTRAQ (multiplexed isobaric tagging technology for relative quantitation) and cytometry by time of flight mass spectroscopy are employed for both qualitative evaluation, such as protein identification and peptide sequencing, and quantitative investigation of proteins and their post-translational alteration [23].

10.6.7 Customized Immunotherapy

Allowing the immune system of patients to fight the disease has emerged as the fifth primary cancer treatment, after radiation therapy, chemotherapy, surgery and antibody-based treatment. This method, known as adoptive cell transfer or ACT, has shown great promise. For example, among the various ACT approaches for glioblastoma multiforme, vaccines based on dendritic cells and T-cells have drawn particular attention due to their ability to penetrate the brain—an immunologically privileged location—and initiate an anti-tumor response [24].

10.6.8 Neurodegenerative Diseases

The way that complicated diseases like Alzheimer's disease (AD) are studied has changed as a result of all these omics technical advancements. AD and other pathologies offer promising environments for examining the connections between various cellular pathways [25].

10.7 CONCLUSION

In summary, omics techniques are instrumental in epigenetic drug discovery as they help identify targets, understand disease mechanisms, discover biomarkers, validate targets, facilitate drug development and screening, enable personalized medicine and monitor treatment efficacy. The main goal of omics technology is to comprehend how complex human disease features are influenced by genes. They play an important role in advancing our understanding of epigenetics and its role in health and disease. Better data analysis tools, sequencing technologies and multi-omics integration are examples of omics innovations. With cost savings and efficiency improvements, high-throughput sequencing methods such as next-generation sequencing have made it possible to conduct large-scale genomic research. These advancements drive the field of omics science toward progressively intricate and important applications.

REFERENCES

[1] Schena M, Shalon D, Davis RW, Brown PO. Quantitative monitoring of gene expression patterns with a complementary DNA microarray. *Science*. 1995;270:467–470. https://www.science.org/doi/10.1126/science.270.5235.467

[2] Narad P, Kirthanashri SV. *Omics Approaches, Technologies and Applications*; Springer; 2019, pp. 1–10.

[3] Watson JD, Crick FH. The structure of DNA. *Cold Spring Harb Symp Quant Biol*. 1953;18:123–131.

[4] McKusick VA. Mendelian inheritance in man and its online version, OMIM. *Am J Hum Genet*. 2007;80:588–604.

[5] Lowe R, Shirley N, Bleackley M, Dolan S, Shafee TJP. Transcriptomics technologies. *PLoS Comp Biol*. 2017;13:5.

[6] Mattick JS, Makunin IVJ. Non-coding RNA. *Human Mol Genetics*. 2006;15(suppl_1):R17–R29.

[7] Biomarkers Definitions Working Group. Biomarkers and surrogate endpoints: Preferred definitions and conceptual framework *Clin Pharmacol Ther*. 2001;69(3):89–95.

[8] Huss R. *Chapter 19—Biomarkers Translational Regenerative*. Medicine Academic Press; 2015, pp. 235–241.

[9] Davis MA, Eldridge S, Louden C. Chapter 10—Biomarkers: Discovery, qualification and application. In *Haschek and Rousseaux's Handbook of Toxicologic Pathology* (3rd ed.). Academic Press; 2013, pp. 317–352.

[10] Doustjalali S. Two-dimensional gel electrophoresis: An overview of proteomic technique in cancer research. *J Proteom Bioinformat*. 2014;7. https://doi.org/10.4172/jpb.1000305

[11] DePrimo SE. Biomarkers. In *Comprehensive Medicinal Chemistry II*. Elsevier; 2007, pp. 69–85. https://doi.org/10.1016/B0-08-045044-X/00076-6

[12] Patel S, Ling J, Kim SJ, Schey KL, Rose K, Kuchtey RW. Proteomic analysis of macular fluid associated with advanced glaucomatous excavation. *JAMA Ophthalmol*. 2016;134(1):108–110.

[13] Clish CB. Metabolomics: An emerging but powerful tool for precision medicine. *Cold Spring Harb Mol Case Stud*. 2015;1(1):a000588. https://doi.org/10.1101/mcs.a000588

[14] Kuehnbaum NL, Britz-McKibbin P. New advances in separation science for metabolomics: Resolving chemical diversity in a post-genomic era. *Chem Rev*. 2013;113:2437–2468.

[15] Gerszten RE, Wang TJ. The search for new cardiovascular biomarkers. *Nature*. 2008;451:949–952.

[16] Patil R, Satpute R, Nalage D. The application of omics technologies to toxicology. *Toxicol Adv*. 2023;5(2):6. https://doi.org/10.53388/TA202305006

[17] Cecchin E, Stocco G. Pharmacogenomics and personalized medicine. *Genes (Basel).* 2020;11(6):679. Published online 2020 Jun 22. https://doi.org/10.3390/genes11060679

[18] Orsini A, Diquigiovanni C, Bonora E. Omics technologies improving breast cancer research and diagnostics. *Int J Mol Sci.* 2023;24:12690. https://doi.org/10.3390/ijms241612690

[19] Wang X, Xu W, Li J, Shi C, Guo Y, Shan J, Qi R. Nano-omics: Frontier fields of fusion of nanotechnology. *Smart Med.* 2023;e20230039. https://doi.org/10.1002/SMMD.20230039

[20] Kim D-H, Kim Y-S, Son N-II, Kang C-K, Kim A-R. Recent omics technologies and their emerging applications for personalised medicine. *IET Syst Biol.* 2017;11(3):87–98. Published online 2017 Apr 19. https://ietresearch.onlinelibrary.wiley.com/doi/epdf/10.1049/iet-syb.2016.0016

[21] Lo YM, Corbetta N, Chamberlain PF et al. Presence of foetal DNA in maternal plasma and serum. *Lancet.* 1997;350:485–487.

[22] Chiu RW, Chan KA, Gao Y, et al. Non-invasive prenatal diagnosis of foetal chromosomal aneuploidy by massively parallel genomic sequencing of DNA in maternal plasma. *Proc Natl Acad Sci.* 2008;105(51):20458–20463.

[23] Ong S-E, Blagoev B, Kratchmarova I, et al. Stable isotope labeling by amino acids in cell culture, silac, as a simple and accurate approach to expression proteomics. *Mol Cell Proteomics.* 2002;1(5):376–386.

[24] Yu JS, Wheeler CJ, Zeltzer PM, et al. Vaccination of malignant glioma patients with peptide-pulsed dendritic cells elicits systemic cytotoxicity and intracranial t-cell infiltration. *Cancer Res.* 2001;61(3):842–847.

[25] Sancesario GM, Bernardini S. Alzheimer's disease in omics era. *Clin Biochem.* 2018;59:9–16.

11 The Role and Mechanisms of Gut Microbiota in Epigenetic Healing

Arshia Razdan,[1] Romi Shreshtha,[1] M V N L Chaitanya,[2] and Smriti Arora[1]
[1]School of Health Sciences and Technology, UPES, Dehradun, Uttarakhand, India
[2]School of Pharmaceutical Sciences, Lovely Professional University, Punjab, India

11.1 INTRODUCTION

Humans are a home to a vast and diverse community of microbes. Most of these are present in the gastrointestinal tract and hence are known as gut microbiota. The human gut microbiota is a collection of microbes that accentuates the gut barrier, affects physiological processes and is linked to neural and behavioral processes in humans [1,2]. These microbes, ranging from fungi to viruses, populate the colon, small intestine and other parts of the GI tract and help in digestion, fermentation, nutrient absorption, regulating and building immunity, increasing metabolism, mental health and detoxification [3,4]. The gut microbiota hosts trillions of microbes, giving it high diversity and species richness. The latter are also the signatures of a healthy and resilient gut microbiome. This state of homeostasis of the gut is called eubiosis; when this state of homeostasis is lost, it is referred to as gut dysbiosis [5,6].

In recent times, the compound interactivity between human health and the gut microbiota has become a compelling focus of scientific exploration [7]. While the gut microbiota shapes human health, it is shaped by diet, which is directly linked to metabolic diseases such as diabetes, dyslipidemia, obesity, cardiovascular diseases and non-alcoholic fatty liver disease [2,8]. Beyond the gut microbiota's well-established impact on digestion and metabolism, it is increasingly acknowledged as central in shaping epigenetic processes [9]. Epigenetics refers to alteration of transcriptional state in response to environmental cues without altering the genome [10]. This recognition has paved the way for fresh perspectives on comprehending the mechanisms that govern cellular healing (Figure 11.1).

As mentioned, epigenetics refers to changes or alterations in the genetic expression or phenotype of the cell that are heritable and occur without altering the original DNA sequence. The modification can occur at the histone level or directly in the DNA or via alteration in noncoding RNA (ncRNA) expression. Epigenetic alterations include methylation of DNA, methylation and acetylation of histones and alterations in ncRNA expression. Regulation of gene expression directly affects epigenetic properties [12].

The objective of this book chapter is to integrate existing knowledge, clarifying how the gut microbiota plays a role in the mechanisms of epigenetic healing and consequently shapes the overall health and resilience of the host organism [13]. Although microbiota-associated epigenetic mechanisms have been delineated in both peripheral tissues and intestinal cells, extended investigation is very important for completely understanding the intricate connection between the host, microbiota and epigenetics [10]. Furthermore, the mechanisms of epigenetic healing are discussed.

DOI: 10.1201/9781032706511-11

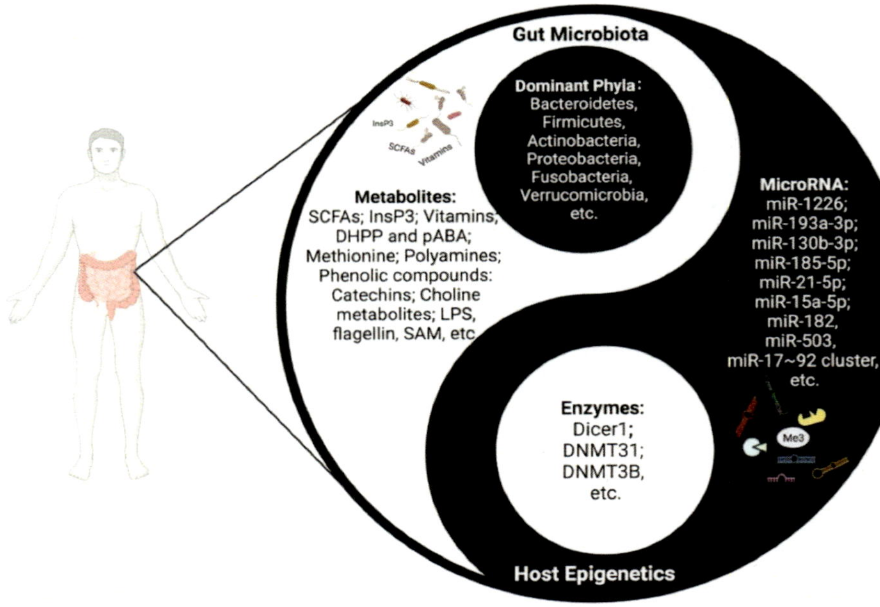

FIGURE 11.1 Interaction of gut microbiota, their metabolites and epigenetics repercussions during diseases.

Source: Adapted from [11].

11.2 EPIGENETIC REGULATION BY GUT MICROBIOTA

Microbes within the gut flora regulate epigenetic alterations including DNA or histone modifications, in addition to regulation of ncRNAs, by producing histone modifiers and microbial signals as metabolites, for instance short-chain fatty acids (SCFAs), indole derivatives, secondary bile acids, trimethylamine N-oxide and polyamines (Figure 11.2). Such metabolites can modulate gene expressions via epigenetic modifications and thus, might have therapeutic as well as disease potential.

SCFAs such as butyrate, propionate and acetate influence gene expression by either inhibiting or promoting epigenetic enzymes. For instance, butyrate produced by *Eubacterium rectale*, *Faecalibacterium parunitzii* and several other gut bacteria inhibits histone deacetylases (HDACs) via conformational changes at its active site. HDACs are enzymes that remove acetyl groups from histones, which leads to more compact chromatin structure, ultimately, making them inaccessible for transcription (Figure 11.3).

The inhibition of HDAC activity by butyrate involves multiple mechanisms. It binds to the zinc (II) ion present in the catalytic site of HDAC, resulting in inability of HDACs to remove acetyl groups from histone, leading to hyperacetylation of histones and increased gene transcription. Another mechanism involves activation of G-coupled receptors by SCFAs that inhibits HDACs. Under normal physiological conditions, butyrate and beta-hydroxybutyrate can activate GPR109a, suppressing pro-inflammatory mediators like TNF-alpha, IL-1 beta and IL-6. GPR109a is a G-coupled receptor that along with GPR43 and GPR41 is a receptor of gut microbial-derived SCFAs. Recent studies are focusing more on treating inflammatory diseases, for instance, inflammatory bowel disease (IBD), through GPR109a activation via butyrate (Figure 11.4). More studies are required to further explore these mechanisms [11,14].

Moreover, butyrate facilitates the polarization of M2 macrophages by inhibiting HDACs deacetylation activity. Along with butyrate, lncRNA–MM2P is a modulator required for macrophage M2

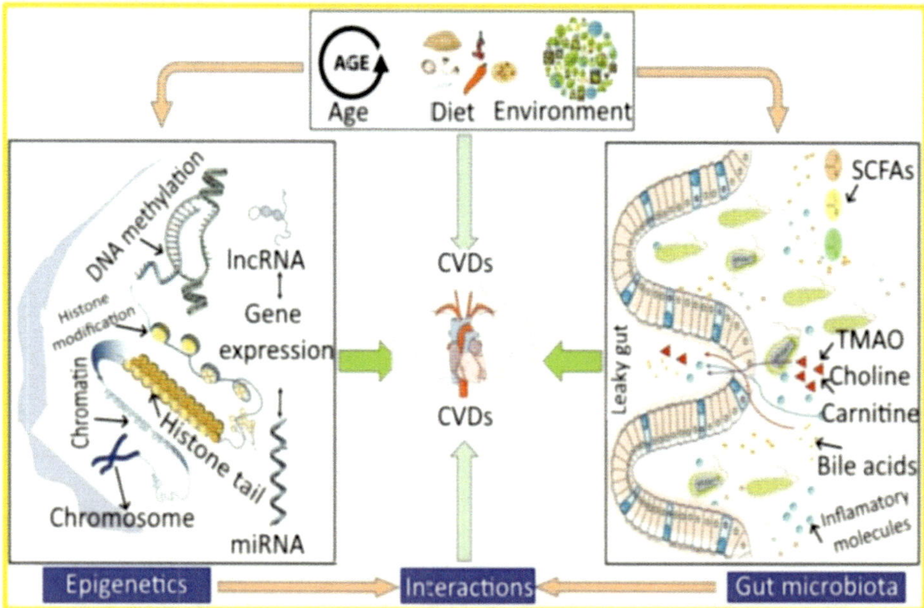

FIGURE 11.2 Gut microbiota and its interactions during cardiovascular diseases. The gut flora, their metabolites, impact epigenetic pathways including miRNA, DNA methylation and histone modification in the host.

Source: Adapted from [21].

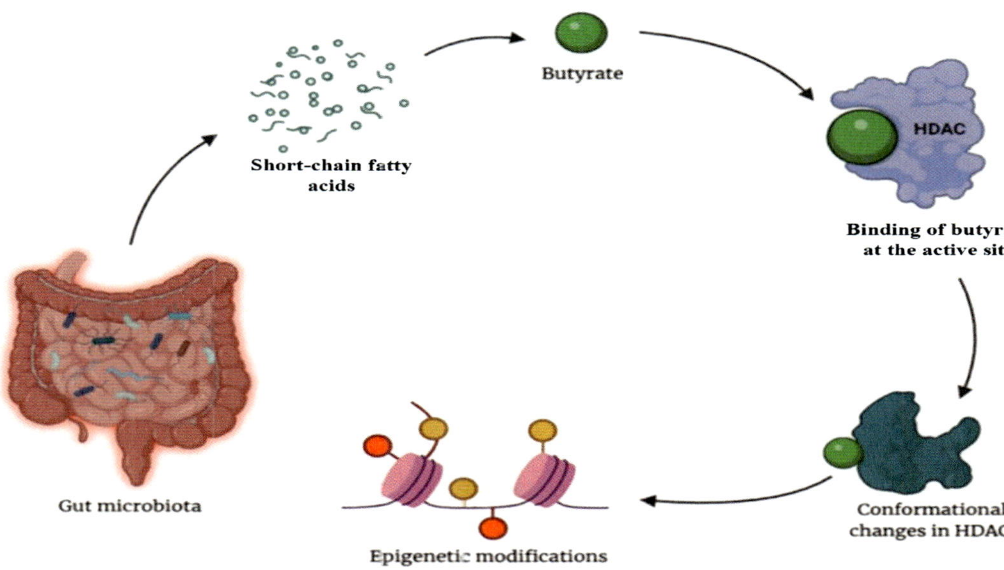

FIGURE 11.3 Butyrate produced as a result of fermentation by gut microbiota exerts a conformational change in the catalytic site of HDACs.

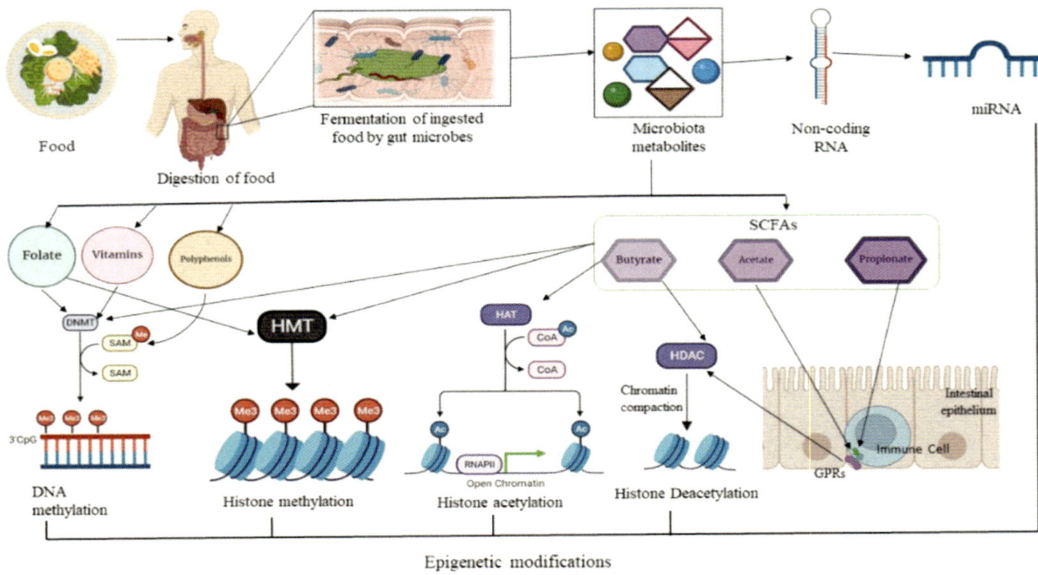

FIGURE 11.4 SCFAs produced by gut microbiota leading to activation of G-coupled receptor and non-coding RNA to exert epigenetic modification and ensuing inflammatory responses.

polarization [15]. The M2 macrophages play a critical role in fibrosis, tissue repair and remodeling, angiogenesis and immunologic tolerance. By targeting HDACs, the patterns of epigenetic modifications and gene expression can be modulated and epigenetic healing processes can be attempted. In addition to these metabolites, commensal gut microbes generate S-adenosylmethionine (SAM); folate and vitamins like B2, B12 and K, which modulates methylation of histones on DNA by providing them with methyl groups [16]. In DNA methylation, methyl groups are covalently added by the DNA methyl transferases (DNMTs) to the cytosine of CpG (cytosine-phosphate-guanine) islands.

Intestinal microbes can change the methylation patterns of cells and thereby regulate cellular expression. For instance, methylation at TLR4 (Toll-like receptor 4) promoter induces anti-inflammatory responses and augments immunity against intracellular pathogens. In some cases, epigenetic changes contribute to the development of disorders including Alzheimer's disease and cardiovascular diseases. The SCFAs produced by the gut bacteria have neuroactive effects along with the ability to cross the blood–brain barrier and modulate epigenetic processes in the CNS, demonstrating the potential role of such metabolites and gut microbes in the development as well as treatment of CNS disorders [17].

The gut microbiome can be manipulated through diet, probiotics, postbiotics, synbiotics, fecal microbiota transplant and genetic engineering [18,19]. Targeting the intestinal microbiota and their metabolites could be an approach for modulating these epigenetic modifications and metabolic dysfunction, augmenting immune responses and mitigating the pathogenesis of various disorders (see Table 11.1). Dysbiosis in the gut microflora affects metabolite production, which in turn affects immune modulation and epigenetic modifications. Loss or imbalance in the gut microbiome is leading to altered mucosal barrier functions and ultimately to inflammatory disorders, altering the microbial ecology; imbalances today are mainly caused by modern diets that are high in fats and calories, low fiber content and poor-quality carbohydrates, [20].

11.3 CONCLUSION

To summarize, with this book chapter, we aimed to shed light on the substantial influence of microbial communities present in human gut on the host's epigenetic framework. By consolidating

TABLE 11.1
Epigenetic Modifications Associated with Gut Microbiota

Epigenetic Factors	Associated Microorganism	Mechanism	Effect on Health/Related Diseases
MicroRNA-1226	*Escherichia coli*	Entry into the cells	Regulates gene transcription and microbe growth [22]
MiR-185-5p, MiR-21-5p, MiR-130b-3p	*Bacteroides eggerthi*	Downregulation of gene expression	Host metabolism pathways such as insulin signaling, degradation of glycerol and fatty acids and lipid metabolism [23]
MicroR-15a-5p	*H. parainfluenzae*	Downregulation of certain genes	Inhibits expression of endogenous UCP2; stimulates insulin secretion and increases ATP [23]
MicroRNA-193a-3p	Metabolic gut microbiota	Downregulation of the gut microbiota in the colon suppresses PepT1 and NF-κB pathway	Reduces inflammation in the colon, protects and restores but homeostasis [24]
MicroRNA enzyme Dicer1	Gut microbiota composition	miRNA reduction	Gut microbiota dysbiosis and aggravation of colitis due to absence of Dicer1 [25]
Methylation promoters DNMT3B and DNMT31	*Desulfovibrio* sp. and *Enterococcus* spp.; *Eubacterium rectale, Faecalibacterium prausnitzii and Lactobacillus*	*Promotes methylation of SFRP2 promoter, decreases Desulfovibrio sp. and Enterococcus spp.; increases E. rectale and Lactobacillus*	Increases risk of colon cancer and intestinal inflammation [26]
76 microRNAs (including miR-182, miR-503, and miR-17~92 cluster)	*Bacteroidetes, Proteobacteria and Firmicutes*	Mediates in transferring signals between immune and stromal cells	Impact on tumor invasion and metastasis [27]

the knowledge that we have acquired, the complex interplay between gut microbiota and epigenetic processes has been emphasized, showcasing their crucial role in shaping cellular healing and, consequently, the overall well-being and adaptability of the host organism.

Our investigation has underscored the significance of comprehending the dynamic relationship between the gut microbiota and epigenetic regulation, evident not only in local intestinal cells but also in peripheral tissues. This awareness opens up novel possibilities for therapeutic strategies, harnessing the potential of the gut microbiota to positively impact epigenetic mechanisms, thereby advancing the field of personalized medicine and tailored interventions for diverse health conditions.

Nevertheless, it's crucial to recognize that our comprehension of the host–microbiota dynamics and their implications for epigenetic healing is still evolving. Further exploration is warranted to decipher the precise mechanisms through which gut microbiota influence epigenetic modifications gives us a broader perspective about the molecular complexities of this symbiotic association. Advanced technologies such as metagenomics and single-cell sequencing holds promise for unraveling the intricacies of microbial interactions and their effects on host epigenetics.

In the quest to unlock the complete therapeutic potential of the gut microbiota in epigenetic healing, collaboration among microbiologists, geneticists, and clinicians will be pivotal. As we navigate this frontier, the convergence of microbiome research and epigenetics holds the potential for groundbreaking discoveries that could revolutionize our approach to health and disease. Linking epigenetics with gut microbiota is a bridge that researchers are still constructing, upon successful completion of which we will have answers to many mysteries about diseases.

REFERENCES

[1] Dinan, T.G., et al., Collective unconscious: How gut microbes shape human behavior. *J Psychiatr Res*, 2015. **63**: pp. 1–9.

[2] Arora, S., et al., The beneficial role of healthy microbiome in metabolic syndrome and cardiovascular health, in *Microbiome, Immunity, Digestive Health and Nutrition*. 2022, Elsevier. pp. 109–124.

[3] Moszak, M., M. Szulińska, and P. Bogdański, You are what you eat-the relationship between diet, microbiota, and metabolic disorders-a review. *Nutrients*, 2020. **12**(4).

[4] Engel, P. and N.A. Moran, The gut microbiota of insects—diversity in structure and function. *FEMS Microbiol Rev*, 2013. **37**(5): pp. 699–735.

[5] Iebba, V., et al., Eubiosis and dysbiosis: The two sides of the microbiota. *New Microbiol*. 2016. **39**(1): pp. 1–12.

[6] Arora, S., S. Puri, and N. Bhambri, A designer diet layout for astronauts using a microbiome mediated approach. *FEMS Microbiol Lett*, 2022. **369**(1).

[7] Guarner, F. and J.R. Malagelada, Gut flora in health and disease. *Lancet*, 2003. **361**(9356): pp. 512–519.

[8] Li, D., et al., Diet-gut microbiota-epigenetics in metabolic diseases: From mechanisms to therapeutics. *Biomed Pharmacother*, 2022. **153**: p. 113290.

[9] Lozupone, C.A., et al., Diversity, stability and resilience of the human gut microbiota. *Nature*, 2012. **489**(7415): pp. 220–230.

[10] Woo, V. and T. Alenghat, Epigenetic regulation by gut microbiota. *Gut Microbes*, 2022. **14**(1): p. 2022407.

[11] Wu, Y., et al., Dissecting the interplay mechanism between epigenetics and gut microbiota: Health maintenance and disease prevention. *Int J Mol Sci*, 2021. **22**(13).

[12] Dupont, C., D.R. Armant, and C.A. Brenner, Epigenetics: definition, mechanisms and clinical perspective. *Semin Reprod Med*, 2009. **27**(5): pp. 351–357.

[13] Caruso, R., B.C. Lo, and G. Núñez, Host-microbiota interactions in inflammatory bowel disease. *Nat Rev Immunol*, 2020. **20**(7): pp. 411–426.

[14] Bridgeman, S., et al., Butyrate lowers cellular cholesterol through HDAC inhibition and impaired SREBP-2 signalling. *Int J Mol Sci*, 2022. **23**(24).

[15] Cao, J., et al., LncRNA-MM2P identified as a modulator of macrophage M2 polarization. *Cancer Immunol Res*, 2019. **7**(2): pp. 292–305.

[16] Bhat, M.I. and R. Kapila, Dietary metabolites derived from gut microbiota: Critical modulators of epigenetic changes in mammals. *Nutr Rev*, 2017. **75**(5): pp. 374–389.

[17] Liu, T., et al., Gut microbes influence the development of central nervous system disorders through epigenetic inheritance. *Microbiol Res*, 2023. **274**: p. 127440.

[18] Bai, X., et al., Engineering the gut microbiome, *Nat Rev Bioeng*, 2023,1: 665–679.

[19] Quigley, E.M.M. and P. Gajula, Recent advances in modulating the microbiome. *F1000Res*, 2020. **9**.

[20] Richards, J.L., et al., Dietary metabolites and the gut microbiota: an alternative approach to control inflammatory and autoimmune diseases. *Clin Transl Immunol*, 2016. **5**(5): p. e82.

[21] Mehta, V., et al., Epigenetics and gut microbiota crosstalk: A potential factor in pathogenesis of cardiovascular disorders. *Bioengineering (Basel)*, 2022. **9**(12).

[22] Liu, S. and H.L. Weiner, Control of the gut microbiome by fecal microRNA. *Microb Cell*, 2016. **3**(4): pp. 176–177.

[23] Assmann, T.S., et al., Comprehensive analysis reveals novel interactions between circulating microRNAs and gut microbiota composition in human obesity. *Int J Mol Sci*, 2020. **21**(24).

[24] Dai, X., et al., MicroRNA-193a-3p reduces intestinal inflammation in response to microbiota via downregulation of colonic pepT1. *J Biol Chem*, 2015. **290**(26): pp. 16099–16115.

[25] Liu, S., et al., The host shapes the gut microbiota via fecal microRNA. *Cell Host Microbe*, 2016. **19**(1): pp. 32–43.

[26] Chen, L., et al., Chemoprevention of colorectal cancer by black raspberry anthocyanins involved the modulation of gut microbiota and SFRP2 demethylation. *Carcinogenesis*, 2018. **39**(3): pp. 471–481.

[27] Yuan, C., et al., Interaction between host MicroRNAs and the gut microbiota in colorectal cancer. *mSystems*, 2018. **3**(3).

12 Prebiotics and Probiotics in Epigenetics Therapies

Saranya Punniyakotti,[1] *Krishna Prabha Naduchamy,*[2] *and Nithya Varadarajan*[1]
[1]Department of Pharmacy Practice, Saveetha College of Pharmacy, SIMATS, Thandalam, Chennai, Tamil Nadu, India
[2]Department of Pharmacology, SRM College of Pharmacy, SRMIST, Kattankulathur, Tamil Nadu, India

12.1 INTRODUCTION

A multitude of microorganisms inhabit the human gastrointestinal (GI) tract, including bacteria, fungus, and protozoa. They maintain symbiotic relationships with host species and play a crucial role in biological processes such as nutrition consumption, infection resistance, immune system maturation, and host metabolism [1,2]. The gut microbiota develops from birth to adulthood and depends on dietary, antibiotic and metabolic profile of the host [3]. The human gut microbiome is mostly consistent in individuals over time, except during certain occurrences such as food poisoning, infection, or overseas travel [4]. The gut microbiota comprises trillions of bacteria including the Firmicutes (*Bacillus*, *Lactobacillus*, *Mycoplasma*, and *Clostridium*) and Bacteroidetes, including *Bacteroides*, are primarily present [5].

Prebiotics are primarily found as non-digestible carbohydrates, including galacto-oligosaccharides, fructo-oligosaccharides (FOS), soybean oligosaccharides, inulin, cyclodextrins, glucooligosaccharides (GOS), xylo-oligosaccharides, lactulose, lactosucrose, and isomalto-oligosaccharides, which serve as nutrients for bacteria in the GI tract [6]. The diversity of gut microbiota is primarily influenced by metabolic disorders such as obesity and diabetes [7] as well as neurological disorders including anxiety [8] and depression [9]. The gut microbiota serves as a crucial regulator in human physiology and pathophysiology, influencing host health and illness [10].

Nondigestible food components support human health by selectively stimulating the growth and activity of specific bacteria in the colon [11]. This aligns with the prevailing concept of prebiotics, which refers to dietary components that pass through traverse the upper GI tract undigested and promote the proliferation of beneficial bacteria in the large intestine [12]. Among active microorganisms, bifidobacteria and lactobacilli are selectively stimulated by specific non-viable substrates, referred to as prebiotics. Gibson and Roberfroid initially defined a prebiotic as a "non-digestible dietary component that positively influences the host by selectively promoting the proliferation and/or activity of specific bacteria in the colon, hence enhancing host health" [11,13,14]. Bifidobacteria and lactobacilli generate lactate and acetate, both of which can enhance health benefits [15].

A variety of substrates influence colonic microbiota including dietary components such as resistant starch and non-starch polysaccharides and plant-derived compounds such as pectin, cellulose, hemicellulose, guar, and xylan. Additionally, carbohydrates such as lactose, raffinose, and FOS undergo microbial digestion in the colon [16,17]. Any dietary element that arrives in the colon unaltered is a possible prebiotic [18].

DOI: 10.1201/9781032706511-12

12.2 EPIGENETICS

Epigenetics refers to modifications in gene expression influenced by environmental or other influences without changing the DNA sequence. Alternatively, it is characterized by the human cells interpreting genes differently due to certain contextual alterations, without impacting the DNA sequence [19]. It indicates the absence of DNA methylation, genomic imprinting silencing of the additional X-chromosome, histone changes, and abnormalities in noncoding RNAs [20].

Fermentation of indigestible carbohydrates mediated by gut bacteria result in SCFAs, which include acetate, butyrate, and propionate, which serve as substrates for glucose metabolism [21]. Genetic factors are among those that contribute to obesity. Leptin, leptin-receptor genes, apolipoprotein A1, and phospholipase D1 expressing genes are mostly implicated in obesity [22].

Modifications at the promoters of tumor suppressor genes and proto-oncogenes are occasionally termed epimutations. In colorectal cancer (CRC), genes such as APC, LKB1, GATA4, p16INK4a, and MLH1 undergo mutations [23]. Epigenetic silencing of specific genes linked to genomic instability [24]. The disparity between the enzymatic activity of histone acetyltransferases (HAT) and histone deacetylases (HDACs) contributes to cancer development. Inactivation of HDAC genes, which are implicated in CRC, impede cell cycle progression [25].

12.3 THE THERAPEUTIC MECHANISM OF PREBIOTICS

Prebiotics exhibit both direct and indirect effects on host cells. In the indirect method, indigestible food is fermented by the host bacteria. This promotes the proliferation of microorganisms that reorganize the gut microbiota [26]. The fermentable prebiotics accentuate the growth of beneficial bacteria like bifidogenic species. In addition to this, the ingestion of inulin significantly increases *Bifidobacterium* and *Lactobacilli* [11,27,28]. Alterations in the gut microbiota confer positive impacts on host health. High *Bifidobacterium* levels correlate with high acetate levels while simultaneously reducing *C. difficile* in the gut and inhibiting the translocation of pathogens from the gut lumen to the bloodstream [28]. In the direct method, prebiotics directly influence cells including epithelial and immunological cells.

SCFAs are principal metabolites produced via the fermentation of prebiotics by gut microbes. These metabolites are released into biological systems such as blood, stomach, lung, and placenta, where they interact with the cells. The active metabolites interact with GPR receptors and initiate the signaling pathway involving AMP-activated protein kinase (AMP-K), mammalian target of rapamycin (mTOR), signal transducer and activator of transcription 3 (STAT3), mitogen-activated protein kinases (MAPKs), and nuclear factor kappa-light chain-enhancer of activated B cells (NF-κβ). SCFAs are transported to cytoplasm by passive diffusion. Subsequently, they initiates gene transcription through acetylation or deacetylation via HAT and HDACs, respectively, modifying metabolism, the cell cycle, and microbial activity (Figure 12.1).

12.4 PREBIOTICS IN EPIGENETIC HEALING

The fundamental concept of prebiotics was initially introduced by Glenn Gibson and Marcel Roberfroid in 1995 [11]. Prebiotics, specifically FOS, lactulose, and GOS, promote the proliferation of microbes that generate lactic acid. FOS promotes the proliferation of bifidobacteria in the GI tract [11]. FOS diminishes the hepatic production of triglycerides [29]. A hypothesis exists suggesting that prebiotics are useful in combating colon cancer [30–33]. Lactic acid-releasing bacteria have antimicrobial properties against intestinal infections of both Gram-positive and Gram-negative types [34]. FOS and inulin were reported to protect against exposure to enteric and systemic pathogens as well as tumor-promoting agents [35].

Obesity is a global epidemiological phenomenon defined by the accumulation of adipose tissue, primarily visceral fat, and is linked to multiple factors including genetics, hormones, and environmental

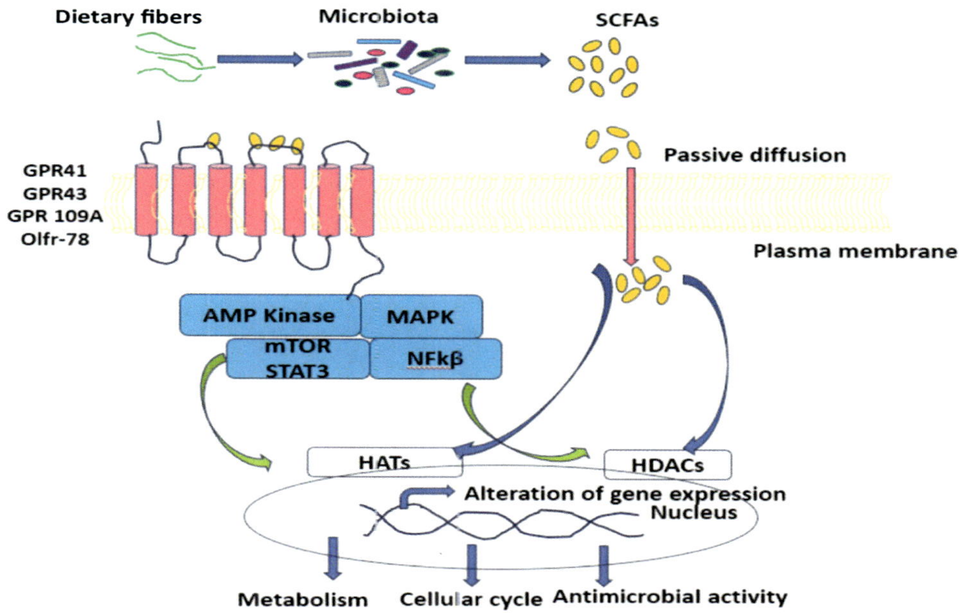

FIGURE 12.1 The indirect effects of prebiotics.

influences [36]. According to WHO, 39% of the current population will become obese in 2035 [37]. It has been managed with the aid of specific gut microbials supplemented by a healthy diet [38,39]. Prebiotic-rich foods lower the risk of obesity by increasing lactobacilli and bifidobacteria [40] and decrease the pathogenic microorganisms Firmicutes and Bacteroidetes in the GI tract [41]. Prebiotics elicit alterations in gut microbiota that lead to enhanced modifications in entero-endocrine cells. These endocrine cells secrete hormones that subsequently modify food consumption and weight growth [42].

Prebiotics were found to improve irritable bowel syndrome (IBS) and Crohn's disease. The outcomes of two distinct clinical trials demonstrated that oligofructose at a dose of 6 g/day for 4 weeks and FOS at a dose of 20 g/day had no effect on patients with IBS,[43,44]. but 5 g/day FOS for 6 weeks and 3.5 g/day GOS for 12 weeks improved IBS symptoms [45,46]. FOS at 15 g/day for 3 weeks increased bifidobacteria in the feces and improved Crohn's disease [47].

In some cases, diet modifies gene expression [48,49]. Metabolites derived from the gut microbiome drive cancer [50]. Populations' shifting toward the Western diet (low plant fibre and high animal fat) from our traditional diet (high plant fibre and low animal fat) is also an important factor for CRC. Bacterial β-glucuronidases aids in treating colorectal cancer by inhibiting glucuronidation of irinotecan.

FOS and GOS suppress necrotizing enterocolitis (NEC), an acute illness prevalent in premature newborns, by promoting *Bifidobacterium* while reducing harmful bacteria [51–53]. Irinotecan, also referred to as camptothecin, is a topoisomerase inhibitor utilized in the treatment of CRC and is inactivated by glucuronidation in the liver. However, the bacterial β-glucuronidases in the gut microbiome impede the glucuronidation of irinotecan, enhancing its chemotherapeutic efficacy in the GI tract (Tables 12.1 and 12.2) [54].

12.5 PROBIOTICS

Probiotics have demonstrated a substantial influence on human health, with prospective therapeutic implications across various diseases and disorders. Specifically, they have been identified as

TABLE 12.1
The Epigenetic Mechanisms of Prebiotics and Their Biological Effects

Substrate	Epigenetic Mechanism	Biological Effect	Reference(s)
Cyclophosphamide, metformin, and berberine		Modifies the microbiome	Viaud et al., 2013 [55] Shin et al., 2014 [58] Zhang et al., 2012 [59]
Cyclophosphamide	Alkylation	Alters the microbiota and stimulates an immune response against cancer	Viaud et al., 2013 [55]
Metformin	High prevalence of genus Akkermansia	Antidiabetic efficacy in type 2 diabetes mellitus	Shin et al., 2014 [58]
Berberine	Alters the microbiome	Mitigated obesity and insulin resistance	Zhang et al., 2012 [59]
Vorinostat (also known as SAHA)		To treat lymphomas	West and Johnstone., 2014 [25]
Romidepsin		To treat lymphomas	West and Johnstone., 2014 [25]
Oligofructose	Attenuates triglyceride levels, adipose tissue mass, and muscle lipid infiltration	Anti-obesogenic	Everard et al., 2011 [60]
α-cyclodextrins	Elevated the expression of PPARγ and PPARα while reducing sterol regulatory element-binding protein-1c (SREBP-1c) and fatty acid synthase.	Anti-obesogenic	Nihei et al., 2018 [61]
Irinotecan, also referred to as camptothecin [11]	Inhibits glucuronidation by bacterial β-glucuronidases	Chemotherapeutic efficacy in the GI tract	Wallace et al., 2010 [56]
	Inactivates HDAC genes	CRC	West and Johnstone., 2014 [25]
Kaempferol	Alters histones, inhibits G9a (HDAC/G9a axis), activates the IRE1-JNK-CHOP pathway; chemopreventive agent for gastric cancer		Kim, Lee, Kim, Cheon, & Ko, 2018 [62]
Alpinetin	DNA methylation, activates AhR, regulates the miR302/DNMT-1/CREB signaling pathway, facilitating Treg differentiation	Improves ulcerative colitis	Lu et al., 2018 [63]
DHCA/Mal-gluc	DNA methylation, histone modifications	Inhibits the CpG-rich IL-6 regions in introns 1 and 3; modulates synaptic plasticity through enhanced histone acetylation of the regulatory sequences of the Rac1 gene; antidepressant	Wang et al., 2018 [64]
PUFAs	DNA and miRNA methylation, alters the miRNA transcriptome; augments promoter methylation of fatty acid desaturase-2	Anti-inflammation	Heberden & Maximin, 2017 [63]
N-acetyl chitooligosaccharides	Histone modifications; attenuates MAPK signaling pathways, particularly via the IL1β-Erk/p38-histone H3 pathway in a type 2 diabetes model in mice	Antidiabetic efficacy	Wu et al., 2017 [66]

(Continued)

TABLE 12.1 *(Continued)*
The Epigenetic Mechanisms of Prebiotics and Their Biological Effects

Substrate	Epigenetic Mechanism	Biological Effect	Reference(s)
mUA	Modulates miR-21 and its downstream targets including PTEN, Akt, and Wnt/β-catenin signaling	Inhibits cell viability	Zhou, Wang, Zheng, & Qiu, 2016 [67]
B-GOS®	Acetate histone modifications	Inhibits brain HDAC activity, inhibits HAT, elevates cortical and hippocampal mRNA, mitigates olanzapine-induced weight gain, enhances cognitive flexibility	Kao, Chan, Anthony, Lennox, & Burnet, 2019 [68]
Acetate	Histone and miRNA changes, inhibits HAT activity, increases hippocampus mRNA levels of HDAC-3 and HDAC-4, activates acetyl CoA carboxylase; reduces hypothalamic AMP-K catalytic activity; induces anorectic neuropeptide expression	Advantages on central and peripheral physiology; Facilitate appetite suppression.	Kao et al., 2019a [68];
Butyrate; D-β-hydroxybutyrate	Histone alterations	Inhibit HDAC; serve as encoding factors for oxidative stress resistance, FOXO3A, and MT2; suppress tumor formation	Ou et al., 2013 [69]; Shimazu et al., 2013 [70]
Propionate	Histone modification and hyperacetylation, inhibits HDAC, activates the AP-1 signaling pathway in epithelial cell lines, regulates Treg cells	Anti-inflammatory properties; Inhibitory impact on tumorigenesis	Louis, Hold, & Flint, 2014 [71] Smith et al., 2013 [72]
Delphinidin	DNA methylation; inhibits DNMTs and HDACs through histone alterations, activating the Nrf2-ARE pathway, which prevents JB6 P+ cell transformation by demethylating 15 CpG sites in the Nrf2 promoter	Chemopreventive agent for skin cancer	Kuo, Wu, Li, Yang, & Kong, 2019 [73]

Notes: B-GOS®, bimunogalacto-oligosaccharide; DHCA, dihydrocaffeic acid; HDAC, histone deacetylase; HAT, histone acetyltransferase; IL-6, pro-inflammatory interleukin 6; Mal-gluc, malvidin-3′-O-glucoside; PUFAs, N-3 polyunsaturated fatty acids; N-acetyl COSs, N-acetyl chitobiose [(GlcNAc)2]; mUA, methylated urolithin A.

regulators of immunological responses, enhancers of intestinal epithelial cell viability, and promoters of barrier function, which may be advantageous in the management of inflammatory bowel diseases (IBDs) [72]. Probiotics also protect against gut inflammation and treat GI illnesses, although the precise mechanisms remain incompletely known. Recent study indicates that the advantageous effects of probiotics may be associated with the modification of microRNAs, which are essential in cellular signaling pathways [73]. These findings offer significant insights into the possible molecular therapeutic implications of probiotics in epigenetic repair. Probiotics provide advantageous impacts on human health when administered in sufficient quantities.

TABLE 12.2
The Epigenetic Mechanisms of Probiotics

Biological Effect	Epigenetic mechanism	Reference(s)
Histone modification alterations	Changes DNA access to transcription machinery, thereby influencing gene expression	Licciardi et al., 2011 [102] M. Hamon et al., 2007 [103]
Genomic imprinting	An epigenetic phenomenon in which specific genes are expressed depending on their parental origin, with considerable ramifications for development and metabolism	Morovic & Budinoff, 2020 [104] Berner, 2010 [105]
Stress response genes	Regulate the expression of genes associated with the body's reaction to stress, potentially impacting mental health and stress-related diseases	Bloemendaal, 2021 [106]
Barrier function genes	Augments the production of genes that preserve the integrity of the intestinal barrier, essential for obstructing pathogens from infiltrating the bloodstream	Paul V Licciardi 2010 [107] Bloemendaal, 2021 [106]
Cell cycle regulation	Affect the epigenetic regulation of cell cycle-related genes, which could have implications for cell proliferation and cancer prevention	R. Patel, 2010 [108]
Detoxification enzymes	Regulates the expression of genes that encode detoxification enzymes, potentially affecting the body's capacity to metabolize and excrete toxins	Bloemendaal, 2021 [106]
Apoptosis	Preventing and controlling cell death	R. Patel, 2010 [108]
Methylation	The methylation of particular genes can adversely impact sperm quality, leading to suboptimal fertilization.	Rotondo, 2021 [109]
Immune responses	Probiotics directly modulate the mucosal immune system, thereby activating both innate and adaptive immune responses.	R. Patel, 2010 [108] Vanderpool, 2008 [75]
Harmful bacteria inhibition	By synthesizing bactericidal agents and competing with infections and poisons for gut epithelium adhesion	Vanderpool [75] 2008
Neurotransmitter activity	Probiotics may impact the capacity of the gut microbiome to modulate neurotransmitter activity in the host.	Snigdha, 2021 [110]
Endogenous sex hormones	Interact with probiotics, specifically estrogens, influencing stress response and cognitive function	Sumon, 2022 [111]
Gender	Can show differing biological effects of probiotics, particularly regarding their influence on the endocrine and central nervous systems	Snigdha, 2021 [110]
Bacterial DNA methylases	Regulate gene expression and control chromosomal DNA replication and repair	Sumon, 2022 [111] Gao, 2023 [112]

(Continued)

TABLE 12.2 (Continued)
The Epigenetic Mechanisms of Probiotics

Biological Effect	Epigenetic mechanism	Reference(s)
Novel epigenetic inhibitors	DNA methyltransferases may serve as potential targets for the development of new epigenetic inhibitors in biological applications.	Ge et al., 2022 [113]
Chromatin remodeling	Probiotics may affect the structure of chromatin, altering gene expression patterns and cell function	P. Licciardi et al., 2011 [102]
NcRNAs and miRNAs	Probiotics modulation impact various biological pathways and cellular processes	Berner, 2010 [105] Malmuthuge, 2021 [114]

Probiotics can mitigate the risk of specific diseases through three overarching, clinically substantiated processes, each demonstrated to play a role in epigenetic control. Probiotics utilize vitamins, proteins, and carbohydrates that the host fails to absorb or digest. This process generates notable secondary metabolites, such as SCFAs, which function as a vital energy source for the host [74]. SCFAs such as acetate, propionate, and butyrate mediate host immune responses influenced by probiotic activity. Finally, numerous probiotics can shape microbiome composition and activity by competitively excluding pathogens. This process often involves adhesion to epithelial cells via pili and antimicrobial mechanisms, both of which are linked to epigenetic regulation.

Epigenetic alterations caused by the microbiota influence human health by changing loci within an individual genome, resulting in diverse phenotypic subpopulations within a clonal culture. The epigenetic elements within the cell are affected by environmental conditions such as temperature, pH, nutrition availability, and the composition of the resident microbiome. Probiotic metabolites, meanwhile, exert an epigenetic influence on the host, affecting critical development during infancy and establishing a foundation for adult health and resilience against conditions such as allergies and autoimmune disorders [75]. The epigenetic alterations prompted by the microbiota are essential in influencing human health and disease vulnerability.

12.5.1 Probiotics and Human Well-Being

Probiotics directly and indirectly impact human health by eliminating toxins, protecting against intestinal bacteria, and enhancing the immune system. Probiotics are recognized for their health advantages. The concept of probiotics has been vital to our comprehension for numerous decades [76]. Probiotic supplements can enhance the efficacy of treatment for some disorders; they contribute to regulating intestinal epithelial homeostasis by improving barrier function, triggering protective responses, modulating the immune system, and competitively adhering to the mucus and epithelium. The gut microflora influences the host's energy balance through diverse mechanisms. Probiotics compete with harmful microbes for adhesion sites in the GI tract. Notably, probiotics seem to offer benefits for premature infants vulnerable to NEC [77].

Karimi et al. and Tijjani et al. both highlighted the potential of probiotics in modulating the intestinal microflora and suppressing inflammatory responses, which can be beneficial in conditions like intestinal pathogenic infections, helicobacter pylori infections, urogenital infection, cancer, obesity, type 2 diabetes, IBD, and lactose intolerance, among others. Probiotic intake strengthens GI health, fortifies the immune system, increases nutritional absorption, alleviates lactose intolerance symptoms, diminishes allergy prevalence in predisposed individuals, and lowers the risk of some cancers. Oyetayo further emphasizes the immunostimulatory effects of probiotics, suggesting their potential as biotherapeutic agents. Probiotics can serve as outstanding vaccine adjuvants [77].

Goldin provides a comprehensive overview of the clinical indications for probiotics, including their effectiveness, in treating and preventing conditions such as diarrhea, food allergies, and urinary tract infections [78]. These studies jointly highlight the therapeutic data advocating for probiotics in human health. Probiotics exhibit preventive effects in illnesses like pouchitis and ulcerative colitis, both IBDs.

12.5.2 Probiotics in Epigenetic Healing

Probiotics have demonstrated efficacy in mucosal illnesses, infection prevention, and immune system regulation along with their many other demonstrated functions [79]. They have also been associated with enhanced healing of intestinal ulcers and infected skin wounds through pathogen displacement and fortification of the epithelial barrier [80]. Probiotics were explored for preventing diarrhea and modulating immunity, with the potential for cancer prevention and cholesterol level reduction [81]. Moreover, the probiotic strain *Lactobacillus rhamnosus GG* demonstrated the ability to inhibit cytokine-induced death in intestinal epithelial cells, indicating a potential function in tissue repair [82]. These data collectively indicate functions for probiotics in epigenetic repair, but additional study is required to comprehensively elucidate their processes and applicability in this setting. Beyond modulating microbiota and inducing epigenetic modifications, probiotics alter gut microbiota composition, inhibit inflammation, and augment immunological responses, all of which can exert epigenetic effects on gene expression and cellular function [83].

The therapeutic potential of butyrate and HDAC inhibitors (HDACIs) is substantial regarding human health. Butyrate is an SCFA that inhibits HDACs to affect gene expression and chromatin remodeling. It has been studied for its therapeutic benefits on immunity, obesity, insulin resistance, cardiovascular disease, and cancer treatment. Butyrate reactivates repressed genes in neoplastic cells, facilitating apoptosis, and possibly acts as a cancer therapy [84].

CRISPR-based approaches can be used for targeted DNA demethylation in probiotics, offering a precise method for manipulating DNA methylation patterns. Nuclease-deactivated Cas9 or base editing enables researchers to selectively affect specific methylation sites of interest while keeping methyltransferases intact. This method shows potential for creating new probiotic strains with targeted health benefits. Overall, the therapeutic implications of butyrate and HDACIs, along with the potential use of CRISPR-based approaches for targeted DNA demethylation in probiotics, represent exciting avenues for leveraging epigenetic mechanisms in probiotics for human health benefits and developing novel therapeutic interventions [84].

12.5.3 Probiotics' Impacts on Epigenetic Healing

The epigenetics of probiotics can be affected by multiple factors, such as environmental conditions, nutrition, lifestyle, stress, antibiotics, and the host's genetic makeup. These variables influence the gut microbiota, leading to health issues such as allergies, IBD, infections, and cancer. Ingested probiotics influence disease risk by modulating epigenetic pathways that regulate immune and inflammatory responses related to epigenetic control. Probiotics generate significant secondary metabolites including SCFAs that supply energy to the host and regulate immunological responses, both linked to epigenetic regulation [76]. The mechanisms that influence probiotics' epigenetics underscore the intricate interaction among the microbiome, host variables, and environmental factors in determining the epigenetic regulation of probiotics and their possible therapeutic effects in a multitude of conditions [85–87].

12.5.3.1 Allergies

Research showed that supplementation with *Lactobacillus reuteri* induced significant changes in DNA methylation, particularly in genes associated with immune response transduction. However, these alterations did not demonstrate a substantial link with the onset of allergy diseases in children. The study's authors proposed that the microbial composition in the GI system and its interaction with the host elicits epigenetic DNA alterations that could transmit to subsequent generations. In the fetal stage, the mother's microbiota and epigenetic makeup could impact fetal immune system development [88]. Consequently, probiotic supplementation has been suggested as a method for the prevention and management of allergies. These findings underscore the need for further research into the role of probiotics in the epigenetic remediation of allergies using advanced technologies [89].

12.5.3.2 Colorectal Cancer

Probiotics significantly contribute to the prevention and treatment of colorectal cancer (CRC) via multiple methods. They decrease tumor incidence, size, and quantity and prevent precancerous lesions in CRC mice models. They accomplish this by inhibiting inflammation, promoting death of nascent tumor cells, reinstating gut barrier integrity, and rectifying microbiota composition [90]. Nonetheless, extensive long-term clinical research remains necessary to validate the preventive effects of probiotics in clinical environments.

In CRC management, probiotics mitigated chemotherapy-induced mucositis, reduced the occurrence of diarrhea and weight loss, and restored intestinal barrier integrity. They modulated gut microbiota, reduced inflammation, and improved immune responses [84]. Additionally, probiotics studied in the perioperative setting for CRC surgery improved barrier function and reduced postoperative infectious complications [85]. Furthermore, probiotics reduced colonic inflammation and enhanced antitumor immunity [91].

12.5.3.3 Neurological and Mental Illnesses

Probiotics demonstrated the ability to restore diminished H3K27me3 levels in the adult hippocampus and partially mitigate memory dysfunction, indicating that memory impairment may be improved by epigenetic modulators or modifications to gut microbiota. Moreover, miRNA generated by gut epithelial cells regulated bacterial gene expression [92]. In addition, gut microbiota affects the epigenetics of neuropsychiatric conditions including anxiety, depression, and neurodegenerative diseases [93].

Furthermore, microbial colonization modulates global histone acetylation and methylation in host tissues, with diet-dependent shifts in microbial composition influencing epigenetics. The direct gut–brain link, along with gut impacts on DNA and histone methylation, offers insight into how the microbiome influences brain development [94]. The findings indicate that utilizing probiotics and dietary modifications to reestablish normal interactions between the central nervous system and the gut may be advantageous in treating brain problems.

The gut microbiome significantly impacts brain function and behavior via metabolite synthesis and epigenetic control. Dysbiosis of the gut microbiota has been associated with neurological and mental illnesses, underscoring the potential for microbiome-based therapies in addressing these conditions. Probiotics may affect epigenetic regulation by changing gut microbiota in relation to neurological and mental illnesses. Probiotic-targeted therapy, combined with dietary modifications, could restore normal gut–brain axis communication and potentially affect epigenetic regulation, presenting an innovative method for addressing neurological and psychiatric disorders by modulating gut microbiota and its epigenetic implications [93].

12.5.3.4 Mood Disorders

Probiotics and metabolites of commensal bacteria affect epigenetic modifications in the gut and regulate mood through controlling DNA methylation, histone acetylation, and redox signaling and synthesizing advantageous chemicals such as SCFAs, B vitamins, and bile salts. *Lactobacillus* and *Bifidobacterium* have demonstrated the ability to affect host physiology and mood via epigenetic regulation. Gut dysbiosis, stress, and antibiotic therapy can perturb the gut microbiota, intensifying mood problems. The gut–brain axis is essential in this bidirectional communication network, with probiotics potentially mitigating depressed symptoms. The influence of dietary nutrients and gut microbiota on epigenetic changes and redox signaling, together with the prospective therapeutic effects of probiotics, fatty acids, choline, and S-adenosylmethionine, is under investigation [95].

12.5.3.5 Type 1 Diabetes

Probiotics and prebiotics provide potential pathways for the prevention and control of Type 1 Diabetes (T1D). Probiotics mitigated insulitis, diminished pancreatic β-cell death, and improved T1D by inhibiting NF-kB activation pathways. These actions altered gene expression patterns via epigenetic mechanisms, ultimately facilitating the repair process in diabetes. Prebiotics enhanced SCFA synthesis, promoted mucin production, and bolstered intestinal integrity while mitigating intestinal inflammation [96].

Prebiotics also enhanced intestinal permeability, diminish inflammation, augment the population of Bacteroidetes and lactic acid-producing bacteria, and improved glycemic regulation, decreasing the likelihood of T1D onset. These methods of action illustrate the potential of probiotics and prebiotics in the prevention and management of T1D by influencing the gut microbiome and immune system to enhance overall metabolic health [96].

12.5.3.6 Type 2 Diabetes

Probiotics have demonstrated advantageous benefits on type 2 diabetes (T2D) by impacting the gut microbiota and immune system, specifically ameliorating obesity, insulin resistance, T2D, and non-alcoholic fatty liver disease (NAFLD) in human studies. Moreover, several probiotics have been linked to improvements in T2D by modulating the immune system, suppressing intestinal inflammation, and stimulating Toll-like receptors to elicit anti-inflammatory actions [97]. Furthermore, probiotics generated SCFAs, which regulate energy homeostasis and interact with cell receptors in peripheral tissues, indicating possible involvement in the prevention and management of T2D [98].

In conclusion, probiotics may beneficially influence the prevention and management of T2D by affecting gut microbiota, modulating the immune system, and producing SCFAs. Additional study is required to comprehensively elucidate the molecular processes of probiotics in the prevention and treatment of type 2 diabetes and to investigate their potential function in epigenetic restoration.

12.5.3.7 NAFLD

Probiotics have demonstrated advantageous benefits in NAFLD by reducing liver lipid levels and altering fecal metabolite concentrations that function as noninvasive and useful biomarkers for assessing treatment response. SCFAs generated by probiotics exert physiological, biochemical, and molecular impacts across various tissues including the liver. They interact with cell receptors in peripheral tissues and regulate energy homeostasis and metabolism, indicating a possible involvement in the prevention and treatment of obesity and related illnesses. These data indicate that probiotics may play a significant role in the epigenetic remediation of NAFLD by influencing metabolite concentrations, SCFAs, and the immune system [99–114].

12.6 CONCLUSION

The microbiota of the GI system plays a crucial role in the host's nutrition and health. The composition of the microbiota is modified by a diet rich in prebiotics and probiotics. Since the 1990s, the importance of both in humans has been intensively studied for understanding the mechanisms of prebiotics and probiotics regarding the diverse health benefits they confer to the host through diet, pharmaceuticals, enzymes, and gut microbiota. This chapter aimed to examine the effects of prebiotics and probiotics on epigenetic healing, substantiated by molecular evidence of their impact on epigenetic pathways. Nonetheless, an opportunity persists to elucidate the therapeutic potential of prebiotics and probiotics in a symbiotic relationship with microbes, examining how it can confer advantages to both individuals and society at large.

REFERENCES

[1] Brestoff, Jonathan R., and David Artis. "Commensal bacteria at the interface of host metabolism and the immune system." *Nature Immunology* 14 no. 7 (2013): 676–684.

[2] Belkaid, Yasmine, and Timothy W. Hand. "Role of the microbiota in immunity and inflammation." *Cell* 157 no. 1 (2014): 121–141.

[3] Reynolds, Lisa A., and Brett B. Finlay. "Early life factors that affect allergy development." *Nature Reviews Immunology* 17 no. 8 (2017): 518–528.

[4] Faith, Jeremiah J., Janaki L. Guruge, Mark Charbonneau, et al. "The long-term stability of the human gut microbiota." *Science* 341 no. 6141 (2013): 1237439.

[5] De Clercq, Nicolien C., Albert K. Groen, Johannes A. Romijn, et al. "Gut microbiota in obesity and undernutrition." *Advances in Nutrition* 7 no. 6 (2016): 1080–1089.

[6] Younis, Kaiser, Syahida Ahmad, and Kausar Jahan. "Health benefits and application of prebiotics in foods." *Journal of Food Processing & Technology* 6 no. 4 (2015): 1.

[7] Patterson, Elaine, Paul M. Ryan. John F. Cryan, et al. "Gut microbiota, obesity and diabetes." *Postgraduate Medical Journal* 92, no. 1087 (2016): 286–300.

[8] Neufeld, Karen-Anne M., Nancy Kang, John Bienenstock, et al. "Effects of intestinal microbiota on anxiety-like behavior." *Communicative & Integrative Biology* 4 no. 4 (2011): 492–494.

[9] Dinan, Timothy G., and John F. Cryan. "Mood by microbe: Towards clinical translation." *Genome Medicine* 8 (2016): 1–3.

[10] Torres-Fuentes, Cristina, Harriet Schellekens, Timothy G. Dinan, et al. "The microbiota–gut–brain axis in obesity." *The Lancet Gastroenterology & Hepatology* 2 no. 10 (2017): 747–756.

[11] Gibson, Glenn R., and Marcel B. Roberfroid. "Dietary modulation of the human colonic microbiota: Introducing the concept of prebiotics." *The Journal of Nutrition* 125 no. 6 (1995): 1401–1412.

[12] Bindels, Laure B., Evelyne M. Dewulf, and Nathalie M. Delzenne. "GPR43/FFA2: Physiopathological relevance and therapeutic prospects." *Trends in Pharmacological Sciences* 34 no. 4 (2013): 226–232.

[13] Roberfroid, Marcel, Glenn R. Gibson, Lesley Hoyles, et al. "Prebiotic effects: Metabolic and health benefits." *British Journal of Nutrition* 104 no. S2 (2010): S1–S63.

[14] Gibson, Glenn R. *Food Science and Technology Bulletin: Fu.* Vol. 2. IFIS Publishing; 2006.

[15] Fukuda, Shinji, Hidehiro Toh, Koji Hase, et al. "Bifidobacteria can protect from enteropathogenic infection through production of acetate." *Nature* 469 no. 7331 (2011): 543–547.

[16] Cummings, J.H., and G.T. Macfarlane. "The control and consequences of bacterial fermentation in the human colon." *Journal of Applied Bacteriology* 70 no. 6 (1991): 443–459.

[17] Salminen, S., C. Bouley, M.C. Boutron, et al. "Functional food science and gastrointestinal physiology and function." *British Journal of Nutrition* 80 no. S1 (1998): S147–S171.

[18] Manning, Thea Scantlebury, and Glenn R. Gibson. "Prebiotics." *Best Practice & Research Clinical Gastroenterology* 18 no. 2 (2004): 287–298.

[19] Berger, Shelley L., Tony Kouzarides, Ramin Shiekhattar, et al. "An operational definition of epigenetics." *Genes & Development* 23 no. 7 (2009): 781–783.

[20] Riscuta, Gabriela, Dan Xi, Dudith Pierre-Victor, et al. "Diet, microbiome, and epigenetics in the era of precision medicine." *Cancer Epigenetics for Precision Medicine: Methods and Protocols* (2018): 141–156.

[21] De Clercq, Nicolien C., Albert K. Groen, Johannes A. Romijn, et al. "Gut microbiota in obesity and undernutrition." *Advances in Nutrition* 7 no. 6 (2016): 1080–1089.

[22] Duranti, Sabrina, Chiara Ferrario, Douwe Van Sinderen, et al. "Obesity and microbiota: An example of an intricate relationship." *Genes & Nutrition* 12 no. 1 (2017): 1–15.

[23] Carmona, F. Javier, and Manel Esteller. "Epigenomics of human colon cancer." *Mutation Research/Fundamental and Molecular Mechanisms of Mutagenesis* 693 no. 1–2 (2010): 53–60.

[24] Akhtar-Zaidi, Batool, Richard Cowper-Sal-Lari, Olivia Corradin, et al. "Epigenomic enhancer profiling defines a signature of colon cancer." *Science* 336 no. 6082 (2012): 736–739.

[25] West, Alison C., and Ricky W. Johnstone. "New and emerging HDAC inhibitors for cancer treatment." *The Journal of Clinical Investigation* 124 no. 1 (2014): 30–39.

[26] Slavin, Joanne. "Fiber and prebiotics: Mechanisms and health benefits." *Nutrients* 5 no. 4 (2013): 1417–1435.

[27] Vandeputte, Doris, Gwen Falony, Sara Vieira-Silva, et al. "Prebiotic inulin-type fructans induce specific changes in the human gut microbiota." *Gut* 66 (2017): 1968–1974.
[28] Hopkins, Mark J., and George T. Macfarlane. "Nondigestible oligosaccharides enhance bacterial colonization resistance against Clostridium difficile in vitro." *Applied and Environmental Microbiology* 69 no. 4 (2003): 1920–1927.
[29] Delzenne, Nathalie M., and Nadine N. Kok. "Biochemical basis of oligofructose-induced hypolipidemia in animal models." *The Journal of Nutrition* 129 no. 7 (1999): S1467–S1470.
[30] Reddy, Bandaru S., Rachid Hamid, and C.V. Rao. "Effect of dietary oligofructose and inulin on colonic preneoplastic aberrant crypt foci inhibition." *Carcinogenesis* 18 no. 7 (1997): 1371–1374.
[31] Bouhnik, Yoram, Bernard Flourie, Michel Riottot, et al. "Effects of fructo-oligosaccharides ingestion on fecalbifidobacteria and selected metabolic indexes of colon carcinogenesis in healthy humans." *Nutrition and Cancer* no. 26 (1996): 21–29.
[32] Buddington, Randal K., Carol H. Williams, Sui-Chi Chen, et al. "Dietary supplement of neosugar alters the fecal flora and decreases activities of some reductive enzymes in human subjects." *The American Journal of Clinical Nutrition* 63 no. 5 (1996): 709–716.
[33] Hylla, Silke, Andrea Gostner, Gerda Dusel, et al. "Effects of resistant starch on the colon in healthy volunteers: Possible implications for cancer prevention." *The American Journal of Clinical Nutrition* 67 no. 1 (1998): 136–142.
[34] Mackey, B.M., and Glenn R. Gibson. "Escherichia coli O157-from farm to fork and beyond." *Journal of General Microbiology* 24 (1997): 55–57.
[35] Buddington, Karyl K., Jillian B. Donahoo, and Randal K. Buddington. "Dietary oligofructose and inulin protect mice from enteric and systemic pathogens and tumor inducers." *The Journal of Nutrition* 132 no. 3 (2002): 472–477.
[36] Bray, Goerge A., Kyoung-Kon Kim, John P.H. Wilding, and World Obesity Federation. "Obesity: A chronic relapsing progressive disease process. A position statement of the World Obesity Federation." *Obesity Reviews* 18 no. 7 (2017): 715–723.
[37] World Health Organization. "Obesity: Preventing and managing the global epidemic: Report of a WHO consultation." (2000).
[38] Morelli, Lorenzo, and Lucio Capurso. "FAO/WHO guidelines on probiotics: 10 years later." *Journal of Clinical Gastroenterology* 46 (2012): S1–S2.
[39] Parnell, Jill A., Maitreyi Raman, Kevin P. Rioux, et al. "The potential role of prebiotic fibre for treatment and management of non-alcoholic fatty liver disease and associated obesity and insulin resistance." *Liver International* 32 no. 5 (2012): 701–711.
[40] Connolly, Michael L., Julie A. Lovegrove, and Kieran M. Tuohy. "In vitro fermentation characteristics of whole grain wheat flakes and the effect of toasting on prebiotic potential." *Journal of Medicinal Food* 15 no. 1 (2012): 33–43.
[41] Parnell, Jill A., and Raylene A. Reimer. "Effect of prebiotic fibre supplementation on hepatic gene expression and serum lipids: A dose–response study in JCR: LA-cp rats." *British Journal of Nutrition* 103 no. 11 (2010): 1577–1584.
[42] Posovszky, Carsten, and Martin Wabitsch. "Regulation of appetite, satiation, and body weight by enteroendocrine cells. Part 2: Therapeutic potential of enteroendocrine cells in the treatment of obesity." *Hormone Research in Paediatrics* 83 no. 1 (2015): 11–18.
[43] Hunter, J.O., Q. Tuffnell, and A.J. Lee. "Controlled trial of oligofructose in the management of irritable bowel syndrome." *The Journal of Nutrition* 129 no. 7 (1999): 1451S–1453S.
[44] Olesen, Merete, and Eivind Gudmand-Høyer. "Efficacy, safety, and tolerability of fructooligosaccharides in the treatment of irritable bowel syndrome." *The American Journal of Clinical Nutrition* 72 no. 6 (2000): 1570–1575.
[45] Paineau, Damien, Flore Payen, Suzanne Panserieu, et al. "The effects of regular consumption of short-chain fructo-oligosaccharides on digestive comfort of subjects with minor functional bowel disorders." *British Journal of Nutrition* 99 no. 2 (2008): 311–318.
[46] Silk, D.B.A., A. Davis, J. Vulevic, G. Tzortzis, and G.R. Gibson. "Clinical trial: The effects of a trans-galactooligosaccharide prebiotic on faecal microbiota and symptoms in irritable bowel syndrome." *Alimentary Pharmacology & Therapeutics* no. 29 (2009): 508–518.
[47] Lindsay, James O., Kevin Whelan, Andrew John Stagg, et al. "The clinical, microbiological and immunological effects of fructo-oligosaccharide in patients with Crohn's disease." *Gut* 55 no. 3 (2005): 348–355.
[48] Cooney, Craig A., Apurva A. Dave, and George L. Wolff. "Maternal methyl supplements in mice affect epigenetic variation and DNA methylation of offspring." *The Journal of Nutrition* 132 no. 8 (2002): 2393–400S.

[49] Waterland, Robert A., and Randy L. Jirtle. "Transposable elements: Targets for early nutritional effects on epigenetic gene regulation." *Molecular and Cellular Biology* 23 no. 15 (2003): 5293–5300.
[50] Camp, J. Gray, Christopher L. Frank, Colin R. Lickwar, et al. "Microbiota modulate transcription in the intestinal epithelium without remodeling the accessible chromatin landscape." *Genome Research* 24 no. 9 (2014): 1504–1516.
[51] Knol, Jan, Günther Boehm, Maura Lidestri, et al. "Increase of faecal bifidobacteria due to dietary oligosaccharides induces a reduction of clinically relevant pathogen germs in the faeces of formula-fed preterm infants." *Acta Paediatrica* 94 (2005): 31–33.
[52] Boehm, G., M. Lidestri, P. Casetta, J. Jelinek, et al. "Supplementation of a bovine milk formula with an oligosaccharide mixture increases counts of faecal bifidobacteria in preterm infants." *Archives of Disease in Childhood Fetal and Neonatal Edition* 86 no. 3 (2002): F178.
[53] Kapiki, Angeliki., Christos Costalos, Christina Oikonomidou, et al. "The effect of a fructo-oligosaccharide supplemented formula on gut flora of preterm infants." *Early Human Development* 83 no. 5 (2007): 335–339.
[54] Wallace, Bret D., Hongwei Wang, Kimberly T. Lane, et al. "Alleviating cancer drug toxicity by inhibiting a bacterial enzyme." *Science* 330 no. 6005 (2010): 831–835.
[55] Viaud, Sophie, Fabiana Saccheri, Grégoire Mignot, et al. "The intestinal microbiota modulates the anticancer immune effects of cyclophosphamide." *Science* 342 no. 6161 (2013): 971–976.
[56] Shin, Na-Ri., June-Chul Lee, Hae-Youn Lee, et al. An increase in the Akkermansia spp. population induced by metformin treatment improves glucose homeostasis in diet-induced obese mice. *Gut* 63 no. 5 (2014): 727–735.
[57] Zhang, Xu, Yufeng Zhao, Menghui Zhang, et al. "Structural changes of gut microbiota during berberine-mediated prevention of obesity and insulin resistance in high-fat diet-fed rats." *PLoS ONE* no. 7 (2012): e42529.
[58] Everard, Amandine, Vladimir Lazarevic, Muriel Derrien, et al. "Responses of gut microbiota and glucose and lipid metabolism to prebiotics in genetic obese and diet-induced leptin-resistant mice." *Diabetes* 60 no. 11 (2011): 2775–2786.
[59] Nihei, Nanako, Hinako Okamoto, Takahiro Furune, et al. "Dietary α-cyclodextrin modifies gut microbiota and reduces fat accumulation in high-fat-diet-fed obese mice." *Biofactors* 44 no. 4 (2018): 336–347.
[60] Kim, Tae Woo, Seon Young Lee, Mia Kim, et al. "Kaempferol induces autophagic cell death via IRE1-JNK-CHOP pathway and inhibition of G9a in gastric cancer cells." *Cell Death & Disease* 9 no. 9 (2018): 875.
[61] Lu, Yuanyuan, Chaonan Fan, Aimin Liang, et al. "Effects of SCFA on the DNA methylation pattern of adiponectin and resistin in high-fat-diet-induced obese male mice." *British Journal of Nutrition* 120 no. 4 (2018): 385–392.
[62] Wang, Jun, Georgia E. Hodes, Hongxing Zhang, et al. "Epigenetic modulation of inflammation and synaptic plasticity promotes resilience against stress in mice." *Nature Communications* no. 9 (2018): 477.
[63] Heberden, C., and E. Maximin. "Epigenetic effects of N-3 polyunsaturated fatty acids." In *Handbook of Nutrition, Diet, and Epigenetics* (pp. 1–15). Cham: Springer International Publishing; 2017. https://doi.org/10.1007/978-3-319-31143-2_45-1
[64] Wu, Xia, Jing Wang, Yuqin Shi, et al. "N-Acetyl-chitobiose ameliorates metabolism dysfunction through Erk/p38 MAPK and histone H3 phosphorylation in type 2 diabetes mice." *Journal of Functional Foods* no. 28 (2017): 96–105.
[65] Zhou, Benhong, Jing Wang, Guohua Zheng, et al. "Methylated urolithin A, the modified ellagitannin-derived metabolite, suppresses cell viability of DU145 human prostate cancer cells via targeting miR-21." *Food and Chemical Toxicology* no. 97 (2016): 375–384.
[66] Kao, Amy Chia-Ching, Ka Wai Chan, Daniel C. Anthony, et al. "Prebiotic reduction of brain histone deacetylase (HDAC) activity and olanzapine-mediated weight gain in rats, are acetate independent." *Neuropharmacology* no. 150 (2019): 184–191.
[67] Ou, Junhai, Franck Carbonero, Erwin G. Zoetendal, et al. "Diet, microbiota, and microbial metabolites in colon cancer risk in rural Africans and African Americans." *The American Journal of Clinical Nutrition* no. 98 (2013): 111–120.
[68] Shimazu, Tadahiro, Matthew D. Hirschey, John Newman, et al. "Suppression of oxidative stress by β-hydroxybutyrate, an endogenous histone deacetylase inhibitor." *Science* no. 339 (2013): 211–214.
[69] Louis, Petra, Georgina L. Hold, and Harry J. Flint. "The gut microbiota, bacterial metabolites and colorectal cancer." *Nature Reviews Microbiology* no. 12 (2014): 661–672.

[70] Smith, Patrick M., Michael R. Howitt, Nicolai Panikov, et al. "The microbial metabolites, short-chain fatty acids, regulate colonic Treg cell homeostasis." *Science* no. 341 (2013): 569–573.

[71] Kuo, Hsiao-Chen Dina., Renyi Wu, Shanyi Li, et al. "Anthocyanin delphinidin prevents neoplastic transformation of mouse skin JB6 P+ cells: Epigenetic re-activation of Nrf2-ARE pathway." *The AAPS Journal* no. 21 (2019): 1–10.

[72] Tijjani, Kabwanga Ismail, Munanura James, and Cansu Altın. "Probiotics and their attributes in human health therapy." *International Journal of Research-Granthaalayah* no. 8 (2020): 158–164.

[73] Vanderpool, Charles, Fang Yan, and Brent D. Polk. "Mechanisms of probiotic action: Implications for therapeutic applications in inflammatory bowel diseases." *Inflammatory Bowel Diseases* no. 14 (2008): 1585–1596.

[74] Dylag, Katarzyna, Magdalena Hubalewska-Mazgaj, Marcin Surmiak, et al. "Probiotics in the mechanism of protection against gut inflammation and therapy of gastrointestinal disorders." *Current Pharmaceutical Design* no. 20 (2014): 1149–1155.

[75] Davoodvandi, Amirhossein, Havva Marzban, Pouya Goleij, et al. "Effects of therapeutic probiotics on modulation of microRNAs." *Cell Communication and Signaling* no. 19 (2021): 1–22.

[76] Karimi, Omid, and Amado Salvador Pena. "Probiotics: Isolated bacteria strain or mixtures of different strains? Two different approaches in the use of probiotics as therapeutics." *Drugs of Today (Barcelona, Spain: 1998)* no. 39 (2003): 565–597.

[77] Oyetayo, Olusegun, and Oyetayo Folake. "Potential of probiotics as biotherapeutic agents targeting the innate immune system." *African Journal of Biotechnology* no. 4 (2005): 123–127.

[78] Goldin, Barry, and S. L. Gorbach. "Clinical indications for probiotics: An overview." *Clinical Infectious Diseases* no. 46 (2008): S96–S100.

[79] Pakdaman, Yasaman. "A review on applications of probiotics in human health and disease." *Journal of Biomedical and Pharmaceutical Research* no. 8 (2019): 55–70.

[80] Lukic, Jovanka, Vivien Chen, Ivana Strahinic. "Probiotics or pro-healers: The role of beneficial bacteria in tissue repair." *Wound Repair and Regeneration* no. 25 (2017): 912–922.

[81] Kaur, Indu Pal, Kanwaljit Chopra, and Amarpreet Saini. "Probiotics: Potential pharmaceutical applications." *European Journal of Pharmaceutical Sciences* no. 15 (2002): 1–9.

[82] Fang, Yan, and D. Brent Polk. "Probiotic bacterium prevents cytokine-induced apoptosis in intestinal epithelial cells." *Journal of Biological Chemistry* no. 277 (2002): 50959–50965.

[83] Mackowiak, Philip A. "Recycling Metchnikoff: Probiotics, the intestinal microbiome and the quest for long life." *Frontiers in Public Health* no. 1 (2013): 52.

[84] Fong, Winnie, Qing Li, and Jun Yu. "Gut microbiota modulation: A novel strategy for prevention and treatment of colorectal cancer." *Oncogene* no. 39 (2020): 4925–4943.

[85] Liu, Z., H. Qin, Z. Yang, et al. "Randomised clinical trial: The effects of perioperative probiotic treatment on barrier function and post-operative infectious complications in colorectal cancer surgery–a double-blind study." *Alimentary Pharmacology & Therapeutics* no. 33 (2011): 50–63.

[86] Chen, Zhao-Fei, Luo-Yan Ai, Ji-Lin Wang, et al. "Probiotics *Clostridium butyricum* and *Bacillus subtilis* ameliorate intestinal tumorigenesis." *Future Microbiology* no. 10 (2015): 1433–1445.

[87] Khani, Soghra, Hamideh M. Hosseini, Mohammad Taheri, et al. "Probiotics as an alternative strategy for prevention and treatment of human diseases: A review." *Inflammation & Allergy-Drug Targets (Formerly Current Drug Targets-Inflammation &Allergy)(Discontinued)* no. 11 (2012): 79–89.

[88] Fiocchi, Alessandro, Michael D. Cabana, and Maurizio Mennini. "Current use of probiotics and prebiotics in allergy." *The Journal of Allergy and Clinical Immunology: In Practice* no. 10 (2022): 2219–2242.

[89] Wu, Zhiwei, Entezar MehrabiNasab, Poonam Arora, et al. "Study effect of probiotics and prebiotics on treatment of OVA-LPS-induced of allergic asthma inflammation and pneumonia by regulating the TLR4/NF-kB signaling pathway." *Journal of Translational Medicine* no. 20 (2022): 130.

[90] Ambalam, Padma, Maya Raman, Ravi Kiran Purama, et al. "Probiotics, prebiotics and colorectal cancer prevention." *Best Practice & Research Clinical Gastroenterology* no. 30 (2016): 119–131.

[91] Chen, Danfeng, Duochen Jin, Shumin Huang, et al. "Clostridium butyricum, a butyrate-producing probiotic, inhibits intestinal tumor development through modulating Wntsignaling and gut microbiota." *Cancer Letters* no. 469 (2020): 456–467.

[92] Xiao, Jie, Tian Wang, Yi Xu, et al. "Long-term probiotic intervention mitigates memory dysfunction through a novel H3K27me3-based mechanism in lead-exposed rats." *Translational Psychiatry* no. 10 (2020): 25.

[93] Kaur, Harpreet, Yuvraj Singh, Surjeet Singh, et al. "Gut microbiome-mediated epigenetic regulation of brain disorder and application of machine learning for multi-omics data analysis." *Genome* no. 64 (2021): 355–371.

[94] Krautkramer, Kimberly A., Julia H. Kreznar, Kymberleigh A. Romano, et al. "Diet-microbiota interactions mediate global epigenetic programming in multiple host tissues." *Molecular Cell* no. 64 (2016): 982–992.

[95] Vitetta, Luis, Matthew Bambling, and Hollie Alford. "The gastrointestinal tract microbiome, probiotics, and mood." *Inflammopharmacology* no. 22 (2014): 333–339.

[96] Al Theyab, Ahmad, Turki Almutairi, Abdulla M. Al-Suwaidi, et al. "Epigenetic effects of gut metabolites: Exploring the path of dietary prevention of type 1 diabetes." *Frontiers in Nutrition* no. 7 (2020): 188.

[97] Panwar, Harsh, Rashmi HogarehalliMallappa, Batish Virender Kumar, et al. "Probiotics as potential biotherapeutics in the management of type 2 diabetes–prospects and perspectives." *Diabetes/Metabolism Research and Reviews* no. 29 (2013): 103–112.

[98] Esteve, Eduardo, Wifredo Ricart, and Jose-Manuel Fernández-Real. "Gut microbiota interactions with obesity, insulin resistance and type 2 diabetes: Did gut microbiote co-evolve with insulin resistance?." *Current Opinion in Clinical Nutrition & Metabolic Care* no. 14 (2011): 483–490.

[99] Plaza-Díaz, Julio, Patricio Solís-Urra, Fernando Rodríguez-Rodríguez, et al. "The gut barrier, intestinal microbiota, and liver disease: Molecular mechanisms and strategies to manage." *International Journal of Molecular Sciences* 21, no. 21 (2020): 8351.

[100] Licciardi, Paul V., Katherine Ververis, and Tom C. Karagiannis. "Histone deacetylase inhibition and dietary short-chain fatty acids." *International Scholarly Research Notices* no. 26 (2011): 869647.

[101] Hamon, Mélanie Anne, Eric Batsché, Béatrice Régnault, et al. "Histone modifications induced by a family of bacterial toxins." *Proceedings of the National Academy of Sciences* no. 104 (2007): 13467–13472.

[102] Morovic, Wesley, and Charles R. Budinoff. "Epigenetics: A new frontier in probiotic research." *Trends in Microbiology* no. 29 (2021): 117–126.

[103] Berner, Carolin, Eva Aumüller, Anne Gnauck, et al. "Epigenetic control of estrogen receptor expression and tumor suppressor genes is modulated by bioactive food compounds." *Annals of Nutrition and Metabolism* no. 57 (2011): 183–189.

[104] Bloemendaal, Mirjam, Joanna Szopinska-Tokov, Clara Belzer, et al "Probiotics-induced changes in gut microbial composition and its effects on cognitive performance after stress: Exploratory analyses." *Translational Psychiatry* no. 11 (2021): 300.

[105] Licciardi, Paul V., Sook-San Wong, Mimi L.K. Tang, et al. "Epigenome targeting by probiotic metabolites." *Gut Pathogens* no. 2 (2010): 1–5.

[106] Patel, Ravi Mangal, and Patricia W. Lin. "Developmental biology of gut-probiotic interaction." *Gut Microbes* no. 1 (2010): 186–195.

[107] Rotondo, John Charles, Carmen Lanzillotti, Chiara Mazziotta, et al. "Epigenetics of male infertility: The role of DNA methylation." *Frontiers in Cell and Developmental Biology* no. 9 (2021): 689624.

[108] Snigdha, Shikha, Kevin Ha, Paul Tsai, et al. "Probiotics: Potential novel therapeutics for microbiota-gut-brain axis dysfunction across gender and lifespan." *Pharmacology & Therapeutics* no. 231 (2022): 107978.

[109] Sumon, Md Afsar Ahmed, Mohammad Habibur Rahman Molla, Israa J. Hakeem, et al. "Epigenetics and probiotics application toward the modulation of fish reproductive performance." *Fishes* no. 7 (2022): 189.

[110] Gao, Qun, Shuwei Lu, Yuwei Wang, et al. "Bacterial DNA methyltransferase: A key to the epigenetic world with lessons learned from proteobacteria." *Frontiers in Microbiology* no. 14 (2023): 1129437.

[111] Ge, Tongxin, Xiang Gu, Renbing Jia, et al. "Crosstalk between metabolic reprogramming and epigenetics in cancer: Updates on mechanisms and therapeutic opportunities." *Cancer Communications* no. 42 (2022): 1049–1082.

[112] Malmuthuge, Nilusha, and L.L. Guan. "Noncoding RNAs: Regulatory molecules of host–microbiome crosstalk." *Trends in Microbiology* no. 29 (2021): 713–724.

[113] Kumar, Ritesh, and Desirazu N. Rao. "Role of DNA methyltransferases in epigenetic regulation in bacteria." *Epigenetics: Development and Disease* no. 61 (2012): 81–102.

[114] Bermudez-Brito, Miriam, Julio Plaza-Díaz, Sergio Muñoz-Quezada, et al. "Probiotic mechanisms of action." *Annals of Nutrition and Metabolism* no. 61 (2012): 160–174.

13 Epigenetic Drug Discovery and Development from Marine Sources

Rashmi Saxena Pal,[1] Yogendra Pal,[2] Shubham Kumar,[1] Sharma Vedika Vijaypal,[2] Nisha Kumari,[1] and Satya Kumar Avula[3]

[1]School of Pharmaceutical Sciences, Lovely Professional University, Phagwara, Punjab, India
[2]CTIPS, Jalandhar, Punjab, India
[3]Nizwa University, Nizwa, Sultanate of Oman

13.1 INTRODUCTION TO EPIGENETICS AND ITS IMPORTANCE

The study of epigenetics focuses on how environmental factors and human behavior can alter our genetic makeup, as in Figure 13.1. While epigenetic modifications can alter how our body interprets a DNA sequence, they are reversible and do not alter our DNA sequence like genetic modifications do. The Greek *epi* translates to "on top of" or "in addition to." (Jackson et al., 2013; Li, 2021; Phillips and Roth, 2019). The term "epigenetics" was first used in 1942 by British embryologist C. H. Waddington to refer to the differentiation of cells during embryonic development (Nicoglou and Merlin, 2017). Since then, the word has expanded to include all alterations in an organism caused by variations in gene expression, independent of variations in the underlying DNA sequences (Cooper et al., 2003).

Within epigenetics, the chemical changes that occur to the DNA letters and the proteins that interact with DNA alter the levels of gene activation and inactivation. During cell division or from one generation to the next, specific epigenetic alterations transfer from parent to daughter cells. An epigenome is the entirety of all epigenetic modifications within a genome (Cooper et al., 2003; Jeffries, 2020; Cavalli and Heard, 2019).

13.1.1 How Does Epigenetics Work?

The final step in translating DNA sequences into proteins is called gene expression; however, the DNA itself does not precisely specify how this is done. The traits that an organism displays, or its phenotype, are different from its genotype, or its set of genes. Herein lays the role of epigenetics (Blackledge and Klose, 2021; Cooper et al., 2003). We refer to variations in gene expression as epigenetic variants. These changes in an organism's epigenetic makeup might be inherited or brought about by external factors. This explains why two people with the same DNA or genes may not necessarily have the same results (Ilango et al., 2020; Gibney and Nolan, 2010).

Since the DNA sequences of every cell in our body are usually identical, someone else must be in charge of determining which phenotype is expressed in each area of our body. Epigenetic changes activate and deactivate genes, which impacts the synthesis of proteins within cells. This ensures that the various cells in our body only manufacture the proteins necessary for their respective roles. Although the DNA sequences of muscle and bone cells are identical, bone cells do not create the proteins needed for muscle growth. Beyond the confined synthesis of proteins appropriate

 DOI: 10.1201/9781032706511-13

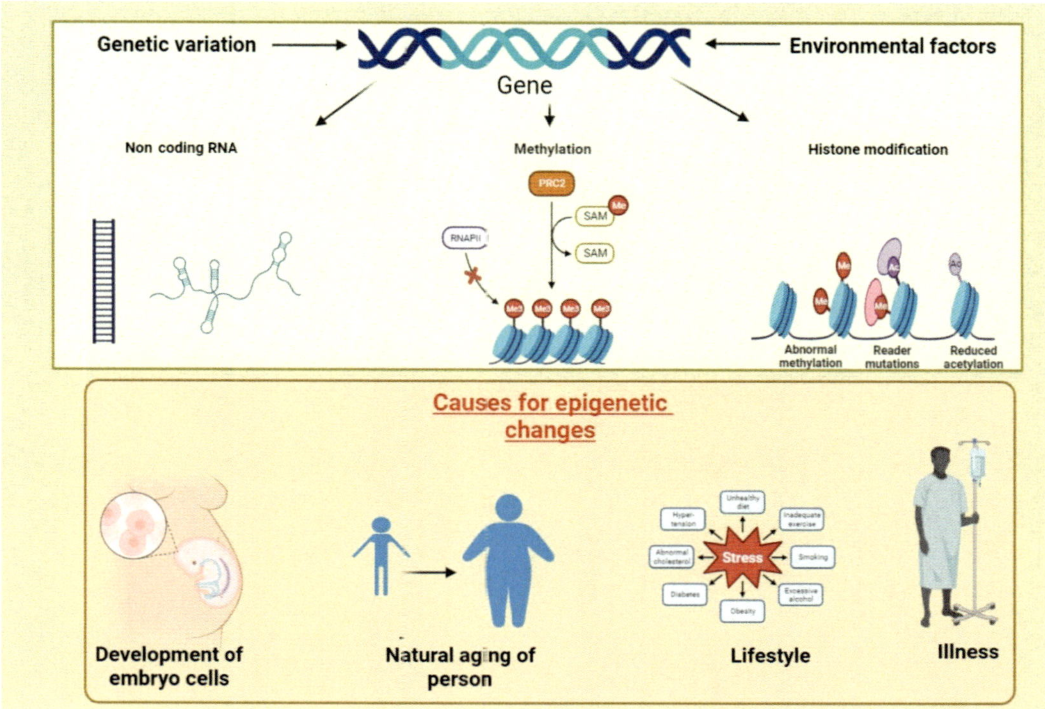

FIGURE 13.1 Causes for epigenetic changes.

for particular cellular tasks, epigenetic changes play a critical role in enabling our fixed DNA to dynamically adapt to our changing surroundings. Changes in epigenetics impact gene expression (Gibney and Nolan, 2010; Lockhart and Winzeler, 2000; Jeffries, 2020).

13.1.2 DNA Methylation

DNA is methylated by introducing a chemical group, usually to specific regions of DNA, as in Figure 13.2, where it obstructs the ability of proteins to bind to DNA and "read" genes. Demethylation eliminates this chemical group. Genes are normally turned "on" by demethylation and "off" by methylation. Methylation adds methyl groups to DNA moleculesr to either suppress or change genes. Generally, they set genes off-kilter, so to speak. Gene activation can also result from demethylation (Venkatesagowda and Dekker, 2021; Moore and Houchen, 2023).

13.1.3 Histone Modification

Proteins called histones aid DNA in the formation of nucleosomes, which eventually transform into chromatin. It controls the accessibility of certain regions of the DNA because it provides structural support to the chromatin. Thus, changes to histones can impact the expression of genes. Histone acetylation is a frequent histone modification (Razin et al., 2007; Saha et al., 2006; Peterson and Laniel, 2004). It involves activating specific genes by adding acetyl groups to histone tails, which increases DNA accessibility (Razin et al., 2007; Shvedunova and Akhtar, 2022). Histone proteins are encircled by DNA. A closely packed set of histones prevents proteins that are meant to read a gene from efficiently accessing the DNA, thereby turning the gene off. More DNA is exposed or not encased in a histone when histones are loosely packed, allowing proteins to read the gene and access it (Shvedunova and Akhtar, 2022; Talukdar and Chatterji, 2023; Azargoonjahromi, 2023).

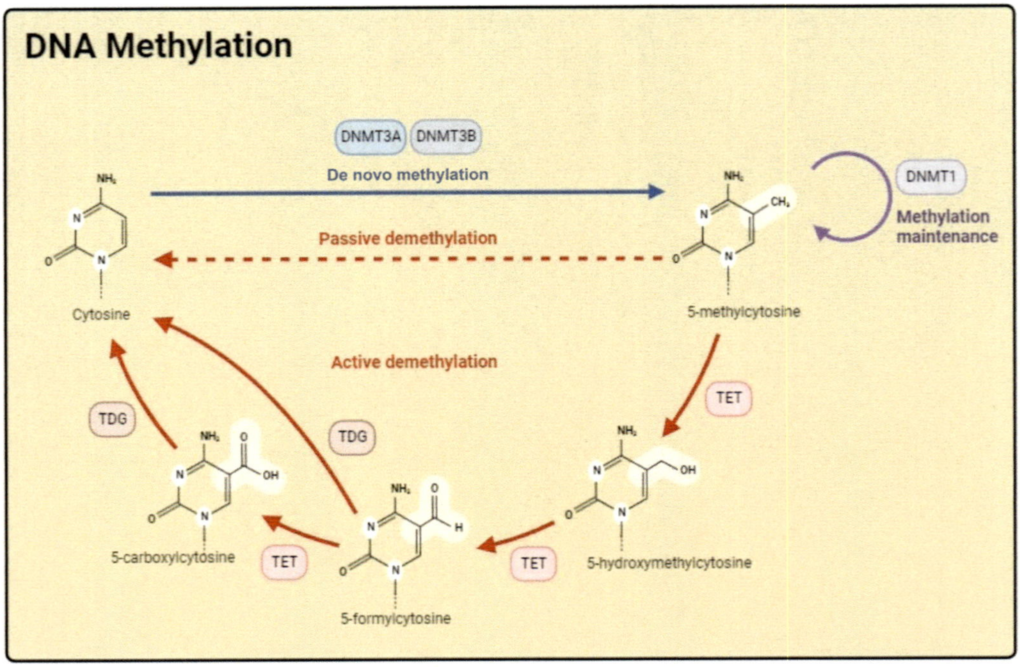

FIGURE 13.2 DNA methylation.

13.1.4 Noncoding RNA

An essential component of an organism's gene expression is RNA. Noncoding RNAs (ncRNAs) are RNA molecules that do not transcribe into proteins. They play a major role in generating differences in the epigenome (Morselli and Dieci, 2022). The instructions for creating both coding and ncRNAs are found in our DNA. Proteins are created using coding RNA. By binding to coding RNA and some proteins to degrade the coding RNA, as in Figure 13.3, so that it can no longer be used to generate proteins, ncRNAs regulate gene expression. Additionally, ncRNAs recruit proteins that change histones to turn genes on or off (Morselli and Dieci, 2022; Dindhoria et al., 2022; Wei et al., 2017; Costa, 2008).

13.2 CAUSES OF EPIGENETIC CHANGES

The mechanisms behind epigenetic variation are now well understood, but what exactly causes epigenetic modifications, particularly in humans?

- Development: Cells in an embryo mature and differentiate into cells with distinct roles based on epigenetic changes.
- Age: Epigenetic modifications take place as a person naturally ages.
- Lifestyle: Epigenetic modifications can be brought about by modifications to an individual's behavior or environment.
- Illness: Our bodies exhibit epigenetic changes as a result of infection, as in Figure 13.4. A person's vulnerability to diseases like cancer may be increased by specific epigenetic variables, and cancer can also bring on epigenetic anomalies (Atlasi and Stunnenberg, 2017; Calvanese et al., 2009; Abdul et al., 2017; El-Sayed et al., 2021).

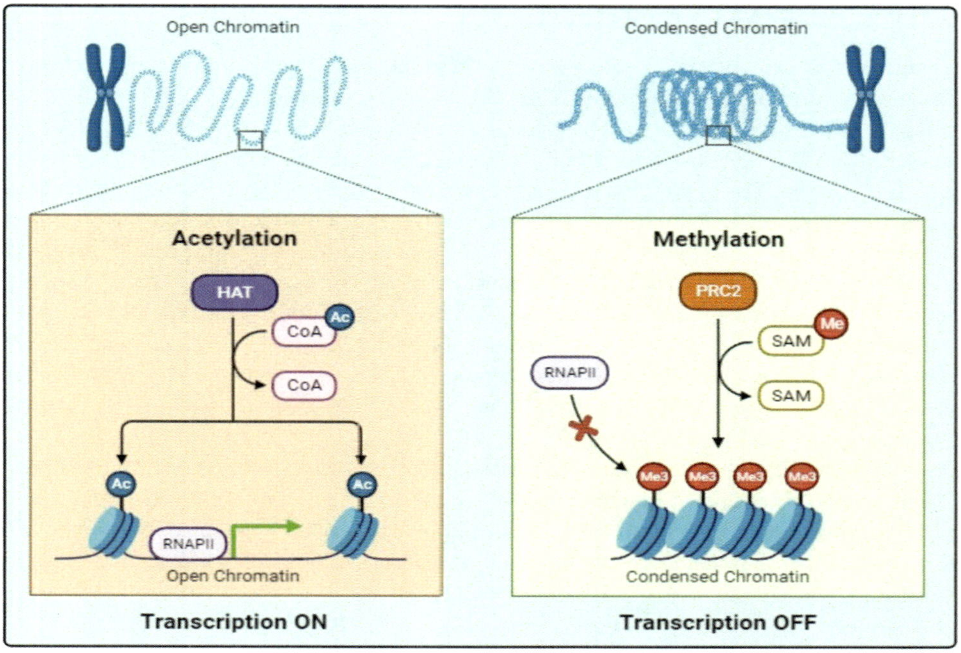

FIGURE 13.3 Regulation of gene expression.

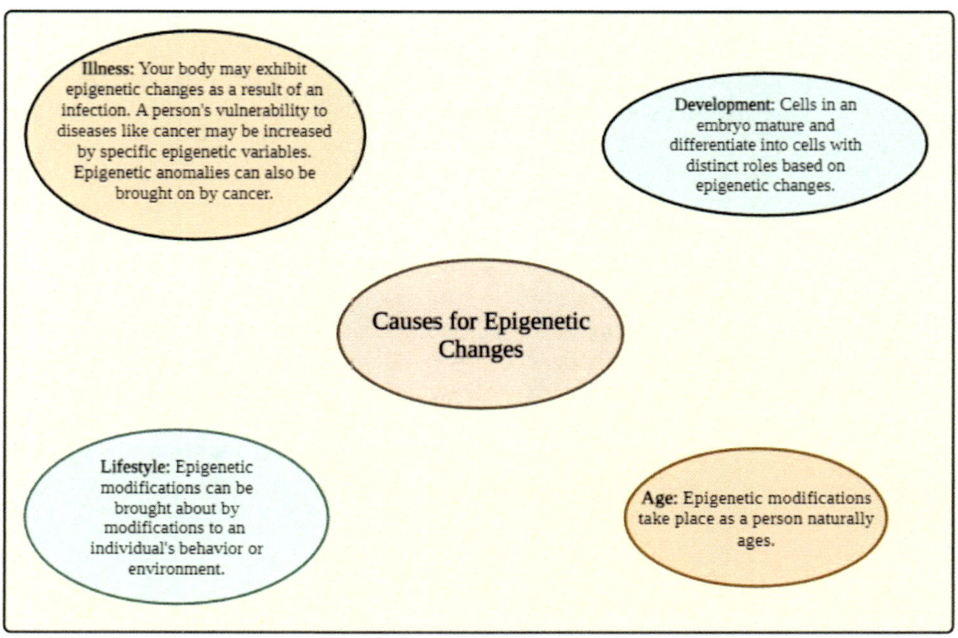

FIGURE 13.4 Causes of epigenetic changes.

13.3 THE IMPORTANCE OF EPIGENETICS

The dynamic process of epigenetic changes reflects the intricate interactions that exist between an organism and its surroundings. For instance, methylation of Lys9 or Lys27 of histone H3 is associated with gene repression, whereas acetylation of lysine residues in histone subunits typically boosts gene transcription. On the other hand, Lys4, Lys36, or Lys79 methylation on H3 is typically linked to gene activity. About 30% of gene expressions are heritable; however, in cells cultured in a lab, this ratio drops to roughly 23%. Numerous organisms, organs, and populations have shown evidence of similar relationships between methylation levels and genetic variants (Migliore and Coppedè, 2022; Miller, 2023; Ajoolabady et al., 2022; Biggar and Li, 2015).

Research in rats has shown that DNA methylation is necessary for hippocampus-dependent contextual learning to form long-lasting fear-related memories. By inhibiting DNA methylation, DNMT inhibitors have the potential to eliminate these memories. DNA methylation plays a major part in the long-term memory storage in cortical brain regions, and histone acetylation and DNA methylation affect the creation of memories (Day and Sweatt, 2011; Sase et al., 2019; Lubin et al., 2011).

A fascinating finding in recent years has been the involvement of methylation in the biological clock. Analysis of 353 epigenetic markers CpG (cytosine-phosphate-guanine) dinucleotide methylation in DNA allowed for predicting tissue aging. Additionally, research on the areas around the Oct4 promoter showed the significance of methylation and acetylation during differentiation (Horvath and Raj, 2018; Unnikrishnan et al., 2019; Bell et al., 2019).

13.3.1 Epigenetics and the Environment

The epigenome refers to the total set of ncRNAs and epigenetic alterations at any given time, including DNA methylation and histone modifications. The epigenome of a cell is dynamic and susceptible to both environmental and genetic influences. Moreover, epigenetic changes have the potential to be reversed, giving the genome flexibility to adapt to changes in the environment brought about by things like diet, stress, toxins, exercise, and medications (Villota-Salazar et al., 2016; Lian et al., 2018; Wattacheril et al., 2023; Anastasiadi and Beemelmanns, 2023).

Food supplies were limited during the winter of 1944–1945 in World War II as a result of German obstructions to food transportation, and the frozen water canals made food delivery by boat unfeasible. Entire communities were impacted by famine and inadequate protein and vitamin intake, but pregnant women were particularly hard hit. Subsequent research on neonates and others experiencing similar circumstances indicated that people who were born after the famine were more vulnerable to certain conditions, including obesity, stress sensitivity, and schizophrenia (Van Der Straeten et al., 2020; Brown, 2011). Separately, folate is an important nutrient for methylation. Through homocysteine remethylation, folate affects the synthesis of methionine by forming 5-methyltetrahydrofolate. According to reports, a folate deficiency or shortage can promote the development of CRC by hypomethylating genomic DNA (Abbasi et al., 2018; Menezo et al., 2022).

One significant environmental factor is stress. Recent research has shown that individuals with post-traumatic stress disorder who experienced childhood abuse display distinct degrees of DNA methylation and gene expressions from those of their non-abused counterparts. Additionally, there is evidence linking maternal stress during pregnancy to psychiatric and neurodevelopmental issues. Fetuses exposed to stress throughout pregnancy demonstrate a susceptibility to later psychiatric problems due to elevated glucocorticoid receptor promoter activity (Watkeys et al., 2018; McCrory et al., 2010; Palma-Gudiel et al., 2015).

13.3.2 Epigenet.ics and Human Diseases

One frequent and extensively utilized method for altering the epigenetic makeup of cells is methylation. Numerous human illnesses, including as various malignancies, autoimmune diseases, and neurological conditions (such as fragile X syndrome, Parkinson's, Alzheimer's, and schizophrenia), have been linked to it. Methylation plays a role in complex disorders, with severity influenced by secondary factors such as age and sex differences (Witoelar et al., 2017; Moosavi and Ardekani, 2016).

13.3.2.1 Cancer

Cancer is significantly impacted by epigenetic changes. Numerous tumor suppressor functions can be rendered inactive by hypermethylation of promoter regions of tumor suppressor genes. Additionally, methylation levels are crucial for cell division, drug resistance, angiogenesis, apoptosis, differentiation, growth factor response, detoxification, and apoptosis (Oleksiewicz and Machnik, 2022; Mesgari et al., 2023).

13.3.2.2 Autoimmune Diseases

Self-tolerance is necessary for the immune system to operate normally and naturally, and a lack of it can lead to autoimmunity. Studies on the concordance of autoimmune diseases in both monozygotic and dizygotic twins have indicated the involvement of epigenetic variables. Failure to maintain epigenetic homeostasis in response to environmental stimuli alters the expression of certain genes in particular differentiated cells, which can lead to deregulated self-tolerance (Fan, 2023; Lleo, 2023; Land, 2023).

13.3.2.3 Neurodegenerative and Psychological Disorders

The specialized sequence segments known as imprinting control regions (ICRs) are responsible for the parental allele-specific gene expressions along imprinted domains. ICRs, which only appear on one of the parental copies, control allele-specific gene expression. Histone methylation and acetylation at ICRs have also been documented, despite the fact that DNA methylation is the epigenetic modification at ICRs that has been studied the most (Matsuzaki et al., 2015; Strogantsev et al., 2015; Kim et al., 2009).

13.4 MARINE SOURCES IN EPIGENETIC DRUG DISCOVERY AND DEVELOPMENT

As a defense mechanism against the outside world, a large proportion of marine animals have evolved to synthesize physiologically active compounds known as secondary metabolites (Benkendorff, 2010; Giordano et al., 2015). The epigenome and the disease-associated epigenetic machinery are modulated by these natural compounds and their derivatives. Seaweeds, corals, and sponges are examples of sessile creatures that are constantly adapting to the biotic and abiotic elements of their environment (Conte et al., 2020). Different types of competition are frequently the outcome of this incredibly dynamic and complicated process, which aims to maintain an ecological niche that is conducive to life.

Epigenetic alterations give marine animals a competitive edge by mediating the creation of complex chemical compounds that have therapeutic consequences, serving as a form of environmental adaptation. By controlling important transcriptional factors implicated in cancer hallmarks through coordinated molecular pathways, bioactive substances can disrupt epigenetic targets and establish tumor microenvironment signaling relationships critical to cancer phenotypes. One crucial way that marine creatures adapt is by producing particular secondary metabolites (Conte et al., 2020; Arora et al., 2020; Norouzitallab et al., 2019).

The last ten years have witnessed a fast expansion in the field of epigenetics due to a move away from lab-based studies and toward real-world implementations, such as fieldwork and experiments. The development of molecular technology has allowed for the affordable analysis of epigenetic changes in a variety of organisms, which has facilitated this progress. This expansion has also been aided by better knowledge of genetic–epigenetic interactions and epigenetic mechanisms in ecologically significant species (Whitehouse, 2021; Li, 2021; Vogt, 2017).

In turn, the marine environment—which is made up of both biotic and abiotic elements—plays a fundamentally selective role in the intrinsic changes that organisms go through. It does this by inducing genetic, phenotypic, and epigenetic alterations that affect species over generations. Because environmental exposures in the marine ecosystem can trigger the reprogramming of epigenetic states, secondary metabolites produced by a multitude of organisms may be promising candidates for novel natural compounds with potential pharmacological activity for the treatment of cancer. Several new marine-derived compounds, some of which have potential anticancer activities, were isolated and characterized using state-of-the-art chromatographic isolation and purification, pharmacological screening, mass spectroscopy and nuclear magnetic resonance (NMR) (Harvey et al., 2014; Kansara et al., 2022; Zompra et al., 2022).

The epigenome of the organism and any possible epigenetic effects are greatly influenced by the chemical makeup of isolated molecules, which reflects a sophisticated and interdependent network of information exchange. The present review delineates the therapeutic potential of marine-derived secondary metabolites and their synthetic derivatives in the context of cancer therapy. Emphasis is placed on their role as epigenetic modulators, which yield modifications that are posttranscriptional, inductive (arising from the organism's metabolism), and induced (arising from changes in the marine environment). We also talk about the difficulties in finding novel marine biocompounds with anticancer properties, both synthetic and natural, considering the great diversity of the species and their surroundings (Whitehouse, 2021; Li, 2021; Vogt, 2017).

But the main driver is the increasing attention that ecologists, evolutionary biologists, environmental scientists, taxonomists, and others are paying to the role that epigenetics plays in the generational variability of phenotype. These mechanisms include developmental effects, life-history features, pathogenicity, aging, and more, and they extend beyond the body's reaction to external stimuli. This demonstrates the essential function of epigenetic control in influencing gene expression and delivering environmental cues to the genome (Burggren, 2016; Jablonski, 2017).

With these important elements in place and the rate of progress continuing, environmental epigenetics is set up for continued rapid growth that shows no signs of abating (Lock, 2015).

Zhao et al. begin their studies with marine algae, providing a thorough examination of the roles played by proteins involved in chromatin remodeling (polycomb and trithorax groups). For their research, they used *Phaeodactylum tricornutum* as a model diatom. The authors thoroughly examined histone marks associated with these proteins throughout the entire genome. Crucially, their research highlights the long history of these regulatory systems (Zhao et al., 2020; Veluchamy et al., 2015; Rastogi, 2016).

13.5 HIGH-THROUGHPUT SCREENING IN EXTRACTION TECHNOLOGIES IN EPIGENETIC DRUG DISCOVERY

High-throughput screening is a widely used technique in chemistry, biology, and materials science and is particularly useful in drug development. A researcher can perform millions of experiments quickly using robotics, data processing/control software, liquid handling devices, and sensitive detectors (Ismail et al., 2018; Prabhu and Urban, 2020).

Using HT-29/DEFB1-luc, a stable reporter cell line expressing the luciferase gene driven by a 3-Kb DEFB1 gene promoter, researchers developed a cell-based high-throughput screening assay to identify the compounds that specifically activate human β-defensin 1 (DEFB1) gene transcription. After 148 small-molecule epigenetic drugs were screened, 28 hits with a strictly standardized mean difference of at least 3.0 were found. After additional screening, 14 compounds produced DEFB1 mRNA in human HT-29 colonic epithelial cells. These epigenetic substances also stimulated the human cathelicidin antimicrobial peptide gene (Lyu et al., 2023; Wei et al., 2019).

It is common to screen in vitro tests against libraries of synthetic or natural compounds to determine the chemical starting points of various medications. These assays can be conducted in microtiter plates and can be biochemical (using purified protein) or cell-based (using primary cells, genetically modified cancer cell lines, or other cells). This allows for testing many samples. Researchers have described assays that monitor activities of histone deacetylase, lysine-specific demethylase, histone methyltransferase, histone acetyltransferase, and bromodomain. A significant number of such assays are available to monitor epigenetic target activity. Prior to screening in a particular therapeutic target, it is crucial that a suitable assay be devised and thoroughly validated (Wang et al., 2007; Ganesan, 2018; Gul, 2017).

13.6 EPIGENETIC MARINE MOLECULES

13.6.1 Polyphenols

One of the primary groups of secondary metabolites derived from marine sources is polyphenols. These compounds are present in a variety of pelagic creatures, with their production varying across taxa based on growth conditions, geographic distribution, and abiotic/biotic variables. The presence of one or more aromatic rings—phenolic acids or polyphenols—allows these molecules to be identified. Their bioactivity has been connected to favorable impacts on human health, including antidiabetic, anti-inflammatory, and anticancer benefits, as well as other enzymatic inhibitory effects. Their capacity to scavenge singlet oxygen and free radicals, as well as to act as reducing and chelating agents, is also a very promising field for cancer study and therapy, given their intriguing epigenetic molecular pathways that regulate DNA damage and gene expression (Hastings et al., 2007; Napiórkowska-Krzebietke et al., 2023; Tatipamula and Kukavica, 2021; Zhang et al., 2021).

13.6.2 Cyclic Peptides

Structured peptides with distinct characteristics that can control cancer's epigenetic pathways. Most of these substances are naturally occurring, and their ring-shaped backbone has been utilized to create innovative synthetic compounds that are more targeted and potent medicines (Sinha et al., 2021). Depsipeptides and cyclic tetrapeptides (CTPs) as epigenetic-like anticancer treatments are the main topics of discussion in this area. The nonribosomal peptides known as depsipeptides have one or more amine bonds substituted with the appropriate ester. These compounds are prevalent in the marine environment and often contain amino acids that are not found in proteins (Conte et al., 2020; Duban et al., 2022).

Their synthesis is quite simple and could result in the creation of multiple structural combinations that are helpful in determining which anticancer drugs are the most potent. CTPs, in contrast to depsipeptides, are incredibly challenging to synthesis because of their intricate structure, which consists of four amino acids joined by eupeptide bonds. Particularly, the l-, d-, and cyclic amino acids included in CTPs lessen their cyclic stress. Numerous biological techniques have been developed alongside in-depth investigations of three-dimensional structures using NMR and X-ray crystallography (Conte et al., 2020; Mota-Martorell et al., 2021).

13.6.3 Alkaloids

Natural substances with a nitrogen–heterocyclic structure are known as alkaloids. Particularly, marine creatures including sponges, green, brown, and red algae, coelenterates, and tunicates are the primary sources of marine alkaloids, which are compounds with an amine nitrogen group and a carbon ring. Acting as anticancer, antiviral, antimalarial, antifungal, and anti-osteoporosis medicines, these metabolites exhibit a variety of characteristics. Marine alkaloids can be employed as lead chemicals for structural alteration or as chemotherapeutics (Conte et al., 2020; Zhou and Huang, 2020).

Among the potential naturally occurring compounds known as brominated alkaloids is isofistularin-3 (Iso-3), which comes from the sponge *Aplysina aerophoba*. This molecule shares structural characteristics with PsA, a well-known derivative of bromotyrosine (Conte et al., 2020; Elmallah and

Micheau, 2019). Cyclostellettamines are a class of chemicals containing a macrocyclic ring of the precursor bispyridinium alkaloid that are obtained from marine sources. Cyclostellettamines A and G, along with dehydrocyclostellettamines D and E, isolated from the sea sponge Xestospongia, functioned as HDAC inhibitors in the myelogenous leukemia K562 cell line (Conte et al., 2020; Bumpers, 2021).

13.7 BIOLOGICAL MECHANISMS OF MARINE EPIGENETIC MOLECULES

Trichostatin A (TSA) is a hydroxamic acid that was first discovered from the soil-dwelling bacterium *Streptomyces platensis*. It has a wide range of reversible HDAC inhibitory actions in addition to antifungal, antibacterial, and antineoplastic properties. Data analysis revealed that BRAF-mutated melanoma cells contained a number of up- and downregulated transcripts, demonstrating that TSA significantly altered the transcriptome. In particular, TSA inhibited the BRAF and HDAC pathways while downregulating the MAPK/MEK/BRAF axis. Class III HDAC enzyme sirtuin 6 is involved in a number of epigenetic-like processes, including blood glucose management, stress tolerance, DNA repair, and gene silencing (Acquah, 2021; Conte et al., 2020; Zhou and Huang, 2020; Zhou et al., 2023).

Azumamides derive from the Japanese marine sponge *Mycale izuensis*; they are a class of CTPs that have five isoforms (A, B, C, D, and E). Azumamides A–E, which are distinguished by four nonribosomal amino acid residues—three of which are D-series amino acids and only one of which is a beta amino acid—were the first cyclic peptides with HDAC inhibitory activity to be isolated from marine animals (Conte et al., 2020; Schnekenburger et al., 2014). Azumamides A–E was initially discovered to be strong HDAC inhibitors in a chronic myeloid leukemia cell line after HDAC activity was assessed in vitro (Cappellacci et al., 2020).

In particular, azumamide A produced lethal effects in chronic leukemia and colon cancer cells while inducing histone H3 acetylation at increasing doses (Schnekenburger et al., 2014). Human cancer cell lines showed an HDAC inhibitory effect with IC_{50} values for azumamide versions B, C, and E. Due to its distinct chemical structure, derivate E—containing a carboxylic acid—exhibited stronger affinity for the HDAC active site's zinc ion than the amide groups in azumamides A, B, and D, making it the most potent inhibitor. Only azumamide E induced p21 overexpression in mouse-induced pluripotent stem cells, a well-known marker that controls cell cycle progression (Ganai, 2023; Newkirk et al., 2009).

Trapoxin, derived from *Helicoma ambiens*, is a homodetic cyclic tetrapeptide HDAC inhibitor produced by fungi. As an epoxyketone, it irreversibly inhibits class I HDACs because of the structural similarities between its ketone's carbonyl group and that of the HDAC substrate, acetyl-L-lysine. An X-ray study revealed that trapoxin A binds to HDAC8, confirming its role as a noncovalent HDAC8 inhibitor. Its α, β-epoxyketone side chain interacts with the zinc-containing active site through a chemical transition. A hybrid substance derived from trapoxin A and TSA, cyclic hydroxamic acid-containing peptide 1 replaces the epoxyketone group with hydroxamic acid (Conte et al., 2020; Janssens et al., 2019).

13.8 MARINE NANOTECHNOLOGY

The process of modifying matter at a size close to the atomic level to create novel structures, materials, and gadgets is known as nanotechnology. Science is advancing thanks to this technology in a number of fields, including manufacturing, consumer goods, energy, materials, and medicine. Under conditions of ocean acidification, copepods exhibit epigenetic markers of trans-generational deficits linked to reproduction. The advancement of nanoscale scientific development has led to several technological improvements that help the marine industry. The international network of ports, ships, and shipping routes in the marine industry is one of the earliest capillary systems for the globalized movement of products, and nanotechnology is widely used in most of the world's economy today (Kaul and Sharma, 2022; Langer et al., 2019; Kong and Ohadi, 2010).

13.9 PHARMACOKINETIC CHALLENGES IN EPIGENETICS

Not all of the documented heritable variance in medication response can be explained by pharmacogenetics, the study of how genetic variations between individuals affect drug reaction. Histone acetylation and DNA methylation are two epigenetic processes that could explain some of the unexplained variations (Vaissière et al., 2008; Ghavifekr Fakhr et al., 2013).

Gene expression is modulated by epigenetic mechanisms that also serve as potential therapeutic targets and affect how nonepigenetic medications work (Kelly et al., 2010). The study of the connection between pharmacological response and epigenetic variability is known as pharmacoepigenetics. A significant portion of this research focuses on epigenetic drugs, which target epigenetic systems and are used to treat diseases such as cancer and immune system problems. Moreover, studies point to an epigenetic component in the traditional pharmacological response; nonetheless, little is known in this field. We suggest that by (1) identifying regions of epigenetic variance, (2) identifying important epigenetic biomarkers within these regions, and (3) mapping these biomarkers to a drug-response phenotype, the link between epigenetics and classical drug response can be investigated using data already accessible. By building on current understanding, this method produces potential pharmacoepigenetics linkages that may be put to the test through experiments (Cascorbi, 2013; Cavalli and Heard, 2019; Ilango et al., 2020; Li, 2021).

Although pharmacoepigenetics research in the cancer setting presents significant challenges, epigenetic pathways play a crucial role in cancer progression. Because the cancer epigenome differs from the somatic epigenome of the host and because cancer cells are constantly changing, it is challenging to detect genetic or epigenetic alterations at the population level in cancer cells. Moreover, monitoring which markers are basal to the patient tissue, cancer specific, or the consequence of a pharmaceutical activity might be difficult in the context of current treatment. The current clinical applications of epigenetic profiles are limited by the tissue specificity of epigenetic signals, which also makes it difficult to gather disease-relevant samples from patients through blood draws. However, the prognosis and responsiveness to cancer drugs are influenced by epigenetic markers experiments (Cascorbi, 2013; Cavalli and Heard, 2019; Ilango et al., 2020; Li, 2021).

It is not only difficult to determine the causal links between changes in the epigenetic state and a particular phenotype in cancer. Epigenetic markers are influenced by medications, illness states, and environmental factors. Drug responsiveness and disease prognosis can also be affected by marker levels. Furthermore, an individual's epigenetic states are dynamic and vary throughout the course of their lifespan. As a result, it's not always easy to pinpoint the origin and significance of links that have been found between epigenetic markers, diseases, and medication reactions. Twin studies are sometimes employed to establish causation, but there is still much to learn about the directionality of the connection between medication response and epigenetics (Cavalli and Heard, 2019; Cascorbi, 2013; Jackson et al., 2013; Kelly et al., 2010).

13.10 OMICS IN EPIGENETIC MARINE DRUG DISCOVERY

Omics—a branch of the biological sciences that includes terms ending in -omics, such as proteomics, phenomics, and genomics—refers to a set of technologies used to increase accuracy and efficacy. Modern omics technologies offer a variety of data and aid in understanding the biological processes underlying drug activity. Drug toxicology is just one of the many biological fields that greatly benefit from this contemporary technology. Better bioinformatics tools and increased availability of sequencing data enable emerging methodologies for multi-omics, which combine data on the genome, the transcriptome (the RNA that codes for the proteins), and the proteome (all the proteins) (Whilde et al., 2017; Matthews et al., 2016; Hasin et al., 2017; Subramanian et al., 2020).

13.11 CONCLUSION

In sum, the study of epigenetics has revealed the complex ways in which human behavior and the environment influence our genetic make-up without changing the underlying DNA sequence. Epigenetic changes, such as DNA methylation, histone modification, and regulation of non-coding RNA, control gene expression and affect an organism's phenotype. Since its introduction in 1942 by C. H. Waddington, the term epigenetics has come to refer to any alteration in an organism resulting from differences in gene expression.

Numerous biological processes, including cellular differentiation during embryonic development and DNA's dynamic adaptation to changing surroundings, depend heavily on epigenetic modifications. Because epigenetic alterations are reversible, there is a unique therapeutic avenue that can be explored for controlling gene expression without changing the DNA sequence. Recognizing the various causes of epigenetic modifications, such as maturation, ageing, way of life, and disease, has revealed their extensive influence on human well-being. Epigenetic modifications affecting susceptibility to diseases and conditions have been related to long-lasting consequences, including stress, nutrition, and even starvation during conflict.

13.12 PROSPECTS

Human diseases including cancer, autoimmune diseases, and neurodegenerative disorders are significantly impacted by epigenetics. By deactivating tumor suppressor genes, methylation—a frequent epigenetic modification—plays a critical role in cancer. Marine sources are a promising avenue for the discovery of new epigenetic medicines because of their enormous biodiversity. Through the modification of epigenetic pathways, compounds produced from marine species, including as polyphenols, cyclic peptides, and alkaloids, show potential anticancer effects. The development of extraction methods combined with high-throughput screening has expedited the search for epigenetic medicines. These methods show promise for tailored therapeutics and precision medicine, especially when combined with the rapidly developing science of nanotechnology. However, further research is required due to difficulties with pharmacokinetics and comprehending the intricate interactions between epigenetics and medication response.

The precision and efficiency of finding epigenetic drugs are improved by integrating omics technologies, such as proteomics and genomics. The prospects for the future of epigenetics point to the creation of targeted treatments that are more selective for cancer cells. The reversibility of epigenetic alterations, which opens new avenues for therapeutic interventions as our understanding grows, heralds a revolutionary era in medicine where genetics and environment meet. Future developments in technology and interdisciplinary cooperation will support the rise of environmental epigenetics, which will ensure that novel medical treatments with significant effects on human health will be driven by epigenetic discoveries.

REFERENCES

Abbasi, I. H. R., Abbasi, F., Wang, L., Abd El Hack, M. E., Swelum, A. A., Hao, R., Yao, J. & Cao, Y. 2018. Folate promotes S-adenosyl methionine reactions and the microbial methylation cycle and boosts ruminants production and reproduction. *AMB Express*, 8, 1–10.

Abdul, Q. A., Yu, B. P., Chung, H. Y., Jung, H. A. & Choi, J. S. 2017. Epigenetic modifications of gene expression by lifestyle and environment. *Archives of Pharmacal Research*, 40, 1219–1237.

Acquah, K. S. 2021. *Exploring the Molecular Diversity and Biomedicinal Potential of Marine Invertebrates and South African Actinomycetes for Tuberculosis Drug Discovery.* https://open.uct.ac.za/items/2aef549e-c19a-4221-9936-1ae67abe66ca

Ajoolabady, A., Aslkhodapasandhokmabad, H., Zhou, Y. & Ren, J. 2022. Epigenetic modification in alcohol-related liver diseases. *Medicinal Research Reviews*, 42, 1463–1491.

Anastasiadi, D. & Beemelmanns, A. 2023. Development of epigenetic biomarkers in Aquatic organisms. *Epigenetics in Aquaculture*, 413–433.

Arora, I., Sharma, M., Sun, L. Y. & Tollefsbol, T. O. 2020. The epigenetic link between polyphenols, aging and age-related diseases. *Genes*, 11, 1094.

Atlasi, Y. & Stunnenberg, H. G. 2017. The interplay of epigenetic marks during stem cell differentiation and development. *Nature Reviews Genetics*, 18, 643–658.

Azargoonjahromi, A. 2023. The role of epigenetics in anxiety disorders. *Molecular Biology Reports*, 1–12.

Bell, C. G., Lowe, R., Adams, P. D., Baccarelli, A. A., Beck, S., Bell, J. T., Christensen, B. C., Gladyshev, V. N., Heijmans, B. T. & Horvath, S. 2019. DNA methylation aging clocks: Challenges and recommendations. *Genome Biology*, 20, 1–24.

Benkendorff, K. 2010. Molluscan biological and chemical diversity: Secondary metabolites and medicinal resources produced by marine molluscs. *Biological Reviews*, 85, 757–775.

Biggar, K. K. & Li, S. S.-C. 2015. Non-histone protein methylation as a regulator of cellular signalling and function. *Nature Reviews Molecular Cell Biology*, 16, 5–17.

Blackledge, N. P. & Klose, R. J. 2021. The molecular principles of gene regulation by Polycomb repressive complexes. *Nature Reviews Molecular Cell Biology*, 22, 815–833.

Brown, A. S. 2011. The environment and susceptibility to schizophrenia. *Progress in Neurobiology*, 93, 23–58.

Bumpers, Q. 2021. *Synthetic Studies Toward the Total Synthesis of Xestocyclamine A*. https://ir.vanderbilt.edu/bitstreams/be9b57ca-80ee-4a34-837c-691699 1f0f34/download

Burggren, W. 2016. Epigenetic inheritance and its role in evolutionary biology: Re-evaluation and new perspectives. *Biology*, 5, 24.

Calvanese, V., Lara, E., Kahn, A. & Fraga, M. F. 2009. The role of epigenetics in aging and age-related diseases. *Ageing Research Reviews*, 8, 268–276.

Cappellacci, L., Perinelli, D. R., Maggi, F., Grifantini, M. & Petrelli, R. 2020. Recent progress in histone deacetylase inhibitors as anticancer agents. *Current Medicinal Chemistry*, 27, 2449–2493.

Cascorbi, I. 2013. Overlapping effects of genetic variation and epigenetics on drug response: Challenges of pharmacoepigenomics. *Pharmacogenomics*, 14, 1807–1809.

Cavalli, G. & Heard, E. 2019. Advances in epigenetics link genetics to the environment and disease. *Nature*, 571, 489–499.

Conte, M., Fontana, E., Nebbioso, A. & Altucci, L. 2020. Marine-derived secondary metabolites as promising epigenetic bio-compounds for anticancer therapy. *Marine Drugs*, 19, 15.

Cooper, T. F., Rozen, D. E. & Lenski, R. E. 2003. Parallel changes in gene expression after 20,000 generations of evolution in Escherichia coli. *Proceedings of the National Academy of Sciences*, 100, 1072–1077.

Costa, F. F. 2008. Non-coding RNAs, epigenetics and complexity. *Gene*, 410, 9–17.

Day, J. J. & Sweatt, J. D. 2011. Cognitive neuroepigenetics: A role for epigenetic mechanisms in learning and memory. *Neurobiology of Learning and Memory*, 96, 2–12.

Dindhoria, K., Monga, I. & Thind, A. S. 2022. Computational approaches and challenges for identification and annotation of non-coding RNAs using RNA-Seq. *Functional & Integrative Genomics*, 22, 1105–1112.

Duban, M., Cociancich, S. & Leclère, V. 2022. Nonribosomal peptide synthesis definitely working out of the rules. *Microorganisms*, 10, 577.

Elmallah, M. I. & Micheau, O. 2019. Epigenetic regulation of TRAIL signaling: Implication for cancer therapy. *Cancers*, 11, 850.

El-Sayed, A., Aleya, L. & Kamel, M. 2021. The link among microbiota, epigenetics, and disease development. *Environmental Science and Pollution Research*, 28, 28926–28964.

Fan, J. 2023. Investigation of DNA methylation in autoimmune diseases with genome-wide association analysis. *International Conference on Modern Medicine and Global Health (ICMMGH 2023)*. SPIE, 295–300.

Ganai, S. A. 2023. *Natural HDAC Inhibitors for Epigenetic Combating of Cancer Progression*. CRC Press.

Ganesan, A. 2018. Epigenetic drug discovery: A success story for cofactor interference. *Philosophical Transactions of the Royal Society B: Biological Sciences*, 373, 20170069.

Ghavifekr Fakhr, M., Farshdousti Hagh, M., Shanehbandi, D. & Baradaran, B. 2013. DNA methylation pattern as important epigenetic criterion in cancer. *Genetics Research International*, 2013.

Gibney, E. & Nolan, C. 2010. Epigenetics and gene expression. *Heredity*, 105, 4–13.

Giordano, D., Coppola, D., Russo, R., Denaro, R., Giuliano, L., Lauro, F. M., Di Prisco, G. & Verde, C. 2015. Marine microbial secondary metabolites: Pathways, evolution and physiological roles. *Advances in Microbial Physiology*, 66, 357–428.

Gul, S. 2017. Epigenetic assays for chemical biology and drug discovery. *Clinical Epigenetics*, 9, 41.

Harvey, B. P., Al-Janabi, B., Broszeit, S., Cioffi, R., Kumar, A., Aranguren-Gassis, M., Bailey, A., Green, L., Gsottbauer, C. M. & Hall, E. F. 2014. Evolution of marine organisms under climate change at different levels of biological organisation. *Water*, 6, 3545–3574.

Hasin, Y., Seldin, M. & Lusis, A. 2017. Multi-Omics approaches to disease. *Genome Biology*, 18, 1–15.

Hastings, A., Byers, J. E., Crooks, J. A., Cuddington, K., Jones, C. G., Lambrinos, J. G., Talley, T. S. & Wilson, W. G. 2007. Ecosystem engineering in space and time. *Ecology letters*, 10, 153–164.

Horvath, S. & Raj, K. 2018. DNA methylation-based biomarkers and the epigenetic clock theory of ageing. *Nature Reviews Genetics*, 19, 371–384.

Ilango, S., Paital, B., Jayachandran, P., Padma, P. R. & Nirmaladevi, R. 2020. Epigenetic alterations in cancer. *Frontiers in Bioscience-Landmark*, 25, 1058–1109.

Ismail, F. M., Nahar, L. & Sarker, S. D. 2018. High-throughput screening of phytochemicals: Application of computational methods. *Computational Phytochemistry*, 165–192.

Jablonski, D. 2017. Approaches to macroevolution: 1. General concepts and origin of variation. *Evolutionary Biology*, 44, 427–450.

Jackson, F. L., Niculescu, M. D. & Jackson, R. T. 2013. Conceptual shifts needed to understand the dynamic interactions of genes, environment, epigenetics, social processes, and behavioral choices. *American Journal of Public Health*, 103, S33–S42.

Janssens, Y., Wynendaele, E., Vanden Berghe, W. & De Spiegeleer, B. 2019. Peptides as epigenetic modulators: Therapeutic implications. *Clinical Epigenetics*, 11, 1–14.

Jeffries, M. A. 2020. The development of epigenetics in the study of disease pathogenesis. *Epigenetics in Allergy and Autoimmunity*, 57–94.

Kansara, K., Bolan, S., Radhakrishnan, D., Palanisami, T., Ala'a, H., Bolan, N., Vinu, A., Kumar, A. & Karakoti, A. 2022. A critical review on the role of abiotic factors on the transformation, environmental identity and toxicity of engineered nanomaterials in aquatic environment. *Environmental Pollution*, 296, 118726.

Kaul, I. & Sharma, J. G. 2022. Nanotechnology for the bioremediation of organic and inorganic compounds in aquatic ecosystem/marine ecosystem. *Journal of Applied Biology and Biotechnology*, 10, 22–33.

Kelly, T. K., De Carvalho, D. D. & Jones, P. A. 2010. Epigenetic modifications as therapeutic targets. *Nature Biotechnology*, 28, 1069–1078.

Kim, J., Samaranayake, M. & Pradhan, S. 2009. Epigenetic mechanisms in mammals. *Cellular and Molecular Life Sciences*, 66, 596–612.

Kong, X. & Ohadi, M. M. 2010. Applications of micro and nano technologies in the oil and gas industry-an overview of the recent progress. *Abu Dhabi International Petroleum Exhibition and Conference*. SPE, SPE-138241-MS.

Land, W. G. 2023. Basic trajectories in autoimmunity. In *Damage-Associated Molecular Patterns in Human Diseases: Volume 3: Antigen-Related Disorders*. Springer.

Langer, J. A., Meunier, C. L., Ecker, U., Horn, H. G., Schwenk, K. & Boersma, M. 2019. Acclimation and adaptation of the coastal calanoid copepod Acartia tonsa to ocean acidification: A long-term laboratory investigation. *Marine Ecology Progress Series*, 619, 35–51.

Li, Y. 2021. Modern epigenetics methods in biological research. *Methods*, 187, 104–113.

Lian, Y., Meng, L., Ding, P. & Sang, M. 2018. Epigenetic regulation of MAGE family in human cancer progression-DNA methylation, histone modification, and non-coding RNAs. *Clinical Epigenetics*, 10, 1–11.

Lleo, A. 2023. Geoepidemiology and the key role of sex chromosomes on autoimmune diseases. In *Principles of Gender-Specific Medicine*. Elsevier.

Lock, M. 2015. Comprehending the body in the era of the epigenome. *Current Anthropology*, 56, 151–177.

Lockhart, D. J. & Winzeler, E. A. 2000. Genomics, gene expression and DNA arrays. *Nature*, 405, 827–836.

Lubin, F. D., Gupta, S., Parrish, R. R., Grissom, N. M. & Davis, R. L. 2011. Epigenetic mechanisms: Critical contributors to long-term memory formation. *The Neuroscientist*, 17, 616–632.

Lyu, W., Deng, Z. & Zhang, G. 2023. High-throughput screening for epigenetic compounds that induce human β-defensin 1 synthesis. *Antibiotics*, 12, 186.

Matsuzaki, H., Okamura, E., Takahashi, T., Ushiki, A., Nakamura, T., Nakano, T., Hata, K., Fukamizu, A. & Tanimoto, K. 2015. De novo DNA methylation through the 5′-segment of the H19 ICR maintains its imprint during early embryogenesis. *Development*, 142, 3833–3844.

Matthews, H., Hanison, J. & Nirmalan, N. 2016. "Omics"-informed drug and biomarker discovery: Opportunities, challenges and future perspectives. *Proteomes*, 4, 28.
Mccrory, E., De Brito, S. A. & Viding, E. 2010. Research review: The neurobiology and genetics of maltreatment and adversity. *Journal of Child Psychology and Psychiatry*, 51, 1079–1095.
Menezo, Y., Elder, K., Clement, A. & Clement, P. 2022. Folic acid, folinic acid, 5 methyl tetrahydrofolate supplementation for mutations that affect epigenesis through the folate and one-carbon cycles. *Biomolecules*, 12, 197.
Mesgari, H., Esmaelian, S., Nasiri, K., Ghasemzadeh, S., Doroudgar, P. & Payandeh, Z. 2023. Epigenetic regulation in oral squamous cell carcinoma microenvironment: A comprehensive review. *Cancers*, 15, 5600.
Migliore, L. & Coppedè, F. 2022. Gene–environment interactions in Alzheimer disease: The emerging role of epigenetics. *Nature Reviews Neurology*, 18, 643–660.
Miller, L. J. 2023. *Catalytic and Binding Mechanisms of Histone H3 Lysine 9 and Lysine 36 Methyltransferases*. The University of Wisconsin-Madison.
Moore, L. L. & Houchen, C. W. 2023. Epigenetic landscape and therapeutic implication of gene isoforms of doublecortin-like kinase 1 for cancer stem cells. *International Journal of Molecular Sciences*, 24, 16407.
Moosavi, A. & Ardekani, A. M. 2016. Role of epigenetics in biology and human diseases. *Iranian Biomedical Journal*, 20, 246.
Morselli, M. & Dieci, G. 2022. Epigenetic regulation of human non-coding RNA gene transcription. *Biochemical Society Transactions*, 50, 723–736.
Mota-Martorell, N., Jové, M., Borrás, C., Berdún, R., Obis, E., Sol, J., Cabre, R., Pradas, I., Galo-Licona, J. D. & Puig, J. 2021. Methionine transsulfuration pathway is upregulated in long-lived humans. *Free Radical Biology and Medicine*, 162, 38–52.
Napiórkowska-Krzebietke, A., Skrzypczak, A. R. & Kicińska, A. 2023. Abiotic–biotic interrelations in the context of stabilized ecological potential of post-mining waters. *Water*, 15, 3328.
Newkirk, T. L., Bowers, A. A. & Williams, R. M. 2009. Discovery, biological activity, synthesis and potential therapeutic utility of naturally occurring histone deacetylase inhibitors. *Natural Product Reports*, 26, 1293–1320.
Nicoglou, A. & Merlin, F. 2017. Epigenetics: A way to bridge the gap between biological fields. *Studies in History and Philosophy of Science Part C: Studies in History and Philosophy of Biological and Biomedical Sciences*, 66, 73–82.
Norouzitallab, P., Baruah, K., Vanrompay, D. & Bossier, P. 2019. Can epigenetics translate environmental cues into phenotypes? *Science of the Total Environment*, 647, 1281–1293.
Oleksiewicz, U. & Machnik, M. 2022. Causes, effects, and clinical implications of perturbed patterns within the cancer epigenome. In *Seminars in Cancer Biology*. Elsevier, 15–35.
Palma-Gudiel, H., Córdova-Palomera, A., Eixarch, E., Deuschle, M. & Fañanás, L. 2015. Maternal psychosocial stress during pregnancy alters the epigenetic signature of the glucocorticoid receptor gene promoter in their offspring: A meta-analysis. *Epigenetics*, 10, 893–902.
Peterson, C. L. & Laniel, M.-A. 2004. Histones and histone modifications. *Current Biology*, 14, R546–R551.
Phillips, N. L. H. & Roth, T. L. 2019. Animal models and their contribution to our understanding of the relationship between environments, epigenetic modifications, and behavior. *Genes*, 10, 47.
Prabhu, G. R. D. & Urban, P. L. 2020. Elevating chemistry research with a modern electronics toolkit. *Chemical Reviews*, 120, 9482–9553.
Rastogi, A. 2016. *Phaeodactylum Tricornutum Genome and Epigenome: Characterization of Natural Variants*. Université Paris sciences et lettres.
Razin, S. V., Iarovaia, O. V., Sjakste, N., Sjakste, T., Bagdoniene, L., Rynditch, A. V., Eivazova, E. R., Lipinski, M. & Vassetzky, Y. S. 2007. Chromatin domains and regulation of transcription. *Journal of Molecular Biology*, 369, 597–607.
Saha, A., Wittmeyer, J. & Cairns, B. R. 2006. Chromatin remodelling: The industrial revolution of DNA around histones. *Nature Reviews Molecular Cell Biology*, 7, 437–447.
Sase, A. S., Lombroso, S. I., Santhumayor, B. A., Wood, R. R., Lim, C. J., Neve, R. L. & Heller, E. A. 2019. Sex-specific regulation of fear memory by targeted epigenetic editing of Cdk5. *Biological Psychiatry*, 85, 623–634.
Schnekenburger, M., Dicato, M. & Diederich, M. 2014. Epigenetic modulators from "The Big Blue": A treasure to fight against cancer. *Cancer Letters*, 351, 182–197.

Shvedunova, M. & Akhtar, A. 2022. Modulation of cellular processes by histone and non-histone protein acetylation. *Nature Reviews Molecular Cell Biology*, 23, 329–349.

Sinha, N. J., Langenstein, M. G., Pochan, D. J., Kloxin, C. J. & Saven, J. G. 2021. Peptide design and self-assembly into targeted nanostructure and functional materials. *Chemical Reviews*, 121, 13915–13935.

Strogantsev, R., Krueger, F., Yamazawa, K., Shi, H., Gould, P., Goldman-Roberts, M., Mcewen, K., Sun, B., Pedersen, R. & Ferguson-Smith, A. C. 2015. Allele-specific binding of ZFP57 in the epigenetic regulation of imprinted and non-imprinted monoallelic expression. *Genome Biology*, 16, 1–18.

Subramanian, I., Verma, S., Kumar, S., Jere, A. & Anamika, K. 2020. Multi-omics data integration, interpretation, and its application. *Bioinformatics and Biology Insights*, 14, 1177932219899051.

Talukdar, P. D. & Chatterji, U. 2023. Transcriptional co-activators: Emerging roles in signaling pathways and potential therapeutic targets for diseases. *Signal Transduction and Targeted Therapy*, 8, 427.

Tatipamula, V. B. & Kukavica, B. 2021. Phenolic compounds as antidiabetic, anti-inflammatory, and anticancer agents and improvement of their bioavailability by liposomes. *Cell Biochemistry and Function*, 39, 926–944.

Unnikrishnan, A., Freeman, W. M., Jackson, J., Wren, J. D., Porter, H. & Richardson, A. 2019. The role of DNA methylation in epigenetics of aging. *Pharmacology & Therapeutics*, 195, 172–185.

Vaissière, T., Sawan, C. & Herceg, Z. 2008. Epigenetic interplay between histone modifications and DNA methylation in gene silencing. *Mutation Research/Reviews in Mutation Research*, 659, 40–48.

Van der straeten, D., Bhullar, N. K., De Steur, H., Gruissem, W., Mackenzie, D., Pfeiffer, W., Qaim, M., Slamet-Loedin, I., Strobbe, S. & Tohme, J. 2020. Multiplying the efficiency and impact of biofortification through metabolic engineering. *Nature Communications*, 11, 5203.

Veluchamy, A., Rastogi, A., Lin, X., Lombard, B., Murik, O., Thomas, Y., Dingli, F., Rivarola, M., Ott, S. & Liu, X. 2015. An integrative analysis of post-translational histone modifications in the marine diatom Phaeodactylum tricornutum. *Genome Biology*, 16, 1–18.

Venkatesagowda, B. & Dekker, R. F. 2021. Microbial demethylation of lignin: Evidence of enzymes participating in the removal of methyl/methoxyl groups. *Enzyme and Microbial Technology*, 147, 109780.

Villota-Salazar, N. A., Mendoza-Mendoza, A. & González-Prieto, J. M. 2016. Epigenetics: From the past to the present. *Frontiers in Life Science*, 9, 347–370.

Vogt, G. 2017. Facilitation of environmental adaptation and evolution by epigenetic phenotype variation: Insights from clonal, invasive, polyploid, and domesticated animals. *Environmental Epigenetics*, 3, dvx002.

Wang, M.-W., Hao, X. & Chen, K. 2007. Biological screening of natural products and drug innovation in China. *Philosophical Transactions of the Royal Society B: Biological Sciences*, 362, 1093–1105.

Watkeys, O. J., Kremerskothen, K., Quidé, Y., Fullerton, J. M. & Green, M. J. 2018. Glucocorticoid receptor gene (NR3C1) DNA methylation in association with trauma, psychopathology, transcript expression, or genotypic variation: A systematic review. *Neuroscience & Biobehavioral Reviews*, 95, 85–122.

Wattacheril, J. J., Raj, S., Knowles, D. A. & Greally, J. M. 2023. Using epigenomics to understand cellular responses to environmental influences in diseases. *PLoS Genetics*, 19, e1010567.

Wei, J.-W., Huang, K., Yang, C. & Kang, C.-S. 2017. Non-coding RNAs as regulators in epigenetics. *Oncology Reports*, 37, 3–9.

Wei, R., Dhawan, P., Baiocchi, R. A., Kim, K. Y. & Christakos, S. 2019. PU. 1 and epigenetic signals modulate 1, 25-dihydroxyvitamin D3 and C/EBPα regulation of the human cathelicidin antimicrobial peptide gene in lung epithelial cells. *Journal of Cellular Physiology*, 234, 10345–10359.

Whilde, J., Martindale, M. Q. & Duffy, D. J. 2017. Precision wildlife medicine: Applications of the human-centred precision medicine revolution to species conservation. *Global Change Biology*, 23, 1792–1805.

Whitehouse, H. 2021. *The Ritual Animal: Imitation and Cohesion in the Evolution of Social Complexity.* Oxford University Press.

Witoelar, A., Jansen, I. E., Wang, Y., Desikan, R. S., Gibbs, J. R., Blauwendraat, C., Thompson, W. K., Hernandez, D. G., Djurovic, S. & Schork, A. J. 2017. Genome-wide pleiotropy between Parkinson disease and autoimmune diseases. *JAMA Neurology*, 74, 780–792.

Zhang, J., Duan, D., Song, Z. L., Liu, T., Hou, Y. & Fang, J. 2021. Small molecules regulating reactive oxygen species homeostasis for cancer therapy. *Medicinal Research Reviews*, 41, 342–394.

Zhao, X., Deton cabanillas, A. F., Veluchamy, A., Bowler, C., Vieira, F. R. J. & Tirichine, L. 2020. Probing the diversity of polycomb and trithorax proteins in cultured and environmentally sampled microalgae. *Frontiers in Marine Science*, 7, 189.

Zhou, S. & Huang, G. 2020. Retracted article: The synthesis and biological activity of marine alkaloid derivatives and analogues. *RSC Advances*, 10, 31909–31935.

Zhou, Y.-J., Lu, X.-F., Chen, H., Wang, X.-Y., Cheng, W., Zhang, Q.-W., Chen, J.-N., Wang, X.-Y., Jin, J.-Z. & Yan, F.-R. 2023. Single-cell transcriptomics reveals early molecular and immune alterations underlying the serrated neoplasia pathway toward colorectal cancer. *Cellular and Molecular Gastroenterology and Hepatology*, 15, 393–424.

Zompra, A. A., Chasapi, S. A., Twigg, M. S., Salek, K., Anestopoulos, I., Galanis, A., Pappa, A., Gutierrez, T., Banat, I. M. & Marchant, R. 2022. Multi-method biophysical analysis in discovery, identification, and in-depth characterization of surface-active compounds. *Frontiers in Marine Science*, 9, 1023287.

14 Mapping Novel Epigenetic Drug Delivery Systems with Artificial Intelligence

Gana Manjusha K, Chiranjeevi P, and Aruna Kumari D
Vignan Institute of Pharmaceutical Technology,
Visakhapatnam, Andhra Pradesh, India

14.1 INTRODUCTION

Epigenetics, the study of heritable changes in gene expression that do not alter the underlying DNA sequence, has had an impact on our understanding of disease processes and treatment options. These modifications, which include DNA methylation, histone modification, and noncoding RNA activity, dynamically regulate gene expression and are influenced by external influences, lifestyle, and disease states. Epigenetics is revolutionizing medication development because it provides new options to target and change these processes for therapeutic benefit [1–5].

14.1.1 Epigenetic Mechanisms in Disease

Epigenetic processes are essential for normal cellular function, but their dysregulation has been linked to illnesses including cancer, neurological disorders, cardiovascular disease, and immune dysfunction. For example, aberrant histone modification or hypermethylation of tumor suppressor genes can result in uncontrolled cellular proliferation in cancer. Several epigenetic medications, including 5-azacytidine, 5-aza-2′-deoxycytidine (decitabine), suberoylanilide hydroxamic acid, valproic acid, and entinostat, have been approved or are in clinical trials [5–9].

To improve the efficacy of these medications, strategies include using improved delivery systems such as nanocarriers, changing the drug's chemical structure into prodrugs, and combining them with other therapeutic substances [10]. The choice of a delivery vector is influenced by the drug's physicochemical properties, including polarity (some epigenetic drugs, such as decitabine and 5-azacytidine, are hydrophilic), molecular weight, pharmacokinetics, biodistribution, effective dosage, and the duration of treatment required at the target site [11–13].

In addition, the pH of the target environment, mechanisms of cellular absorption, and the optimum site for intracellular drug release are also important considerations. Surface alterations to delivery methods can improve cellular contacts or pre-uptake interactions, such as those with blood components during intravenous administration or gut tissues and fluids. The successful integration of personalized medicine with improvements in omics technologies and patient data is intrinsically related to the effective use of artificial intelligence (AI), notably machine learning. These tools assist researchers and doctors in managing and evaluating big data from large, complex datasets [14,15]. Machine learning (ML) is an area of AI that focuses on designing and training algorithms that learn from and act on data [16].

14.2 EPIGENETIC MECHANISMS AND THERAPEUTIC TARGETS

Epigenetic treatment seeks to fix abnormal gene expression by targeting reversible changes to DNA, RNA, or histones. These treatments use tiny chemicals or nucleic acids to change DNA methylation,

DOI: 10.1201/9781032706511-14

histones, or RNA-mediated gene control. However, the clinical success of epigenetic therapies is dependent not only on the therapeutic agents' potency but also on their ability to reach the target site; drug delivery systems (DDSs) are important in this context (Table 14.1).

Drug delivery uses a variety of nanocarriers, including polymeric nanoparticles, liposomes, dendrimers, nanogels, and biological vectors. Additional options for improving epigenetic medication delivery include creating prodrugs and combining pharmacological compounds to produce synergistic therapeutic effects (Figure 14.1).

14.2.1 Nanoparticles

Nanoparticles (NPs) used for medication delivery are typically less than 300 nm in diameter [27]. They can be constructed of natural or synthetic polymers such polyethylene glycol (PEG), polylactic acid (PLA), polyglycolic acid (PGA), poly(lactic-co-glycolic acid) (PLGA), dextran, and chitosan. NPs are typically transformed into monomers or low-molecular-weight fragments that are then eliminated by natural biological systems. For example, PLGA degrades into lactic acid and glycolic acid: Lactic acid is metabolized by the tricarboxylic acid (TCA) cycle and ejected as carbon dioxide and water; glycolic acid can either follow the TCA cycle or be removed by the kidneys. [28]

NP formulation and production parameters can be tuned to regulate important properties such as molecular weight, lactide-to-glycolide ratio, hydrophobicity, crystallinity, and degradation kinetics. These variables determine the drug release profile, which can be tailored for first-order or triggered release mechanisms. However, NPs confront obstacles such as phagocytosis, which can

TABLE 14.1
Epigenetic Targets in Drug Development

Epigenetic Mechanism	Description	Key Enzymes/Components	Therapeutic Implications
DNA methylation	Adding methyl groups to cytosine residue to influence gene silencing, imprinting, and X-chromosome inactivation	DNMTs target CpG (cytosine-phosphate-guanine) islands near gene promoters	Regulates gene expression; potential target for reversing abnormal gene silencing in diseases like cancer [17–19]
Histone acetylation	Post-translational modification that decreases histone–DNA interaction by reducing histone's positive charge, increasing DNA accessibility and promoting gene expression	Histone acetyl transferases add acetyl groups; histone deacetylases (HDACs) remove acetyl groups	HDAC inhibitors used in cancer therapy to restore gene expression and counteract transcriptional suppression [20,21]
Histone methylation	Adding methyl groups to amino acids like lysine and arginine, which can either promote or inhibit gene expression depending on the methylation pattern	Histone methyltransferases add methyl groups; histone demethylases remove methyl groups	Modifies chromatin to either activate or repress transcription; potential target for gene-specific regulation [20–22]
RNA-mediated regulation	Gene regulation through small RNAs, including microRNAs (miRNAs), which modulate gene expression by interacting with RNA targets in the cytoplasm or chromatin	miRNAs regulate proliferation, apoptosis, and development, whereas X-inactive specific transcript inhibits X-chromosome activity	Therapeutic use in targeting specific miRNAs to modulate pathways in cancers, neurological disorders, and metabolic diseases [23–26]

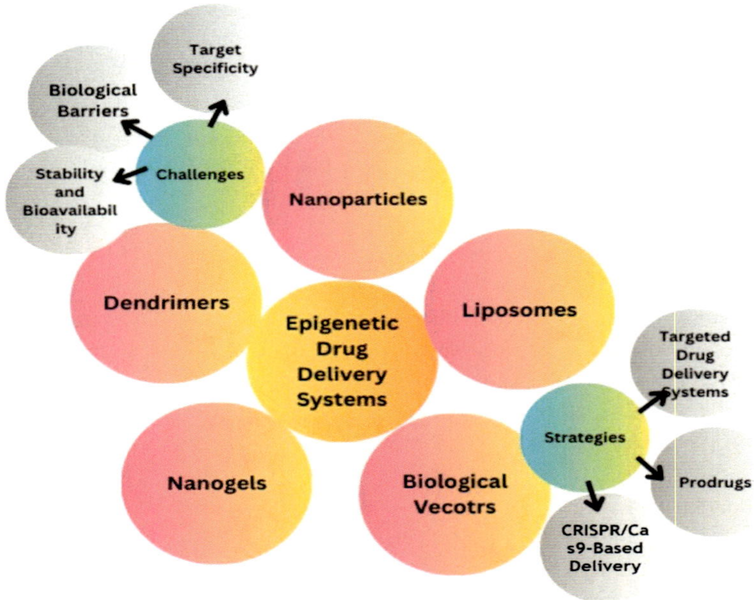

FIGURE 14.1 Epigenetic drug delivery systems.

shorten their circulation duration and cause tumor accumulation [29]. Surface modifications such as PEGylation and the addition of targeted ligands are frequently used to overcome this issue. The therapeutic effectiveness of DDSs based on nanoparticles is ultimately increased by these changes, which also increase cellular absorption, blood circulation half-life, and biodistribution.

14.2.2 Liposomes

Lipid bilayers around an aqueous core form liposomes, which are self-assembling structures. By altering the lipid type or ratio, their characteristics, such as surface charge, functioning, and size, can be improved. PEG is commonly used to modify the surfaces of liposomes to increase their stability and slow down the reticuloendothelial system's (RES) quick clearance [30].

Liposomes have the unique capacity to encapsulate both hydrophilic and hydrophobic medicines. Drugs that are hydrophilic are kept in the aqueous core, whereas those that are hydrophobic are kept between lipid bilayers. However, compared with alternative delivery techniques, liposomes have lower drug-loading capabilities. They also confront obstacles such as circulatory instability and fast burst medication release [30].

Novel liposomal designs are being investigated for overcoming these limitations. For instance, phospholipase A2, an enzyme that is overexpressed in malignancies, can break down the lipid components in non-targeted liposomes made by LiPlasome Pharma. These innovations enable the drug to be broken down and released selectively at the site of the tumor. Another area of investigation is amphoteric liposomes for nucleic acid delivery, such as RNA sequences that function through epigenetic processes [30].

14.2.3 Dendrimers

Nanoscale globular macromolecules with a diameter of 5–10 nm are called dendrimers. They can be altered in shape and characteristics to provide controlled drug release and longer circulation periods. These molecules are constructed with a central core around which a highly branching architecture is created by gradually adding layers of multifunctional monomeric components [31]

The ability of amphiphilic dendrimers to self-assemble into micelles with hydrophilic surface groups makes them effective for targeting compounds and administering a variety of medications and imaging agents. For instance, phase 1 clinical trials for the dendrimer-based nanoparticle CALAA-01 are now underway. This transferrin–cyclodextrin–siRNA complex has demonstrated potential in the treatment of solid tumors by targeting the M2 subunit of ribonucleotide reductase [32].

However, problems persist, such as steric hindrance inside the dendrimer structure, which can restrict drug loading capacity. On the other hand, this characteristic promotes cascade release mechanisms, allowing for sequential drug delivery—for example, delivering an epigenetic medication followed by a chemotherapeutic agent [33]. Further research on dendrimer biodistribution is required to improve tissue localization and therapeutic efficacy.

14.2.4 Nanogels

Hydrogel-based nanoparticles known as nanogels can be made to release a therapeutic payload in response to certain stimuli, including biomolecular recognition, magnetic fields, temperature, and pH [34]. For instance, Vijayaraghavalu and Labhasetwar used NIPAM to develop biodegradable nanogels. These nanogels are designed to hold epigenetic medications like the potent DNMT1 inhibitor decitabine.

Decitabine degrades readily in acidic environments, such as tumor sites, and has a brief half-life (10–35 minutes). The stability and effectiveness of decitabine are enhanced by nanogel encapsulation. Studies have demonstrated that decitabine-loaded nanogels enhance DNMT1 depletion, prolong G2/M phase cancer cell arrest, and significantly enhance its antiproliferative properties against leukemia cells, doxorubicin-resistant breast cancer, and decitabine-resistant melanoma [35].

These findings imply that nanogels may boost decitabine's therapeutic efficacy, particularly in solid tumors, by providing sustained release and minimizing fast elimination.

14.2.5 Biological Vectors

Engineered bacteria, such as the "nanocell," a bacterial cell that has had its DNA removed to inhibit multiplication and mutation, are examples of biological delivery systems. These nanocells, which are about 300 nm in size, can hold molecules with varied solubility and charge, including medications like doxorubicin, paclitaxel, and siRNAs [36].

While these systems provide novel drug delivery alternatives, they are coupled with potential hazards including as immunological responses triggered by the presence of bacterial lipopolysaccharides. More progress is necessary to eliminate these dangers and improve the efficacy of biological vectors in therapeutic applications.

14.3 CHALLENGES IN EPIGENETIC DRUG DELIVERY

Epigenetic treatments, which aim to treat diseases by modifying the epigenome (for example, DNA methylation and histone modification), confront a number of obstacles in terms of drug delivery. These include biological barriers, ensuring target specificity, and increasing medication stability and bioavailability.

14.3.1 Biological Barriers

14.3.1.1 The Blood–Brain Barrier

The distribution of epigenetic medications is significantly hampered by the blood–brain barrier, especially for conditions affecting the central nervous system. The tightly controlled barrier restricts the entry of therapeutic molecules, making it difficult for epigenetic medicines to reach their targets in the brain [37].

14.3.1.2 Cell Membrane Permeability

Low membrane permeability is a feature of many epigenetic therapies, including DNA methyltransferase (DNMT) and histone deacetylase (HDAC) inhibitors. This reduces their ability to enter cells and alter the epigenome [38].

14.3.1.3 Efflux Mechanisms

P-glycoproteins and other efflux pumps can remove drugs from cells before they reach their target, reducing their efficacy [39].

14.3.1.4 Tumor Heterogeneity

Tumors in cancer therapy have diverse epigenetic patterns. This makes it difficult to give a consistent therapeutic dose throughout the tumor population, as some cells do not respond to treatment due to variances in epigenetic alterations [40].

14.3.2 Target Specificity

In nonspecific binding, epigenetic medications, especially tiny compounds, might attach to off-target proteins or DNA sections, causing unexpected effects and harm. High specificity for desirable epigenetic modifications (such as methylated DNA or acetylated histones) is critical to avoiding systemic adverse effects [41].

14.3.3 Epigenetic Modulators

The intricate nature of the epigenome, with overlapping changes and interactions, makes it difficult to develop medications that target specific epigenetic markers while ignoring others. For example, DNMT inhibitors can accidentally influence non-target genes or pathways [42]. In addition, because epigenetic alterations vary between people and even between cell types in the same person, highly tailored medication delivery systems need to be developed that can target unique epigenetic abnormalities in individual patients or cell types [43].

14.3.4 Stability and Bioavailability

Epigenetic medicines may deteriorate or undergo metabolic transformation in bodily fluids due to their chemical instability. DNMT inhibitors such as 5-azacytidine and decitabine have short half-lives, which limits their therapeutic window[44]; they also degrade in acidic environments, such as the tumor microenvironment, reducing their potency. It is critical to formulate these medications with stable delivery methods to ensure that they remain active at the target site [45].

Epigenetic medicines' limited water solubility makes them challenging to formulate into effective DDSs. This causes difficulties in delivering therapeutic concentrations to tissues including tumors, especially when taken orally [46]. More broadly, systemic administration of epigenetic medicines shows low bioavailability due to rapid circulation clearance, poor gastrointestinal absorption, and biological obstacles. Achieving sustained, high local concentrations at the target site is critical for successful therapy [47].

14.4 STRATEGIES TO OVERCOME THESE CHALLENGES

Using NPs such as liposomes, dendrimers, and polymeric NPs allow for circumventing biological barriers, increasing stability, and achieving customized distribution. These systems can be used to encapsulate epigenetic medicines, protecting them from degradation while also increasing their ability to traverse barriers and reach specific target tissues or cells [48].

Targeted DDSs are another solution: Modifying drug carriers, such as via PEGylation or ligand-based targeting, can enhance epigenetic drug targeting to specific cells or tissues. Ligands are targeted that attach to receptors that are overexpressed on target cells [49]. Prodrugs enhance the stability and bioavailability of epigenetic drugs by converting them from inactive to active molecules at the target location. Lipophilic prodrugs, for example, can increase cell membrane penetration and stability before being metabolized at the target site [50]. Additionally, advanced formulations such as NPs, nanogels, and injectable depot formulations improve the solubility and stability of epigenetic medicines, leading to increased therapeutic effectiveness and prolonged release [51].

14.5 EMERGING NOVEL DDSS IN EPIGENETICS

Recent breakthroughs in drug delivery technologies have resulted in considerable improvements to epigenetic medication delivery and controlled release. These innovative DDSs seek to address the issues associated with epigenetic therapies, such as low bioavailability, stability, and specificity. The following are the primary emerging systems in epigenetic drug delivery.

14.5.1 Systems Using NPs

NPs' small size, biocompatibility, and ability to encapsulate both hydrophilic and hydrophobic drugs have made them a promising delivery system for epigenetic drugs. It is possible to develop NPs to target particular tissues or cells, enhance cellular absorption, and enhance the pharmacokinetics of epigenetic medications. Liposomes, polymeric NPs, and gold NPs are some of the most used vehicles for epigenetic medications [52]. By allowing the controlled release of epigenetic modulators, nanoparticles lessen side effects and increase the effectiveness of treatment [53].

14.5.2 Lipid-Based Carriers

The ability of lipid-based carriers such as liposomes, solid-lipid NPs, and nanostructured lipid carriers to transport biologics and small compounds, including epigenetic medications, has drawn much interest. Liposomes, in particular, are frequently employed for drug administration due to their biocompatibility, capacity to encapsulate both hydrophilic and hydrophobic medicines, and potential surface modification to improve targeting. Lipid-based systems are also used in epigenetics to administer DNA methylation inhibitors and HDAC inhibitors, which improves their stability and half-life in vivo [54].

14.5.3 Polymer-Based Delivery Systems

Polymer-based drug delivery methods, such as biodegradable polymers, are being studied for their ability to enable regulated and sustained release of epigenetic medicines. To encapsulate epigenetic medications including DNMT and HDAC inhibitors and small interfering RNAs (siRNAs) that target epigenetic regulators, polymeric nanoparticles, micelles, and hydrogels have been developed [55]. These systems provide benefits such as increased drug stability, lower systemic toxicity, and the capacity to target specific tissues or cells based on surface changes.

14.5.4 CRISPR/Cas9 Delivery for Epigenetic Modulation

CRISPR/Cas9 gene-editing technology has transformed the study of epigenetics by allowing for precise epigenome modifications. By modifying DNMT and histone alterations, CRISPR/Cas9 can provide novel therapeutic options for diseases brought on by epigenetic changes. However, effective delivery of CRISPR/Cas9 components (such as Cas9 proteins and guide RNAs) to specific cells is still a big challenge. Various delivery techniques such as lipid NPs, viral vectors, and polymer-based systems are being developed to improve the safe and effective transport of CRISPR/Cas9 components to cells [56]. These methods have the ability to both cure and fix underlying epigenetic abnormalities in vivo.

14.6 AI IN DDSS

Technologies like AI and machine learning (ML) are becoming more and more crucial for developing and improving DDSs. Researchers can use AI to improve the effectiveness of drug delivery formulations, anticipate drug–target interactions, and evaluate vast datasets to reveal unexpected insights in epigenetic studies. The following are some of the important areas where AI is helping to develop DDSs.

14.6.1 AI in Formulation Design

AI methods like deep learning and neural networks have been applied to the development and improvement of drug formulations. These algorithms can analyze enormous datasets to predict drug candidates' physicochemical qualities, find the best drug delivery carriers, and simulate drug release profiles in a variety of biological contexts. AI-based systems can help identify the best drug carrier materials, determine optimal dosages, and anticipate drug pharmacokinetic characteristics [57]. Optimizing formulations for improved therapeutic effects and cutting down on the time and cost needed to develop novel DDSs are two benefits of using AI in formulation design.

14.6.2 Predicting Drug–Target Interactions with ML

The ability of ML algorithms to anticipate drug–target interactions, which are essential for both drug discovery and the development of effective DDSs, has shown significant promise. ML algorithms use complicated datasets such as genetic, proteomic, and chemical data to anticipate how a medicine will interact with its target receptor or enzyme. These models aid in identifying possible medication candidates that would go undetected using typical experimental procedures. In the area of epigenetics, ML is used to anticipate interactions between epigenetic medicines and their targets, such as enzymes involved in DNA methylation or histone modification [58]. This can hasten the discovery of new epigenetic medicines and increase the specificity of medication delivery.

14.6.3 Big Data Analysis in Epigenetic Studies

The enormous amount of data created by genomic, transcriptomic, proteomic, and epigenomic studies creates both opportunities and constraints. Big data analytics, powered by AI and ML, has enabled academics to extract important insights from these complex datasets. Big data approaches are utilized in epigenetic research to assess DNA methylation patterns, histone changes, and other epigenetic markers that influence disease development. AI-powered methods can also help integrate multi-omics data to better understand how the epigenome functions in diseases including cancer, neurological conditions, and uncommon genetic issues. AI can discover new epigenetic biomarkers for illness diagnosis, predict therapy outcomes, and tailor treatment regimens by analyzing massive datasets.

14.7 COMBINING AI WITH DDSS BASED ON EPIGENETICS

The combination of AI and epigenetic DDSs represents a watershed moment in personalized medicine. AI-powered techniques can help with the development, optimization, and practical use of epigenetic medicines, especially in addressing hurdles like targeted delivery, patient specificity, and continuous monitoring. The following are key areas in which AI is playing an important role in optimizing and enhancing epigenetic medication delivery.

14.7.1 AI for Optimizing Delivery Mechanisms

One of the key issues with developing DDSs is ensuring that therapeutic chemicals are delivered to the appropriate tissue or organ in an efficient and controlled manner. AI algorithms can improve delivery mechanisms by simulating drug interactions at the molecular level and predicting the ideal formulation for controlled release. AI can be used to develop nanoparticles or lipid-based carriers for epigenetic medications that target DNA methylation or histone modification enzymes, increasing stability and bioavailability while decreasing off-target effects [59]. Furthermore, AI models can predict how drug carriers interact with the cellular milieu, improving their release patterns and ensuring that the medications are exclusively activated in the intended tissues [60].

14.7.2 Predictive Modeling for Patient-Specific Therapy

Predictive modeling is made possible by AI's capacity to analyze vast volumes of data, which is highly beneficial for individualized care. Based on genetic, epigenetic, and environmental factors, predictive modeling can be utilized to anticipate patient-specific responses to specific epigenetic therapies, for instance by combining multi-omics data (genomics, transcriptomics, proteomics, etc.). This tailored approach can help guide the development of targeted medicines, ensuring that epigenetic medications are administered at the appropriate dosage and time for maximal therapeutic impact while reducing potential negative effects [57]. This precision medicine strategy is especially important for complex illnesses like cancer and neurological conditions, where each patient's unique epigenetic modifications play a vital part in the course of the illness.

14.7.3 AI in Monitoring and Feedback

Effective monitoring and feedback are required to guarantee that epigenetic DDSs work as intended, especially for medicines requiring long-term administration. AI technology can be used to continuously monitor a patient's response to epigenetic medications using real-time data analysis to spot any symptoms of side effects or treatment failure. Wearable sensors and mobile health applications can collect data on drug efficacy, patient symptoms, and biomarkers, which AI algorithms can then use to provide insights into the medicine's performance [61]. Furthermore, AI can help clinicians change treatment regimens based on ongoing input via feedback loops to improve the delivery system itself, altering parameters like drug release rates or carrier qualities based on the patient's reaction. This dynamic, adaptive approach to drug administration ensures that medicines stay effective while reducing side effects over time.

14.8 REGULATORY AND ETHICAL CONSIDERATIONS

As AI integration with epigenetic medicine delivery systems advances, additional obstacles emerge both regulatory and ethical. AI-powered medicines require complex data analysis, patient-specific treatment algorithms, and unique medication delivery methods, all of which must be thoroughly tested to verify their safety and efficacy. Furthermore, there are significant ethical concerns about the use of AI in personalized medicine and epigenetic editing (Table 14.2).

AI mapping of recent developments in new medication delivery methods has completely changed the field of customized medicine. The development of epigenetic medicines has been greatly hastened by AI's capacity to evaluate intricate datasets, forecast therapeutic effects, and optimize treatment regimens. These developments make it possible to precisely target gene expression, which

TABLE 14.2
Regulatory and Ethical Considerations for AI in Epigenetic DDSs

Category	Considerations	Details
Regulatory challenges	Complexity of AI models	Frequently functioning as "black boxes," ML models make it challenging for authorities to examine how decisions are made. AI transparency is essential for acceptance.
	Data quality and standardization	Ensuring that the data used to train AI models is reliable, unbiased, and consistent across clinical settings to prevent incorrect predictions or unforeseen consequences
	Long-term safety monitoring	Tracking how epigenetic treatments affect gene expression over the long run, developing monitoring procedures, and identifying accountable parties for handling side effects
	Approval pathways	Lack of defined pathways for AI-driven epigenetic medicines; adaptive pathways and real-world evidence are being explored but require further development.
	Recent regulatory efforts	The FDA is working on frameworks for AI in medical devices, such as the Software Precertification Program and AI-based software as a medical device that could inform epigenetic therapy approvals.
Ethical concerns	Patient consent and autonomy	Patients must be informed about how their data is used, the AI models employed, and potential risks and benefits. Patients may struggle to grasp the implications of epigenetic changes for themselves and future generations.
	Privacy and data security	Genetic data must be securely stored and managed to prevent misuse, identity theft, or genetic discrimination. Large datasets used in AI-based therapies require robust data protection measures.
	Bias and inequality	Historical biases in training data can lead to unequal therapeutic outcomes, potentially worsening health disparities and limiting the effectiveness of therapies across diverse populations.
	Epigenetic editing and "designer babies"	Concerns about using epigenetic therapies for non-medical purposes, such as selecting traits or creating "designer babies." Raises ethical issues about genetic modification and societal impacts.
	Long-term impact on human evolution	Potential hereditary modifications could affect future generations, raising concerns about unintended consequences and the ethics of altering the human epigenome.
Ethical frameworks and governance	Guidelines for ethical use	Frameworks should address patient consent, data privacy, AI model bias, and epigenetic technology regulation while fostering continuous public discourse to ensure societal benefits and respect for human dignity.

could provide answers for difficult illnesses like cancer, neurological conditions, and genetic abnormalities.

However, there are also particular ethical and regulatory issues with the use of AI in epigenetic medicine such as the complexity of AI models, the need for high-quality training data, long-term safety monitoring, and the absence of established approval pathways; these issues highlight the pressing need for tailored regulatory frameworks. Ethical concerns, including patient consent, data privacy, bias in AI algorithms, and the implications of hereditary modifications, demand robust governance to ensure equitable and secure application of these technologies. Addressing these concerns is essential for fostering trust, inclusivity, and societal acceptance.

14.9 FUTURE DIRECTIONS AND INNOVATIONS

To fully exploit the potential of AI-driven epigenetic therapeutics, future studies should concentrate on

- Ensuring that the AI models are transparent and explainable to meet regulatory standards and increase clinician confidence.
- Standardizing data collection and usage; ensuring varied, unbiased, and high-quality datasets for more reliable and applicable AI predictions.
- Developing regulatory frameworks to approve and monitor AI-based epigenetic therapeutics throughout time.
- Enhancing ethical governance by establishing explicit standards for patient consent, privacy protection, and equitable access while minimizing abuse concerns.
- Exploring innovative applications such as including preventive medicine and genomic insights while adhering to ethical standards.

REFERENCES

[1] McGowan, P. O., Sasaki, A., D'Alessio, A. C., Dymov, S., Labonté, B., Szyf, M., et al. (2009). Epigenetic regulation of the glucocorticoid receptor in human brain associates with childhood abuse. *Nature Neuroscience*, *12*(3), 342–348. https://doi.org/10.1038/nn.2270

[2] Huehn, J., Polansky, J. K., & Hamann, A. (2009). Epigenetic control of FOXP3 expression: The key to a stable regulatory T-cell lineage? *Nature Reviews Immunology*, *9*(2), 83–89. https://doi.org/10.1038/nri2504

[3] Mastronardi, F. G., Noor, A., Wood, D. D., Paton, T., & Moscarello, M. A. (2007). Peptidyl arginine deiminase 2 CpG island in multiple sclerosis white matter is hypomethylated. *Journal of Neuroscience Research*, *85*(9), 2006–2016. https://doi.org/10.1002/jnr.21356

[4] Adcock, I. M., Ito, K., & Barnes, P. J. (2005). Histone deacetylation: An important mechanism in inflammatory lung diseases. *COPD: Journal of Chronic Obstructive Pulmonary Disease*, *2*(4), 445–455. https://doi.org/10.1080/15412550500374243

[5] Buckstein, R., Yee, K., & Wells, R. A. (2011). 5-Azacytidine in myelodysplastic syndromes: A clinical practice guideline. *Cancer Treatment Reviews*, *37*(2), 160–167. https://doi.org/10.1016/j.ctrv.2010.06.007

[6] Ravandi, F., Alattar, M. L., Grunwald, M. R., Rudek, M. A., Rajkhowa, T., Richie, M. A., et al. (2013). Phase 2 study of azacytidine plus sorafenib in patients with acute myeloid leukemia and FLT-3 internal tandem duplication mutation. *Blood*, *121*(23), 4655–4662. https://doi.org/10.1182/blood-2013-01-475252

[7] Greenberg, P. L., Garcia-Manero, G., Moore, M., Damon, L., Roboz, G., Hu, K., et al. (2013). A randomized controlled trial of romiplostim in patients with low-or intermediate-risk myelodysplastic syndrome receiving decitabine. *Leukemia & Lymphoma*, *54*(2), 321–328. https://doi.org/10.3109/10428194.2012.728291

[8] Thomas, X., Dmoszynska, A., Wierzbowska, A., Kuliczkowski, K., Mayer, J., Shelekhova, T., et al. (2011). Results from a randomized phase III trial of decitabine versus supportive care or low-dose cytarabine for the treatment of older patients with newly diagnosed AML. *Journal of Clinical Oncology*, *29*(15 Suppl), 6504.

[9] Rangwala, S., Duvic, M., & Zhang, C. (2012). Trends in the treatment of cutaneous T-cell lymphoma: Critical evaluation and perspectives on vorinostat. *Blood Lymph Cancer: Targets and Therapy, 2*, 17–27. https://doi.org/10.2147/BLC.S31898

[10] el Bahhaj, F., Dekker, F. J., Martinet, N., & Bertrand, P. (2014). Delivery of epidrugs. *Drug Discovery Today, 19*(9), 1337–1352. https://doi.org/10.1016/j.drudis.2014.05.007

[11] Zhao, M., Rudek, M. A., He, P., Hartke, C., Gore, S., Carducci, M. A., et al. (2004). Quantification of 5-azacytidine in plasma by electrospray tandem mass spectrometry coupled with high-performance liquid chromatography. *Journal of Chromatography B, 813*(1), 81–88. https://doi.org/10.1016/j.jchromb.2004.03.009

[12] Hua, W., Ieradi, T., Lesslie, M., Hoffman, B. T., & Mulvana, D. (2014). Development and validation of a HILIC-MS/MS method for quantification of decitabine in human plasma by using lithium adduct detection. *Journal of Chromatography B, 969*, 117–22.

[13] Adjei, I. M., Sharma, B., & Labhasetwar, V. (2014). Nanoparticles: Cellular Uptake and Cytotoxicity. In *Nanomaterial* (pp. 73–91). Springer.

[14] Brannon-Peppas, L., & Blanchette, J. O. (2012). Nanoparticle and targeted systems for cancer therapy. *Advanced Drug Delivery Reviews, 64*, 206–212. https://doi.org/10.1016/j.addr.2011.08.004

[15] Rauschert, S., Raubenheimer, K., Melton, P. E., & Huang, R. C. (2020). Machine learning and clinical epigenetics: A review of challenges for diagnosis and classification. *Clinical Epigenetics, 12*, 51. https://doi.org/10.1186/s13148-020-00842-4

[16] Toh, T. S., Dondelinger, F., & Wang, D. (2019). Looking beyond the hype: Applied AI and machine learning in translational medicine. *EBioMedicine, 47*, 607–615. https://doi.org/10.1016/j.ebiom.2019.08.027

[17] Jones, P. A., & Takai, D. (2001). The role of DNA methylation in mammalian epigenetics. *Science, 293*(5532), 1068–1070.

[18] Jones, P. A. (2012). Functions of DNA methylation: Islands, start sites, gene bodies, and beyond. *Nature Reviews Genetics, 13*(7), 484–492.

[19] Moore, L. D., Le, T., & Fan, G. (2013). DNA methylation and its basic function. *Neuropsychopharmacology, 38*(1), 23–38.

[20] Bannister, A. J., & Kouzarides, T. (2011). Regulation of chromatin by histone modifications. *Cell Research, 21*(3), 381–395.

[21] Richon, V. M., Sandhoff, T. W., Rifkind, R. A., & Marks, P. A. (2000). Histone deacetylase inhibitor selectively induces p21WAF1 expression and gene-associated histone acetylation. *Proceedings of the National Academy of Sciences, 97*(18), 10014–10019.

[22] Ng, H. H., & Adrian, B. (1999). DNA methylation and chromatin modification. *Current Opinion in Genetics & Development, 9*(2), 158–163.

[23] Avner, P., & Heard, E. (2001). X-chromosome inactivation: Counting, choice and initiation. *Nature Reviews Genetics, 2*(1), 59–67.

[24] Lee, Y., Kim, M., Han, J., Yeom, K. H., Lee, S., Baek, S. H., et al. (2004). MicroRNA genes are transcribed by RNA polymerase II. *The EMBO Journal, 23*(20), 4051–4060.

[25] John, B., Enright, A. J., Aravin, A., Tuschl, T., Sander, C., & Marks, D. S. (2004). Human microRNA targets. *PLoS Biology, 2*(11), e363.

[26] Brennecke, J., Stark, A., Russell, R. B., & Cohen, S. M. (2005). Principles of microRNA–target recognition. *PLoS Biology, 3*(3), e85.

[27] Alexis, F., Pridgen, E. M., Langer, R., & Farokhzad, O. C. (2010). Nanoparticle technologies for cancer therapy. In *Drug Delivery* (pp. 55–86). Springer.

[28] Crotts, G., & Park, T. G. (1998). Protein delivery from poly (lactic-co-glycolic acid) biodegradable microspheres: Release kinetics and stability issues. *Journal of Microencapsulation, 15*(6), 699–713.

[29] Wang, A. Z., Langer, R., & Farokhzad, O. C. (2012). Nanoparticle delivery of cancer drugs. *Annual Review of Medicine, 63*, 185–198.

[30] Gabizon, A., & Martin, F. (1997). Polyethylene glycol-coated (pegylated) liposomal doxorubicin. *Drugs, 54*(4), 15–21.

[31] Kaminskas, L. M., McLeod, V. M., Porter, C. J., & Boyd, B. J. (2012). Association of chemotherapeutic drugs with dendrimer nanocarriers: An assessment of the merits of covalent conjugation compared to noncovalent encapsulation. *Molecular Pharmaceutics, 9*(3), 355–373.

[32] Oh, Y. K., & Park, T. G. (2009). siRNA delivery systems for cancer treatment. *Advanced Drug Delivery Reviews, 61*(10), 850–862.

[33] Gillies, E. R., & Frechet, J. M. (2005). Dendrimers and dendritic polymers in drug delivery. *Drug Discovery Today, 10*(1), 35–43.

[34] Zha, L., Banik, B., & Alexis, F. (2011). Stimulus responsive nanogels for drug delivery. *Soft Matter, 7*(13), 5908–5916.
[35] Vijayaraghavalu, S., & Labhasetwar, V. (2013). Efficacy of decitabine-loaded nanogels in overcoming cancer drug resistance is mediated via sustained DNA methyltransferase 1 (DNMT1) depletion. *Cancer Letters, 331*(1), 122–129.
[36] MacDiarmid, J. A., Mugridge, N. B., Weiss, J. C., Phillips, L., Burn, A. L., Paulin, R. P., et al. (2007). Bacterially derived 400 nm particles for encapsulation and cancer cell targeting of chemotherapeutics. *Cancer Cell, 11*(5), 431–445.
[37] Gonzalez, A., et al. (2018). Blood-brain barrier: Challenges and opportunities in nanomedicine. *International Journal of Pharmaceutics, 550*(1), 89–97.
[38] Toniatti, C., et al. (2019). Challenges in targeting the epigenome in drug discovery. *Nature Reviews Drug Discovery, 18*(11), 857–871.
[39] Duan, J., Zhang, Y., et al. (2015). Targeting the epigenetic landscape in cancer therapy. *Therapeutic Advances in Medical Oncology, 7*(3), 100–114.
[40] Saito, Y., et al. (2020). Overcoming tumor heterogeneity: Strategies for enhancing targeted drug delivery. *Molecular Therapy, 28*(8), 1420–1432.
[41] Kouzarides, T. (2007). Chromatin modifications and their function. *Cell, 128*(4), 693–705.
[42] Gupta, S., Sharma, M., et al. (2017). Targeting epigenetic modifications: A new frontier in cancer treatment. *Frontiers in Oncology, 7*, 101.
[43] Sharma, S., et al. (2020). Epigenetic therapies in cancer: A detailed review. *Journal of Cancer Research and Clinical Oncology, 146*(6), 1459–1468.
[44] Yuan, Y., et al. (2019). Epigenetic drugs in cancer therapy: Challenges and opportunities. *Cancer Research, 79*(6), 1159–1168.
[45] Manjunath, A., et al. (2020). Epigenetic therapies: Current status and future directions. *Pharmacology & Therapeutics, 210*, 107529.
[46] Cheng, Y., Zhan, S., et al. (2021). Challenges in the development of epigenetic therapies for cancer. *Frontiers in Pharmacology, 12*, 683767.
[47] Morris, G., et al. (2021). Advances in drug delivery strategies for epigenetic therapy. *Journal of Controlled Release, 335*, 101–115.
[48] Raza, M. A., et al. (2019). Nanoparticle-based drug delivery in the treatment of epigenetic disorders. *Nanomedicine: Nanotechnology, Biology, and Medicine, 21*, 102062.
[49] Elzoghby, A. O., Samy, W. M., et al. (2012). Liposomes in drug delivery: A review of current applications. *International Journal of Nanomedicine, 7*, 1–10.
[50] Choi, Y., Kang, S., et al. (2019). Prodrug strategies for the delivery of small molecules and biologics in cancer therapy. *International Journal of Nanomedicine, 14*, 2927–2945.
[51] López, J. A., et al. (2017). Nanocarrier-based drug delivery systems for epigenetic therapies. *Current Pharmaceutical Design, 23*(28), 4107–4118.
[52] Zhang, L., et al. (2018). Nanoparticles for drug delivery: A comprehensive review. *Nanomedicine, 13*(15), 1987–1999.
[53] Huang, X., et al. (2019). Nanoparticles in drug delivery and cancer therapy. *Journal of Controlled Release, 240*, 199–217.
[54] Zhao, L., et al. (2018). Lipid-based drug delivery systems for cancer epigenetic therapy. *Journal of Drug Targeting, 26*(9), 804–814.
[55] Batrakova, E. V., et al. (2019). Polymeric drug delivery systems: Current status and future challenges. *Advanced Drug Delivery Reviews, 146*, 56–72.
[56] Kim, H. K., et al. (2020). CRISPR/Cas9-based epigenetic modulation: From research to therapy. *Nature Reviews Molecular Cell Biology, 21*(10), 634–650.
[57] Wang, Z., et al. (2020). Predictive modeling for patient-specific therapy: Integrating AI and epigenomics. *Trends in Pharmacological Sciences, 41*(6), 438–451.
[58] Liu, Y., et al. (2020). Machine learning for predicting drug-target interactions: A review. *Journal of Biomedical Informatics, 103*, 103380.
[59] Gajjar, K., et al. (2020). AI in optimizing nanoparticle-based drug delivery systems for epigenetic modulation. *Pharmaceutics, 12*(3), 259.
[60] Deng, Z., et al. (2021). Artificial intelligence in optimizing the design of drug delivery systems: Focus on epigenetic drugs. *Journal of Controlled Release, 330*, 99–110.
[61] Vermesan, O., et al. (2020). AI in monitoring drug delivery: Application in epigenetic therapy. *Journal of Personalized Medicine, 10*(2), 102.

15 Recent Developments in Epigenetics-Based Personalized Medicine

Deepti Katiyar[1], *Shreya Dixit*[2], *Rajnandani Singh*[2], *Rashmi Saxena Pal*[3], *and Pradeep Singh*[4]
[1]Department of Pharmacognosy, KIET School of Pharmacy, KIET Group of Institutions, Delhi-NCR, Ghaziabad, Uttar Pradesh, India
[2]KIET School of Pharmacy, KIET Group of Institutions, Delhi-NCR, Ghaziabad, Uttar Pradesh, India
[3]School of Pharmaceutical Sciences, Lovely Professional University, Phagwara, Punjab, India
[4]Department of Pharmacy, Debre Tabor University, Ethiopia

15.1 INTRODUCTION

Waddington was ahead of his time when he hypothesized in 1942 that disorders in the body's tissues and organs could be attributed to changes in the epigenome. The area of cancer epigenetics evolved with the first observations of aberrant DNA methylation in cancer in the last decade of the 1980s (Feinberg & Tycko 2004). Toward the middle of the 1990s, DNA methylation supplanted mutations as an essential contributor to oncogenesis. Almost every significant cancer type possessed genes with elevated degrees of DNA methylation, or hypermethylation, with the afflicted genes and regions frequently correlating to the malignancy category (Esteller 2008; Esteller & Kulis 2012). Additional illnesses such as autoimmune diseases, allergies and asthma, diabetes (type 2), metabolic diseases, autism, and neurodegenerative and mental illnesses have also been linked to epigenetic modifications (Szyf 2009; Best & Carey 2010).

Furthermore, the evolutionary origins-of-disease hypothesis postulates that gestation, when the epigenome is largely formed, is the moment of origination for many epigenetic disorders (Thornburg et al. 2010). Numerous techniques have been devised to discover epigenetic alterations in biological specimens in order to improve the detection and diagnosis of epigenetic illnesses. Since methylation changes correlate with disease status, especially in cancer, DNA methylation identification is presently the greatest putative gate indicator for epigenetic modifications in disease. A range of qualitative and quantitative approaches can be used to evaluate DNA methylation indicators, which are persistent in separated DNA and may additionally demonstrate whether a pharmacological therapy is successful or not (Heyn & Esteller 2012). Additionally, preparations derived from noninvasively collected materials, such as feces, serum, and urine, can be used to identify disease-specific DNA methylation patterns and changes that correspond to bladder (Kim & Kim 2009) and colon (Kim et al. 2010) cancer.

Histone alterations are just as significant when seen in relation to the DNA sequence with which they are bound. Therefore, it is necessary to separate complete histone/DNA complexes and entire cells from tumor material (Esteller 2008). The majority of techniques for identifying DNA methylation in medical samples rely on amplifying a bisulfite-treated DNA fragment using the polymerase chain reaction (PCR) (Chaudhry 2010). With this approach, a methylated or unmethylated cytosine

 DOI: 10.1201/9781032706511-15

inside a certain DNA sequence can be identified. Since the bisulfite method can be employed to determine DNA methylation for targeted loci linked to a particular disease, it has given rise to several patent filings. Bisulfite treatments can also be employed in conjunction with next-generation sequencing to assess global DNA methylation alterations that happen within the context of illnesses, providing valuable insights into epigenetics and disease (Esteller 2011).

While epigenetic modifier genes are commonly altered in malignancies, conventional sequencing of DNA can additionally be helpful for comprehending epigenetic modifications in cancer. These alterations may account for certain documented alterations to the epigenome (Lee et al. 2012). As epigenetic alterations are reversed, the enzymes responsible for producing these are viewed as promising candidates for pharmaceutical intervention (Esteller 2011). Reversing an alteration that is intrinsically linked to an illness ought to in principle improve indications and lead to effective treatments. Four epigenetic medications have currently been granted approval by the FDA for the treatment of illness: a pair that block histone deacetylation and another couple that inhibit DNA methylation (Szyf 2009). The fact that epigenetic medications regulate genes so broadly and can therefore result in serious adverse reactions is a major warning indicator. Though more target-specific pharmaceuticals have been invented, petitions for patents continue to be filed for the therapeutic administration of epigenetic enzymes such as histone deacetyltransferase inhibitors, both alone and in conjunction with additional drugs (Szyf 2009).

In addition to offering molecular hints about the genesis of diseases, epigenetic disease correlations may serve as biomarkers for diagnosis. Because epigenetic signatures are specific to phases of development and tissues, there is a lot to discover in creating biomarkers that take advantage of these distinct features (García-Giménez 2015). Moreover, the reversibility of epigenetic modifications has generated substantial curiosity regarding the creation of medications with epigenetic mechanisms of action (Szyf 2009; Hunter 2015). Cancer has produced the greatest corpus of research on disease epigenetics to date. Numerous epigenetic discoveries have been reported since the beginning of the 1980s, when the first connections among DNA methylation and cancer were made. These discoveries implicate various parts of the epigenetic mechanism.

15.1.1 Crucial Epigenetic Biomarkers in Disease Diagnosis

A biomarker is a detectable biochemical substance present in tissues, blood, or bodily fluids. It's associated with the progression of diseases like cancer and include DNA, RNA, proteins, and epigenetic changes. It can be expected that epigenetic biomarkers may be useful as a restorative target in different disease. Several epigenetic therapies have been authorized for the treatment of malignancies such as big B-cell lymphoma, certain kinds of leukemia and myelodysplastic syndrome, and cutaneous T-cell lymphomas, with more cancer treatments now being researched (Hillyar et al. 2020). Epigenetic biomarkers for cancer, migraine, autoimmune disorders (systemic autoimmune rheumatic diseases, thyroid autoimmune diseases, Sjogren's syndrome and type 1 diabetes), cardiovascular diseases, asthma, immunosuppression, rheumatoid arthritis, and neurological disorders such as Alzheimer's disease and Parkinson's disease (Figure 15.1) are discussed in the following sections.

15.2 CANCER

A precise tumor biomarker revealed cancer presence, predicted outcomes, and helped direct specific treatments. Epigenetics investigates heritable modifications in gene expression that happen without changes to the DNA sequence. Epigenetic processes can modify gene expression (Tiffon et al. 2018).

Epigenetic modifications encompass DNA methylation, histone modifications that induce chromatin remodeling, and noncoding RNAs (ncRNAs) governing gene expression at post-translational stages. Epigenetic changes, unlike genetic alterations, are reversible, offering potential as effective targets for treating diseases. Epigenetic therapies work by restoring gene expression in cells where it has been suppressed through epigenetic modifications, essentially reprogramming cancer cells.

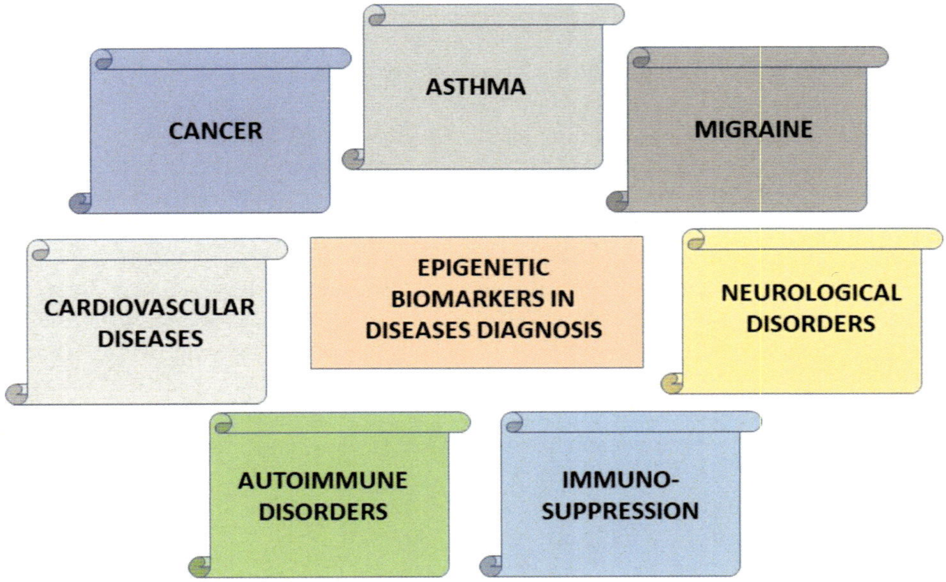

FIGURE 15.1 Epigenetic biomarkers used in disease diagnosis.

Epigenetic changes, like abnormal DNA methylation and histone acetylation at gene promoter areas, are among the first steps in the development of cancer. These changes lead to the suppression of specific genes that play roles in apoptosis, cell cycle regulation, and DNA repair. Epigenetic biomarkers have the potential to serve as predictive indicators for anticipating therapeutic drug responses For instance, the connection between methylation of the KEAP1 gene and improved overall survival. This could potentially act as a biomarker indicating resistance to taxane-based chemotherapy treatments (Berdasco 2019). We next discuss the various epigenetic biomarkers for a range of cancers.

15.2.1 Renal Cancer

Neoplasia of the kidney, renal pelvis, or ureter (ICD-9 189 and ICD-10 C64-C66) is known as renal cancer, and it is the sixth most prevalent cancer in men in Europe. Of all cancers, it accounts for 29,600 new cases annually. In addition to certain environmental and occupational factors, risk factors include obesity, tobacco use, and a diet deficient in vegetables. About 25% of individuals have locally advanced renal carcinoma, 45% have localized disease, and 30% have metastatic disease with renal carcinoma.

15.2.1.1 DNA in Biofluid Methylation as Renal Cell Carcinoma Biomarkers

DNA methyltransferase (DNMTs) enzymes add a methyl group (-CH3) at the carbon-5 position of cytosine residues in the context of cytosine-phosphate-guanine (CpG) dinucleotides, creating DNA methylation in mammalian cell cytosine 5′-methyl (5mC). The most well-researched epigenetic mechanisms involve biological processes such as gene expression regulation in animal development. Promoter CpGs, typically found in normal cells, are linked to active gene expression and remain hypomethylated during differentiation (Turajlic et al. 2018a; Turajlic et al. 2018b). However, the promoter CpGs are frequently hypermethylated in cancers such as kidney cancer.

15.2.1.2 Biofluid MicroRNAs as Renal Cell Carcinoma Biomarkers

MicroRNAs (miRNAs) are a class of short ncRNAs with a length of 18 to 25 nucleotides that control the expression of target genes by binding to the complementary 3′UTR of mRNA and preventing the protein's translation or encouraging its destruction (Stirzaker et al. 2014) Evidence is mounting that miRNAs are involved in a variety of functions relevant to cancer formation and progression, such as drug and radiation response, invasion, angiogenesis (Nilsen 2007), cell proliferation (Pillai et al. 2007; Fang et al. 2011), apoptosis (Chen et al. 2018; Chen et al. 2018; Xiao et al. 2019), metastasis (Jia et al. 2018; Sharifi & Moridnia 2017), and opposition (Fan et al. 2018; Xiao et al. 2017; Kabir et al. 2018; Ma & Dong 2010). Multiple independent investigations found that the expression of miRNAs was abnormal in renal cell carcinoma (Schaefer et al. 2010; Lichner et al. 2014). Additionally, miRNAs can be found in a variety of biofluid sources such as urine, saliva, plasma, and serum (Weber et al. 2010).

15.2.1.3 Biofluid Long Noncoding RNAs as Renal Cell Carcinoma Biomarkers

Long noncoding RNAs (lncRNAs) are a family of single RNAs with a length more than 200 nucleotides but no capacity to code for proteins (Peng et al. 2017). They regulate gene expression through chromatin-modifying factor recruitment or transcriptional control. Additionally, they influence post-transcriptional regulation by interacting with miRNAs, mRNAs, protein complexes, and other factors (Qu et al. 2016). LncRNAs contribute to renal cancer development (Bhan et al. 2017; Liu et al. 2018); similar to how miRNAs work, lncRNAs contribute to the emergence and spread of cancer, regulating migration, senescence, apoptosis, the cell cycle, and other processes drug resistance, invasion, and other topics (Liu et al. 2016; Lin et al. 2016; Yuan et al. 2017).

15.2.1.4 Other Noninvasive Epigenetic Cancer Detection Methods

The presence of circulating nucleosomes and their modifications, as well as other ncRNAs such P-Element induced wimpy testis-interacting RNAs (piRNAs), may also operate as non-invasive biomarkers for cancer (Xiao et al. 2017) while other ncRNAs are far less suitable (McAnena et al. 2017; Sole et al. 2019; Dietrich et al. 2014).

15.2.2 Gastric Cancer

Around 4.1 million individuals are examined with gastrointestinal (GI) cancers every year, leading to approximately 3 million cancer-related deaths worldwide resulting from late disease detection. Pancreatic (338,000) and esophageal (456,000) cancers are less frequent, yet their low overall survival rates (330,000 and 400,000 cancer-related deaths) still classify them as significant contributors to cancer fatalities. Early-stage GI cancers often lack symptoms, leading to late-stage diagnoses with limited treatment choices and unfavorable patient outcomes. Aberrant promoter DNA methylation, detectable in body fluids like bile, feces, and blood, is gaining significant interest as an indicator for noninvasive cancer biomarkers. These markers could aid in early detection, prognosis, and predicting treatment outcome (Portela et al. 2010). While gastric cancer occurrence has decreased in many Western nations, it still ranks among the main causes of cancer-related fatalities in Asia (Ferlay et al. 2013). In nations like Korea and Japan, where gastric cancer (GC) rates are notably elevated, screening initiatives involving methods like endoscopy and barium-meal have been introduced (Leung et al. 2008).

15.2.2.1 Histone Modifications

Many pathophysiological cellular processes, including inflammation, carcinogenesis, and the epithelial–mesenchymal transition, are impacted by histone post-translational modifications in GC, which primarily include acetylation, methylation, phosphorylation, and ubiquitination. The appearance and advancement of other gastrointestinal tumors, such as esophageal, colorectal, and hepatocarcinoma liver cancer, have been linked in recent years to some new modifications, including succinylation, sumoylation, butyrylation, and crotonylation. These discoveries offer fresh perspectives into the workings of these tumors as well as possible treatments for cancer diagnosis and treatment (Tang et al. 2022).

15.2.2.2 DNA Hypermethylation

DNA hypermethylation with abnormalities typically occurs in the promoter region of GC tumor suppressor genes such as p16, RASSF1A, and hMLH1. By decreasing the binding of transcription factors, hypermethylation prevents gene transcription, hindering DNA readability and ultimately leading to gene silencing. In particular, 5-fluorurazil chemo-resistance in GC was induced and the cell cycle in G1 phase was inhibited by changes in methylation in the p16 promoter. Infection by *Heliobacter pylori* and Epstein–Barr virus are recognized as significant gastric cancer risk factors. These infections are linked to heightened levels of promoter DNA methylation (Maekita et al. 2006).

15.2.3 Esophageal Cancer

Diagnosis of esophageal cancer commonly occurs in advanced stages, presenting limited treatment avenues. This cancer often originates from the precancerous Barrett's esophagus (Barrett et al. 1999), and assessing its risk in people with Barrett's esophagus is currently a subject of ongoing investigation. Numerous extra genes have been proposed as potential biomarkers for forecasting disorder advancement in individuals with Barrett's esophagus (Clement et al. 2006b). A biomarker set comprising eight genes—CDKN2A, RUNX3, HPP1, NELL1, TAC1, SST, AKAP12, and CDH13—was confirmed through a multicenter validation (Jin et al. 2009).

15.2.3.1 DNA Methylation

Despite numerous DNA methylation biomarkers proposed for esophageal cancer, only a handful have been investigated non-invasively in blood samples (Kawakami et al. 2000). There is a limited availability of DNA methylation biomarkers with predictive significance for esophageal cancer, frequently studied in small sample sets (Eads et al. 2001). In a recent investigation, CDO1 methylation emerged as an autonomous prognostic indicator for esophageal squamous cell carcinoma patients (multivariate HR = 2.00, P = 0.03) in an analysis of 169 cancer samples (Ushiku et al. 2017).

DNA methylation biomarkers found in precancerous or initial stages of cancer might aid in earlier detection, potentially lowering mortality rates. Nonetheless, currently, only a limited subset of the recognized biomarker possibilities for esophageal cancer have been assessed in noninvasive samples. DNA methylation takes place on CpG islands, which are frequently found close to transcriptional regulatory regions. Research has demonstrated a strong correlation among the carcinogenesis of esophageal cancer and the DNA methylation that takes place here. In particular, changed DNA methylation in CpG islands causes chromosomal rotation, silencing, and loss of oncogene expression that results in the growth of tumors. Numerous related genes, including p16, E-cadherin, DNMT1, and MTHFR, have had their methylations studied in relation to esophageal cancer (Liu et al. 2023).

15.2.3.2 Histone Modification

Histone methylation and acetylation play a significant role in the pathophysiology of human esophageal cancer. Histone arginine methyltransferases and lysine demethylases, which can control gene expression and preserve chromatin structure at the corresponding sites, dynamically regulate the majority of the lysine and arginine residues of H3 and H4 that contain histone methylation sites (Liu et al. 2023).

15.2.3.3 NcRNA Regulation

Research has increasingly revealed the role of ncRNAs in tumor growth, invasion, and metastasis; they influence not only the diagnosis but also the course of treatment and prognosis of esophageal cancer. Using miR-455-3p antagonists, Liu A et al. (2017) discovered that they could significantly increase the sensitivity of chemotherapy in patients with esophageal cancer (Liu et al. 2017).

15.2.4 Colorectal Cancer

The third most frequent malignancy and the fourth leading cause of cancer-related mortality is colorectal cancer (CRC). The majority of CRC cases are found in Western nations, and the disease is becoming more common every year. The likelihood of having CRC is between 4% and 5%, and personal characteristics or behaviors including age, a history of chronic illness, and lifestyle choices are linked to the risk of acquiring CRC.

The first stool DNA test for detecting average-risk CRCs to receive FDA approval was Cologuard, in 2014. It is a highly sensitive assay that targets multiple molecular targets in stool DNA (mt-sDNA), including N-Myc downstream-regulated gene 4 protein (NDRG4), the methylation status of bone morphogenetic protein 3 (BMP3), and the KRAS mutation, fecal immunochemical testing, and β-actin for CRC screening. Unlike other screening techniques, the mt-sDNA test detects DNA methylation and mutations across premalignancy to advanced stages of carcinogens, using fecal hemoglobin as a single marker. Fecal hemoglobin alone, however, is more trustworthy because precancerous lesions might not bleed in late stages (Duran et al. 2020; Kim & Park 2018; Ferrari et al. 2021; Wang et al. 2021).

In 2016, the FDA authorized the first blood-based screening test, Epi proColon®. The test is based on the qualitative examination of methylated SEPT9 (septin 9) detected by PCR. Both tissue and blood have much higher levels of methylated SEPT9 than usual. Samples from CRC patients have better specificity than samples from Cologuard® and other screening techniques, but the test is less sensitivity (Lamb & Dhillon 2017). Single-target stool DNA testing is available for CRC screening through both Epi proColon® and Cologuard®. The FDA has authorized EarlyTectTM-Colon Cancer (Genomictree, South Korea). Colosafe® (Creative Biosciences China) was authorized as a CRC screening test by the China National Medical Products Administration (Ferrari et al. 2021; Wang et al. 2021). Some of the biomarkers for colorectal cancer are discussed below.

15.2.4.1 Syndecan-2

The transmembrane protein syndecan-2 (SDC2) is involved in angiogenesis, cell migration, and proliferation (Oh et al. 2017). The amount of SDC2 methylation was measured in two investigations employing tissue and stool (Wang et al. 2021, Niu et al. 2017, Oh et al. 2017). Park et al. (2018) used tissue and bowel lavage, and three research groups (Han et al. 2019; Wang et al. 2020; Su et al. 2021) focused solely on stool DNA methylation. In one study, samples were paired. Stages I and II had sensitivity ranges of 83.3%–91.4% and 89.6%–100%, respectively. Although early SDC2 methylation evaluation is advised, the test appeared to be more sensitive for detecting stage III/IV CRC (Niu et al. 2017).

15.2.4.2 SEPT9

As a member of the septin family and a GTP-binding protein, SEPT9 is a part of the cytoskeleton. As a result, it is involved in vital activities including cell division. The SEPT9 promoter's island hypermethylation results in altered expression, which eradicates its tumor suppressor action, which aids in the development of cancer (Sun et al. 2020).

A promising single biomarker is mSEPT9, and it is being investigated fairly widely. Six research groups tested with plasma (Shayan et al. 2013; Fu et al. 2018; Xie et al. 2018; Yang et al. 2019; Sun et al. 2019; Ma et al. 2019), and one used paired samples to investigate identical twins' plasma and feces (Liu et al. 2020; Houshmand et al. 2017). The researchers commonly used Epi proColon® to assess plasma mSEPT9. Plasma samples with methylated SEPT9 were more sensitive than recognized tumor protein indicators like FOBT, CEA, and CA19-9. The combination of the mSEPT9 and FOBT assessments produced the highest sensitivity (100%) for stage I CRC. Additionally, stool samples have higher sensitivity for mSEPT9 than plasma samples, despite having similar specificity.

15.2.4.3 BMP3

Osteoblasts and osteocytes release BMP3, which regulates bone mass (Kokabu & Rosen 2018) and is crucial for cellular growth and development. Additionally, BMP3 promoter hypermethylation results in its inactivation, which has consequences for carcinogenesis (Wen et al. 2019). The methylation status was highest in stage IIA CRC (66.6%) in both of the studies under review, suggesting a function for BMP3 in early detection, although both studies revealed a decline in overall sensitivity and specificity. Additionally, the scientists concluded that BMP3 is not adequate as a sole biomarker; rather, it should be evaluated in conjunction with other methylation genes (Houshmand et al. 2017; Ashoori et al. 2018)

15.2.4.4 NDRG4

A study using tissues from NDRG4 wild-type, heterozygous, and knockout mice and humans revealed that the enteric nervous system (ENS) is the only place where NDRG4 is expressed explicitly in enteric neurons. Although NDRG4 has primarily been studied in the brain and heart; it is expressed throughout the body in nervous system structures (Vaes et al. 2017). An investigation into whether the ENS, via NDRG4, influences intestinal tumorigenesis recently revealed that NDRG4 knockdown in CRC models and in an indirect co-culture of primary ENS cells was associated with the development of enlarged intestinal adenomas; the organoid growth was enhanced by the Ndrg4/ENS cell secretome. The ENS, via loss of Ndrg4, showed involvement in cancer development (Vaes et al. 2021).

15.2.4.5 Alpha 1-Antitrypsin

Alpha 1-antitrypsin (A1AT) is mentioned in a different study as a possible biomarker for the early diagnosis of CRC. Both CEA and A1AT plasma concentrations were assessed in healthy people and in CRC patients. A1AT plasma level and tumor incidence were associated; A1AT assessment demonstrated higher sensitivity and specificity for early stage than CEA CRC detection. When inflammation occurs, especially when it is localized at hepatocytes, cell lines that have colonic tumors express this gene as well. Some research found increased A1T1 in pancreatic, breast, and liver tumors (Jaberie et al. 2020). In a literature review, one study evaluated CLIP4, three examined TWIST1, and two examined LINE-1 as single biomarkesr (Suehiro et al. 2018). Other researchers used a panel of methylation biomarkers to evaluate C9orf50 along with CLIP4 and KCNQ5, along with TWIST1, or only with KCNQ5 (Cao et al. 2021).

15.2.5 Breast Cancer

Breast cancer is the most common cancer among women. It has recently been increasing in incidence, and it has now surpassed lung cancer as the most diagnosed cancer worldwide. According to the 2020 global cancer statistics from the International Agency for Research on Cancer and the American Cancer Society, breast cancer accounted for about 11.7% of new cancer cases, followed closely by lung cancer at 11.4%. Other prevalent cancers included colorectal, prostate, and stomach cancers (Sung et al. 2021).

In breast cancer, there are two main types: noninvasive (limited to epithelial components, known as carcinoma in situ) and invasive carcinoma. Invasive carcinoma occurs when abnormal cells proliferate in breast tissue, causing them to penetrate from duct walls into the surrounding stroma (Makki 2015).

In one study on the presence of p16 promoter hypermethylation in breast cancer, p16 served as both a prognostic and a predictive indicator for the efficacy of hormonal therapy (Goyal et al. 2019). In simpler terms, when the p16 gene is overly methylated and specific other genes like BRCA1, BRCA2, ERα, and RARβ2 show similar methylation patterns, it might indicate a pre-cancerous condition. This could potentially help identify women who need close monitoring for breast cancer even before visible signs appear (Thomas & Marcato 2018).

Recent reports indicate a growing count of identified genes with abnormal methylation in breast cancer. Aberrant DNA methylation is a promising biomarker for liquid biopsies due to its early appearance, cancer specificity, stability, and presence in bodily fluids. Its strong stability and detectability in circulating cell-free tumor DNA make it an attractive option for a quick, accurate, affordable, and noninvasive breast cancer test (Mancilla et al. 2020).

15.2.5.1 DNA Methylation and DNA Methyltransferase Inhibitors

DNA methylation, facilitated by DNMT enzymes like DNMT1, DNMT3A, and DNMT3B, is a key epigenetic regulation mechanism. This process involves adding a methyl group to cytosine within CpG dinucleotides, which plays a roles in transcription, chromatin remodeling, gene imprinting, and suppressing repeat elements. In breast cancer, global demethylation in tumors is observed, but specific CpG island promoters can become methylated, leading to gene repression (Skvortsova et al. 2019). DNMT inhibitors like decitabine and 5-azacytadine are approved for hematological malignancies but not yet used in solid malignancies (Silverman et al. 2002; Fenaux et al. 2009; Griffiths et al. 2020).

15.2.5.2 Chromatin Remodeling, Super-Enhancers, Bromodomain, and Extra-Terminal Motif Inhibitors

Chromatin remodelers play crucial roles in altering chromatin structure, influencing DNA damage repair, recombination, replication, and transcriptional control. Mutations in chromatin remodeler genes, particularly in SWI/SNF subunits, are prevalent in various cancers, including breast cancer and was observed, SWI/SNF complexes, operating in an ATP-dependent manner, remodel chromatin to facilitate transcription and transcription factor binding. Notably, SWI/SNF is essential for transcriptional activation by nuclear receptors in breast cancer, highlighting its significance in gene regulation in this context (Morel et al. 2020).

15.2.6 Lung Cancer

Lung cancer is the second most common cancer worldwide. Around 13% of individuals diagnosed with lung cancer manage to survive beyond a five-year period. The development and advancement of lung cancer are influenced by a combination of genetic, epigenetic, and environmental factors. Epigenetic changes, like the modification of DNA methylation in gene promoters causing gene suppression, are frequent occurrences in lung cancer. These changes occur in specific nuclear locations and chromosome regions due to exposure to environmental risk elements like smoking, drugs, and chronic inflammation (Tomasetti et al. 2017)

A new technique called a ‘liquid biopsy’ has become available, allowing us to study the genetic changes in tumors without invasive procedures. This helps us track tumor evolution and find cancer biomarkers in fluids. Changes in the epigenetic profile can offer valuable information about the origin and progression of cancer. These alterations often happen before visible tissue changes, making them useful as biomarkers for diagnosing cancer and assessing risk (Liloglou et al. 2014).

Epigenetic abnormalities provide adaptable and reversible opportunities for treating cancer. Changes like overexpression, mutation, or rearrangement are being increasingly identified in genes that govern the epigenome. DNA methylation and miRNAs have rapidly come up as valuable biomarkers in bodily fluids, with the potential to assist in the medical management of lung cancer. Methylation occurs early in the development of lung tumors, and numerous genes with abnormal methylation patterns have been discovered in lung cancer (Liloglou et al. 2014).

DNA methylation changes can serve as diagnostic indicators for specific tumor types or stages, highlighting the promise of using DNA methylation as a molecular marker. For instance, the methylation of the p16 promoter region is suggested as a potent biomarker for identifying early lung cancer and tracking prevention trials (Belinsky et al. 1998). DNA methylation abnormalities were identified through either direct methods or by using drugs to reverse the silencing of genes that had been turned off due to epigenetic changes in lung cancer cells.

Discovering new biomarkers is crucial for predicting outcomes and tumor responses accurately in lung cancer treatment. Epigenetic biomarkers like DNA methylation and microRNA expression are especially promising due to their distinct characteristics, making them valuable for predicting prognosis. Many studies have extensively investigated and documented these biomarkers in the context of lung cancer (Rauch et al. 2007; Rauch et al. 2008).

15.3 MIGRAINE

The CALCA gene produces calcitonin gene-related peptide (CGRP), and by alternative splicing, calcitonin as well. Normal CALCA expression is only found in neurons and endocrine cells; glial cells do not express the gene. The CGRP receptor, which is made up of the calcitonin receptor-like receptor (CRLR) and RAMP1 genes, regulates CGRP activity. Increased CGRP levels have been linked to the etiology of migraines. A number of studies show that CGRP is involved in the pathophysiology of migraine.

Environmental variables like stress, food, and toxins can change the epigenome. Growth, development, and illness risk are all affected by altered epigenome. The interaction of the epigenetic environment and genetic background appears to be more pertinent at crucial developmental stages, such those in childhood, where the plasticity of the epigenome is increased. Numerous research projects on human cohorts and animal models have demonstrated connections between epigenetic modifications in the progeny and the environment during pregnancy, such as maternal and fetal toxicant exposures (Alavian-Ghavanini & Rüegg 2018). DNA methylation has been the subject of the majority of studies to far, although the consequences on other epigenetic alterations, such histone modifications, caused by environmental variables additionally developing are miRNA expression

15.3.1 The JNK Signalling Pathway in Epigenetic Regulation

The brain exhibits considerable hyperexcitability in migraine patients, responding abnormally to both internal and external stimuli and perhaps precipitating attacks. Oral and nasal trigeminal nerve fibers are susceptible to a variety of environmental stimuli, including poisons and chemical irritants. Phosphorylation of the inducible transcription factor by JNK, the ability of c-Jun to trigger gene transcription is largely determined by it. In reaction to injuries in the nerve fibers as well as infection and inflammation of the nervous system, phosphorylation is the final signal in a series of signals. In vitro histone acetylation in response to a neurotoxic stimulus (mustard oil) and the function of the JNK/c-Jun cascade play a regulatory role in controlling histone H3 acetylation in rat trigeminal neurons. This was the first study to provide solid evidence of JNK's involvement in the epigenetic modulation of histones in peripheral trigeminal neurons in response to chemical environmental stimuli (Wu et al. 2008).

15.3.2 CGRP, RAMP1, and Migraine

The pathophysiology of migraines is heavily influenced by CGRP (Edvinsson 2019; Durham 2008) The CGRP is encoded by the alternative splice variant of the CALCA gene encodes both CGRP and the hormone calcitonin. (Tendl et al. 2013). Trigeminal neurons have high levels of CGRP and its receptor expression (Malhotra 2016; Edvinsson & Warfvinge 2019); CGRP functions as a powerful vasodilator when it causes neurogenic inflammation by activating adenylyl cyclase in vascular smooth muscle (Malhotra 2016).

Intravenous CGRP induces migraine-like symptoms; during acute episodes, heightened CGRP blood levels are detected in chronic migraine sufferers as well as during active migraine attacks versus sufferers of episodic migraine (Edvinsson 2007; Russo 2015; Cernuda-Morollon et al. 2013). The components of membrane heterodimer complexes, or CGRP receptors, include RAMP1, an auxiliary protein, and calcitonin receptor-like receptor (CLR). The RAMPs are a small family of three single transmembrane proteins that modify the functioning, pharmacology, and cell trafficking of CGRP family peptide receptors. CLR and RAMP1 must combine to generate a complex for the CGRP peptide receptor to function. The expression of these peptide receptors in the trigeminal system is still unclear, and their potential functional roles have not yet been determined, owing to the intricacy of this system. Numerous pertinent areas, including the periaqueductal gray, area postrema, pontine raphe nucleus, spinal trigeminal nucleus, and spinal cord, were found to express CLR and RAMP1 mRNA and protein (Edvinsson et al. 2020; Barwell et al. 2012; Gingell et al. 2019).

15.4 AUTOIMMUNE DISORDERS

An immune reaction to antigenic elements of the host itself (autoantigens) defines autoimmune disorders. Systemic autoimmune illnesses and organ-specific autoimmune disorders are the two basic categories that can be identified. While in organ-specific diseases the immune response is focused on a particular organ, in systemic diseases, the immune system assaults its own antigens in multiple organs in a generalized fashion.

15.4.1 Systemic Autoimmune Rheumatic Diseases

A class of uncommon chronic inflammatory disorders known as systemic autoimmune rheumatic diseases (SARDs) includes Sjogren's syndrome (SS), systemic lupus erythematosus (SLE), systemic sclerosis (SSc), and rheumatoid arthritis (RA). The loss of self-tolerance leading to aberrant humoral and cell-mediated reactions against self-components is known as autoimmunity. Studies have increasingly shown that epigenetically changed immune cells and molecules, such as CD4+ T-cells and costimulatory molecules, are major contributors to SARDs, impairing the efficiency and control of the cellular immune system and control (Jeffries & Sawalha 2015; Strickland & Richardson 2008; Brooks et al. 2010; Meda et al. 2011). Epigenetic changes that take place across these autoimmune diseases can provide important information about their pathogenesis and management. Many researchers have therefore investigated the epigenetic processes involved in the most significant epigenetic changes in SARDs control of autoimmunity. Historically, epigenetic considerations for the key associated kinds of RA, SLE, SSc, and SS, among other autoimmune diseases, have been studied to determine a close relationship between new insights about epigenetics and systemic autoimmunity SARDs' regulatory framework.

15.4.2 Systemic Lupus Erythematosus

The autoimmune condition most closely associated with epigenetic alteration is SLE. It is more common in women and is mostly brought on by T lymphocyte dysregulation, making it a challenging and complicated condition to treat. Autoantibodies against nuclear antigens can develop in any organ system or tissue, including the kidney and blood arteries. Epigenetic changes like DNA methylation and histone modifications can control the expression of certain genes in mature T cells. Many genes, including CD11a, perforin, CD70, and CD40LG, were discovered to be hypomethylated in the T cells of SLE patients (Richardson 2007). Early research on SLE CD4+ T cells showed that DNMT inhibitor therapy resulted in a conversion to autoreactivity and an induced lupus-like syndrome (Richardson 2007; Yung et al. 1995; Quddus et al. 1993; Richardson 1986). Only cloned and polyclonal human and murine CD4+ cells have displayed auto-reactivity; CD8+ cells have not, and the cause of this is yet unknown (Richardson 1986; Richardson et al. 2019).

15.4.3 Autoimmune Thyroid Diseases

Autoimmune thyroid disorders (AITDs) are regulated by B and T lymphocytes. Flajani–Basedow–Graves disease are the two primary clinical manifestations of AITD, Hashimoto's thyroiditis (HT) and Graves' disease (GD) (Lee et al. 2015; McLachlan & Rapoport 2014). The pathogenesis of AITD's epigenetic pathways is methylation of DNA. In-depth examinations of lncRNA and histone alterations have been conducted in AITDs, the clinical relevance of epigenetics modulation is still difficult to ascertain. Patients with an AITD exhibited widespread DNA hypomethylation, affecting genes crucial for immune function and immune cell activation. This dysregulation can result in an autoimmune attack on thyroid tissues. (Cai et al. 2015; Limbach et al. 2016). A genome-wide analysis identified more than 200 genomic sites, including ICAM1, that were hyper- and hypomethylated in GD patients. These sites play a role in regulating antigen processing and presentation, along with the DNMT1 and MECP2 (Fodor et al. 2017). In GD patients, CD4+ and CD8+ cell profiling revealed that highly methylated ICAM1, CD247, and CTLA4 gene sites are linked to T cell receptor signaling, leading to overexpression.

15.4.4 SS

Dry eyes and mouth are symptoms of SS, an autoimmune condition of unclear origin that affects the salivary and tear glands. SS, an autoimmune condition that affects about four million people in the US, is fairly common. This autoimmune disease is accompanied by blood autoantibody synthesis that is directed against different body tissues, causing an inflammatory reaction. Reduced DNA methylation of immune cells is one of the most researched epigenetic markers in SS, similar to other autoimmune illnesses.

15.4.5 Type 1 Diabetes

Type 1 diabetes (T1D) is a chronic autoimmune condition that causes both cell death and a significant inflammatory response. Diabetes is one of the leading causes of death due to its increasing prevalence throughout the world. In more detail, the destruction of the insulin-producing pancreatic cells by the body's immune system lowers and eventually eliminates the pancreas' capacity to produce insulin. The illness does not yet have a cure. Although exogenous insulin is still an essential treatment, the damaging process that leads to T1D is believed to involve an adaptive immune response. Recent research has highlighted the role of epigenetics in T1D, with some studies producing favorable insights (Olsson et al. 2014).

In the past ten years, a few studies have revealed that noncoding RNAs, histone changes, and DNA methylation all play significant roles in T1D. SNP-CpG methylation patterns were once more found to be potentially relevant for the genetic connection of insulin expression and/or secretion in human pancreatic islets by genome-wide DNA methylation analysis (Rui et al. 2016). More specifically, it is known that potential genes including GPX7, GSTT1, and SNX19 have a direct effect on a number of biological processes involving proliferation and death in pancreatic cells. The relationship between epigenetic alterations in insulin secretion and diabetes risk was further supported by a study conducted in mice models (Wilson et al. 2009).

15.4.6 Alterations in Immune System Epigenetics

Numerous recent findings unequivocally show that major immune cell activities, including hematopoietic lineage selection, antigen-receptor rearrangement, allelic exclusion, and immunological responses to infections, are epigenetically controlled (Aune et al. 2009; Bergman & Cedar 2010; Cedar & Bergman 2011; Fer nández-Morera et al. 2010; Janson et al. 2009; Lal Bromberg 2009; Lee et al. 2006; Pang et al. 2009; Sawalha 2008; Vanden Berghe et al. 2006; Wilson et al. 2005; Allan et al. 2012). For instance, epigenetic mechanisms control T helper cell development and the constancy of their lineage (Wilson et al. 2009; Allan et al. 2012); exposing dendritic cells to oxidized phospholipids alters a cell's phenotype through epigenetic mechanisms (Blüml et al. 2009), and epigenetic mechanisms regulate the B cell's ability to produce antibodies (Subrahmanyam et al. 2012). These epigenetic-controlled immunological developmental processes may be changed to encourage autoimmune disease.

15.5 CARDIOVASCULAR DISEASE

The missing heritability factors of complex diseases including atherosclerosis, hypertension, metabolic syndrome, and diabetes that have not yet been explained by genetic studies of sequence variation have been theorized to be due to epigenetic alterations (Slatkin 2009; Manolio et al. 2009). In a recent study, the inheritance of single-nucleotide polymorphisms (SNPs) close to known imprinted genes was followed to assess the impact of parental origin on illness connection. These findings revealed six SNPs where risk was affected by a gene's parental origin. One of these SNPs that was connected to type 2 diabetes mellitus also had a direct correlation with methylation status. These results imply that more non-sequence dependent variants contribute to heritable features.

15.5.1 The Advanced Biomarker cTn

One biomarker has been proven to be effective for both diagnosis and prognostication (Eur Heart 2000; Alpert et al. 2000; Antman et al. 1996; Ohman et al. 1996): cTn. The biomarkers for cTn, assuming superior analytical techniques, whether for cTnI or cTnT, are extremely sensitive and specific for cardiac damage.

15.5.2 Emerging Markers

A marker must make a diagnosis or describe the prognosis in order to effectively and usefully direct therapy. There are two new markers in this regard: C-reactive protein (CRP) and the natriuretic peptides BNP (B-type natriuretic peptide) and NT-proBNP. The liver produces the acute-phase CRP. Interleukin 6 is the stimulus that is closest to it. Recently, it was discovered that CRP has a unique receptor (Jaffe et al. 2000).

The fluctuation of CRP levels is a subject of debate (Devaraj et al. 2005; Pearson et al. 2003). Some have found that levels vary depending on gender and ethnicity (Rifai & Warnick 2004). The test should be repeated at least two weeks later if the patient has an elevation, is critically unwell, or has an infarction that is progressing (Khera et al. 2005). Acute disease is usually the cause of values higher than 10 mg/l (Pearson et al. 2003). Values of 3 mg/l and 1 mg/l are related to increased risk and lower risk, respectively (Pearson et al. 2003). The intermediate range is defined as 1 to 3 mg/l (Pearson et al. 2003). There are numerous CRP assays (called high-sensitive CRP or hsCRP), and most provide comparable results.

15.5.3 Developing Markers

- *sCD40 Ligand* sCD40 is a signaling protein that shows how platelets and inflammation interact with the plaque (Rifai & Ridker 2001). Patients with ACS have been demonstrated to have elevated sCD40 ligand. Furthermore, studies have indicated that it may be significant for these patients' prognosis.
- *Myeloperoxidase* Myeloperoxidase is a byproduct of degranulation produced by white blood cells (Rifai & Ridker 2001). Even when results are derived from the coronary effluent that drains a non-culprit coronary channel, elevations happen. Research shows that systemic levels can help predict outcomes for patients who come to the ER with chest discomfort, although studies have frequently only included the admission troponin level or high cTn threshold values in their comparisons. A commercial assay has just become accessible.

- *Ischemia-Modified Albumin* Ischemia-modified albumin has been promoted as a means of identifying ischemia in the absence of necrosis (Rifai & Ridker 2001). The idea is that weight has a metal effect on the amounts of BNP. The binding site on the amino terminus of albumin is damaged by ischemia. It is marketed for its negative predictive value. It has been difficult to validate ischemia-modified albumin because of the lack of a gold standard for ischemia. In addition, the assay is cumbersome to use.
- *Pregnancy-Associated Plasma Protein-A* Pregnancy-associated plasma protein-A, a protein that belongs to the insulin-like growth factor family, is secreted during neovascularization and could therefore be a sign of impending plaque rupture. It has no correlation with any of the heart damage markers that are currently accessible. It appears to identify patients who are at risk for subsequent occurrences in certain trials, but there isn't a standardized assay available. Since phospholipids are released when choline is released, it is possible that choline is a signal of ischemia and/or necrosis. The marker may help predict patients' outcomes in cases of ACS, according to a number of studies (Rifai & Ridker 2001).
- *Placental Growth Factor* Placental growth factor is a member of the vascular endothelial growth factor family, which some claim to be predictive in acute coronary syndrome (ACS) patients but not others. Some of the positive investigations have not used acceptable cTn cutoffs. The assay for placental growth factor is not standardized (Rifai & Ridker 2001).
- *Citrate C* After tubular reabsorption, this biomarker, a low-molecular-weight basic protein (13 kDa), is freely filtered and metabolized. It appears to be less affected by age, gender, and muscle mass than serum creatinine. The US Food and Drug Administration (FDA) has approved an analytically reliable test that demonstrated effectiveness in predicting outcomes in heart failure and ACS (Apple et al. 2005; Krauser et al. 2005), which makes sense given that renal prognosis is significantly influenced by function. It appears that the marker is affected by inflammation, which could make its use more difficult.
- *Protein That Binds Fatty Acids* Since fatty acid binding protein is quickly released following myocardial infarction, it has been suggested as a possible substitute for myoglobin. Although it has reportedly improved), it still lacks cardiac specificity. The FDA has not yet authorized any tests (Sarnak et al. 2005).

15.6 IMMUNOSUPPRESSION

Sepsis is marked by an initial strong inflammatory reaction involving cell activation and a cytokine storm. Alongside this, there's a prolonged compensatory anti-inflammatory response called immunological tolerance, which can result in weakened immune function. From a clinical perspective, this condition leads to the failure of multiple organs, ultimately causing the patient's demise. While the exact mechanisms driving sepsis are not completely known, there's compelling evidence that epigenetic alterations such as DNA methylation and changes in histone structures play a role in regulating the inflammatory response during sepsis (Falcão-Holanda et al. 2021).

Sepsis is a critical condition where the body's response to an infection harms its own tissues. Those who survive sepsis can experience weakened immunity, raising the likelihood of mortality. Studying the immune system's exposure to epigenetic alterations during and after sepsis has led researchers to explore novel biomarkers for identifying septic patients experiencing immunoparalysis. This avenue of research could provide fresh epigenetic-oriented approaches for predicting both immediate and lasting negative outcomes associated with this critical condition (Beltran-Garcia et al. 2020).

An investigation explains the primary epigenetic mechanisms responsible for changes in the innate and adaptive immune reactions seen in sepsis and septic shock. It also delves into how these alterations lead to states of immunosuppression, highlighting potential epigenetic markers that might enhance sepsis treatment and enable the anticipation of immunosuppression among post-septic individuals (Beltran-Garcia et al. 2020). Epigenetics is a broad concept that covers all the mechanisms controlling how genes are expressed, without making changes to the fundamental DNA sequence of an organism (Delcuve et al. 2009). The realm of epigenetics has grown extensively in the field of immunology, as a growing body of research suggests that epigenetic mechanisms perform a role in controlling both beneficial immune reactions and harmful conditions like immunosuppression and autoimmunity.

15.6.1 RA

RA is a condition where the immune system attacks the body's joints, causing intense inflammation and damage within a setting of widespread inflammation (Cristina Moran-Moguel et al. 2018). However, not all the varied clinical traits in RA can be accounted for by genetic variation or environmental factors alone. Conversely, epigenetic factors, particularly DNA methylation and histone modification, have significant roles in the development of RA. These modifications are closely linked to RA's progression and could serve as valuable biomarkers for tracking disease advancement and predicting drug responses (Guo et al. 2020).

A recent focus on the role of epigenetic modifications in RA sparked significant research. Exploring the use of epigenetic methods has become a cutting-edge area in understanding the development of RA. This review emphasizes the last five years' worth of research into RA's pathogenesis through epigenetic modifications, offering fresh approaches and tactics for diagnosing and treating the condition (Ciechomska et al. 2019). RA affects around 1% of the global population; there is currently, and its underlying development mechanisms remain unclear. Biomarkers can aid in enhancing diagnosis, prognosis, and treatment choices for improved efficacy; genes, SNPs, and epigenetic markers like HDACs can all be valuable in the handling and treatment of RA (Sharma et al. 2017).

Rheumatoid factor (RF) and anti-CCP are biomarkers linked to RA and connected with the disease. Remarkably, anti-CCP can be identifiable in the blood of RA patients, even years before clinical symptoms of the disease become apparent (Sharma et al. 2017). As RA advances, CRP and ESR lose sensitivity as predictors. In contrast, levels of acute phase serum amylase-A (A-SAA) can indicate both local and systemic inflammation in RA (Connolly et al. 2012). Additionally, A-SAA has the potential to serve as a valuable biomarker for gauging the therapeutic response to disease-modifying antirheumatic drugs (Hwang et al. 2016). RF is valuable for identifying a specific disease early on, but not for gauging its progression. On the other hand, anti-CCP is effective for predicting both the initial and advancing phases of the disease (Nell et al. 2005)

SNPs are heralded as a novel category of biomarkers for RA. Traditional biomarkers exhibit enhanced effectiveness when combined with SNPs, providing a comprehensive insight into RA prognosis and early detection. Moreover, SNPs play a role in anticipating treatment outcomes. Ultimately, the selection of effective prognostic and diagnostic biomarkers contributes to early intervention, improved disease control, and reduced treatment expenses (Mowla et al. 2017).

15.6.2 Neurological Disorders

A neurological disorder is any disorder of the nervous system. Structural, biochemical, or electrical abnormalities in the brain, spinal cord, or other nerves can result in a range of symptoms.

Genetic mutations can result in disrupted epigenetic processes, contributing to specific neurodevelopmental disorders. Modified epigenetic patterns can directly link to mutations in genes responsible for these disorders (Urdinguio et al. 2009). Epigenetics' significance in human diseases began in oncology and has now expanded to encompass areas like neurodevelopment and neurodegenerative disorders.

15.6.3 Alzheimer's Disease

Alzheimer's disease is a complex age-related neurological condition involving various biological levels like molecular, genetic, epigenetic, cellular, and brain networks. These changes result in mechanisms such as protein buildup, neuroinflammation, synaptic issues, and neurodegeneration, ultimately causing cognitive and behavioral decline (Zhao et al. 2016). Indeed, the ε4 isoform of apolipoprotein E has been associated with an increased risk of Alzheimer's disease. This heightened risk is attributed, in part, to its diminished capacity to crosslink amyloid-beta (Aβ) proteins, which perform a significant role in the development of Alzheimer's pathology (Gunzburg et al. 2007).

In the context of late-onset Alzheimer's disease and the clearance of Aβ, the main enzyme responsible for breaking down Aβ in the brain (Iwata N et al. 2000), neprilysin (also referred to as discoidin domain receptor tyrosine kinase 1 or DDR1), experiences an increase in methylation levels within cerebral endothelial cells following exposure to elevated concentrations of Aβ (Chen et al. 2009). A decrease in neprilysin activity associated with aging, which is inversely linked to the levels of Aβ, has been identified. This decline could impact the development of Alzheimer's disease (Hellstrom et al. 2008).

In the context of the amyloid hypothesis for Alzheimer's disease, there are additional epigenetic factors at play. Apart from Aβ peptide formation, the cleavage of amyloid precursor protein (APP) also produces an APP C-terminal peptide known as AICD.

Certainly, high-throughput screening has revealed genes that undergo epigenetic regulation in Alzheimer's disease even though they may not directly align with the amyloid hypothesis. Notably, two genes, S100A2 (a part of the calcium-binding S100 protein family) and SORBS3 (a cell-adhesion protein containing sorbin and SH3 domains), exhibit significant differences in DNA methylation levels between Alzheimer's disease and control cases (Al-Mahdawi et al. 2008). While S100A2 has been recognized as a promoter of metastasis in non-small-cell lung cancer, its involvement in the development of Alzheimer's disease is still not understood. S100B, a different member of the S100 protein family, functions as a neurotrophic and pro-survival factor for neurons and could play a role in the development of Alzheimer's disease (Druse et al. 2007).

15.6.4 Parkinson's Disease

Parkinson's disease (PD) stands as the second most prevalent neurodegenerative condition. Its exact cause remains uncertain, and there is currently no treatment that can modify the progression of the disease. It is classified as a multifactorial disorder because its development is influenced by both genetic and environmental factors (Angelopoulou et al. 2022). This complex neurodegenerative disorder manifests as symptoms like resting tremor, muscle rigidity, bradykinesia (slowed movement), and difficulty maintaining posture (Selvakumar et al. 2022).

Various environmental factors can lead to changes in cellular activities, like gene expression. Recent findings suggest that activities such as smoking, drinking coffee, exposure to pesticides, and contact with heavy metals like manganese, arsenic, and lead might impact the risk of developing PD, possibly through epigenetic changes (Angelopoulou et al. 2022). Neuroinflammation is also a key contributor to the development and advancement of neurodegeneration in various conditions,

PD included. The primary reason for the progression of PD is believed to be the degeneration of dopaminergic neurons in the substantia nigra (Selvakumar et al. 2022).

There is a strong need for biomarkers in PD to enable early detection, predict disease advancement, and gauge treatment effectiveness. The epigenome, a partially changeable set of epigenetic marks influenced by genetics and PD-related environmental factors, shows great potential for creating molecular biomarkers. These marks could potentially serve as a unified indicator of susceptibility to PD. Epigenetic irregularities have been observed in PD and other neurodegenerative conditions linked to aging (Jakubowski et al. 2017).

In the 2010s, researchers pinpointed mutations in six specific genes that lead to PD development: SNCA (α-synuclein), PARK2 (parkin), PINK1 (PTEN-induced kinase protein 1), UCHL1 (ubiquitin carboxyl-terminal hydrolase isozyme L1), DJ1 (protein DJ-1), and LRRK2 (leucine-rich repeat serine/threonine-protein kinase 2) (Urdinguio et al. 2009). Numerous other potential biomarker types for PD have been evaluated, encompassing clinical, imaging, pathological, biochemical, and genetic factors, but none have reached a level of excellence or even satisfaction. Nonetheless, promising recent advancements provide optimism that PD's early detection might become feasible in the upcoming years (Peplow et al. 2022).

The discovery of dependable biomarkers for PD could be the outcome of achievements in neuroimaging, such as positron emission tomography (PET), single-photon emission CT (SPECT), and MRI along with advancements in novel biochemical approaches. Illustrations of these techniques involve assessing aggregated α-synuclein using real-time quaking-induced conversion and protein misfolding cyclic amplification, as well as the creation of biomarkers based on epigenetics (Peplow et al. 2022).

Biomarkers could aid in confirming a PD diagnosis, comprehending disease advancement and its associated characteristics, as well as tracking treatment outcomes. Because initial clinical symptoms often overlap, distinguishing PD from other neurological conditions becomes challenging. PD might be mistaken for conditions like progressive supranuclear palsy, multiple system atrophy, dementia with Lewy bodies, Parkinson's disease dementia, and other unconventional parkinsonian disorders (Koga et al. 2015).

PET scan radiotracers release electrons and positively charged antiparticles (positrons) that share a comparable mass to electrons. SPECT can also utilize radioligands for dopamine D2 receptors, which function as dopamine antagonists. Both PET and SPECT scans are capable of identifying the initial stages of PD and the decline of dopaminergic neurons through the utilization of radiotracers for 3D imaging. Most radiotracers are noninvasive radiopharmaceuticals with a brief half-life, leading to their decay shortly after imaging is done. Furthermore, PET and SPECT scans' 3D images display certain brain functions, whereas MRI is limited to observing brain anatomy and structure (Histed et al. 2015).

15.6.5 Asthma

Asthma is a multifaceted, diverse, and long-lasting inflammatory respiratory condition characterized by various clinical presentations resulting from a range of triggers and underlying physiological processes (Alizadeh et al. 2017).

Asthma, a persistent airway inflammation condition, is shaped by the interaction of genetic and environmental elements, a process increasingly recognized to be modulated by epigenetic mechanisms. Irregular DNA methylation, changes in histone modifications, distinct microRNA patterns, and other adjustments in chromatin structure collectively coordinate an intricate reprogramming of early-life immune T-cell responses, dendritic cell behavior, macrophage activation, and a disruption of the airway epithelial barrier. These factors collectively determine the risk and severity of asthma in later life (Meda et al. 2011).

The rising occurrence of asthma over recent decades has highlighted the influence of epigenetics in asthma development. Environmental factors act as initiating signals through epigenetic processes. Three epigenetic mechanisms, DNA methylation, alterations in histone structure, and the involvement of small noncoding RNAs, govern immune responses and the expression of inflammatory genes in asthma (Alizadeh et al. 2017).

Previous research has demonstrated that epigenetic processes control a multitude of genes, including those associated with inflammatory and immune responses. The primary epigenetic mechanisms presently under investigation for understanding the causes and progression of asthma are DNA methylation, histone modifications, and noncoding RNAs (Edris et al. 2019).

Histone modifications encompass posttranslational changes like acetylation, phosphorylation, methylation, and ubiquitination of core histones, which form the building blocks of chromatin as octameric proteins. Research exploring histone modifications in relation to asthma is less prevalent in comparison to studies on DNA methylation (Tessarz & Kouzarides 2014).

MiRNAs are a category of small ncRNAs intricately involved in diverse biological activities, including cell signaling, biochemical pathways, and the development of tissues and organs. These miRNAs are recognized for their role in governing the inherent functions of immune cells and their growth in response to pathological stimuli (Ivey & Srivastava 2015).

15.7 EPIGENETICS-RELATED INTELLECTUAL PROPERTY RIGHTS

The investigation of epigenetics provides insight into the molecular processes that control the expression of genes (Mancilla et al. 2020; Cavalli & Heard 2019). An increasing number of social researchers, bioethicists, and legal experts have expressed their opinions on the possible effects on society as well as medicine of advances in science and technology in the field of epigenetics during the past ten years (Müller et al. 2017; Chiapperino 2018; Huang & King 2018; Dupras et al. 2019). According to positive evaluations, the discipline is thought to provide a greater comprehension of the biosocial and molecular processes that lead to illnesses and wellness disparities, as well as legislation that might be beneficial for their prevention (Rozek et al. 2014; Loi et al. 2013; Dupras et al. 2014). The application of epigenetic techniques to the identification, management, and outlook of disease is another potential approach (Feinberg 2018; Berdasco 2019; Oh & Petronis 2021).

Simultaneously, challenges related to ethics, law, and society (ELSI) were also identified. These difficulties range from worries about the confidentiality of people participating in epigenetic investigations (Dyke et al. 2015; Erwin 2015; Dupras & Bunnik 2021) to the possibility of prejudice because of a person's epigenetic characteristics (Taki & de Melo-Martin 2021). When conducting epigenetic studies, investigators might question which findings—whether accidental or medically significant—should be shared with the study's participants and give rise to ELSI (Dyke et al. 2019).

ELSI can also occur during information translation. For instance, media reports have raised questions regarding the frequently hurried, deceptive, or inaccurate dissemination of scientific findings in epigenetics (Juengst et al. 2014; Dyke et al. 2019). Throughout the last 25 years, multiple nations have implemented regulations aimed at preventing genetic discrimination due to the potential threats associated with the improper utilization of particular epigenetic knowledge (Joly et al. 2020). Nevertheless, it appears that these regulations are not providing a structure for the application of epigenetic knowledge since they concentrate on "genetic characteristics" and "the outcomes of genetic tests" (Dupras et al. 2018, Dupras, Beauchamp, & Joly, 2020).

Given the growing excitement in employing epigenetic techniques by a number of beneficiaries, particularly forensic scientists, life insurance firms, and immigration authorities (Miller & Freeman 2018; Shapo & Masar 2020; Abbott 2018b; Vidaki & Kayser 2018; Shabani et al. 2018; Dupras, Saulnier, & Joly 2019; Abbott 2018a; Ihar 2020), the aforementioned oversight gap may be problematic. Concerns about the scientific reliability of investigations that are promoted and offered on the internet, along with the degree to which directly to consumers epigenetic testing organizations adhere to best practice guidelines and legislation on healthcare equipment created specifically for the customer genomics industry, are additionally being addressed by the rapidly expanding customer epigenetics industry (Dupras, Joly, & Rial-Sebbag, 2020; Knoppers et al. 2021). It is complicated to determine the extent to which epigenetic studies and advances necessitate novel or modified regulations prohibiting possible inappropriate utilization, and if so, how those regulations ought to be created.

First, epigenetics demands reflection on the physiological features that render genetic variations morally dubious (and deemed deserving of legal safeguards) and the degree to which these characteristics are shared by epigenetic variations. Given, for example, the associations between particular genetic and epigenetic variants, the pre-birth programming of numerous epigenetic variants, as well as the continued stability of numerous of these modifications through an individual's entire life, a certain population might assume that epigenetic variants are "similar enough" to genetic mutations to warrant legal safeguards. Given the flexibility and reversibility of certain epigenetic variations as well as the level of prospective personal ownership across epigenetic programming, a few may contend that epigenetic variants are significantly unique as well as do not, consequently, require such safeguards (Vears & D'Abramo 2018; M'hamdi et al. 2018; Dubois et al. 2019; Dupras & Ravitsky 2016).

Second, there are significant theoretical and linguistic difficulties that policymaking inherently confronts. In fact, it's hard to say what kinds of biological variations ought to be governed as "epigenetic" variants and categorized as appropriately. While discussing epigenetic variants, some individuals may be comprehensive and include gene expression controlling pathways like transcription factors and RNA interference, whereas other individuals might restrict the discussion to the methylation of DNA and histone modifications (Henikof & Greally 2016; Greally 2018).

The distinction between DNA methylation and histone modifications ought to be considered "similarly epigenetic" or because their physiological features differ sufficiently to prevent policymakers from conflating the two is up for debate. Contrary to popular belief (Wu & Morris 2001), DNA methylation does alter the linear DNA sequence when it converts cytosines into methylcytosines, as noted by Lappé and Landecker (Lappé & Landecker 2015). Therefore, it is possible to conceptualize DNA methylation as a component of genetics from the standpoint of policy interpretation, and as such, it would be covered by laws prohibiting genetic discrimination. With histone alterations, this wouldn't be the situation. In the last decade, a number of patents related to epigenetics have been granted (Table 15.1).

15.8 CONCLUSION

Epigenomic therapy is experiencing rapid advancement due to multiple variables. The initial factor is the significant overall extent of fascination with the discipline. Next is the continuous process of characterizing standard epigenomes in order to facilitate and expedite scientific endeavors. The final factor is the accessibility of potent techniques for characterizing epigenomes, like next-generation DNA sequencing. The fact that epigenetic changes are cell-specific adds an additional dimension of complication. Although this is advantageous in numerous capacities and may offer exact understanding into the origin cell's functioning, it additionally presents certain sample collection challenges, especially in the context of clinical studies. With numerous medications in advanced clinical studies, we anticipate the release of multiple novel epidrugs in the following five years. Possibly the biggest development in this field is expected to be the arrival of epidrugs for purposes other than cancer treatment.

TABLE 15.1
Patents Related to Epigenetics (2013–2023)

Patent No	Description	Assignee	Inventor	Date
US20130129668	Evaluation of the individual's diagnosis, prognosis, or reactions to therapies for conditions like osteoarthritis or rheumatoid arthritis employing epigenetics techniques, compositions, and kits	Regents of the University of California (Oakland, CA, USA)	Firestein G, Nakano K	5/23/2013
US20130283404	Compositions and techniques for the identification, tracking, management, and/or prevention of autoimmune diseases, such as lupus, that are predicated on a causal relationship with different epigenetic markers, such as chromosome demethylation, lupus marker overexpression, or PKCd nitration in response to oxidative stress	Regents of the University of Michigan (Ann Arbor, MI, USA)	Richardson BC, Hewagama A, Gorelik G	10/24/2013
US20130323742	Techniques and equipment for dividing, concentrating, and/or identifying molecules based on variations in their binding affinities to a probe; these can be applied in the domains of oncology and epigenetics to identify the existence of particular biomolecules or biomolecules that have been differentially modified, to provide diagnosis for conditions like genetic disorders in fetuses, or to identify biomarkers in cancer, organ failure, disease states, or infection.	University of British Columbia (Vancouver, British Columbia, Canada)	Marziali A, Pel J, Thompson JD, Shibahara G	12/5/2013
KR101363412	Technique for controlling the activity of the domain-rearranged methyltransferase protein in plants; beneficial for enhancing the characteristics of food, forage, floriculture, horticultural, and energy crops; also accessible for a variety of medical uses, including the differentiation of cancer cells and neural cells, which are primarily involved in processing epigenetic data	SNU R&DB Foundation (Seoul)	Shin CS, Park JJ, Hwang DG	3/19/2014
US20140080162	Techniques for testing an enzyme's activity that rely on measuring its fluorescence lifetime	Almac Sciences (Edinburgh)	Cotton G, Dunsmore C	3/20/2014
US20140170228	Treatments for proliferative diseases (like cancer) that involve giving a patient a therapeutic amount of a taxane in the form of nanoparticles as a component of the initial treatment and then giving them an operational dose of another representative that alters a cell's epigenetics as part of the subsequent treatment	Abraxis BioScience (Los Angeles)	Desai NP, Soon-Shiong P	6/19/2014
US20140212873	Examination for epigenetic patterns linked to a pathologic situation in both individuals and communities. The FMR1 gene contains the following sites from which epigenetic patterns are revealed: FREE3; intron 2; an intron, intron-exon boundary and/or splicing region downstream of intron 2; and a location inside the FREE2 portion of intron 1 when combined with a full mutation	La Trobe University (Melbourne, Victoria, Australia), Murdoch Children's Research Institute (Parkville, Victoria, Australia)	Godler D	7/31/2014
CN104069482	Using the transcription factor CREB to control the expression of the KDM3A (JHDM2A) gene and lower pig lipid deposition. Pig KDM3A gene expression is upregulated via CREB overexpression, increased CREB phosphorylation, or a combination of phosphorylated CREB and pig CREB-binding protein. This lowers pig lipid deposition and regulates lipid–metabolism pathways from an epigenetic perspective.	China Agricultural University (Beijing)	Hu X, Li Y, Li N	10/1/2014

CN104152407	Histone deacetylase inhibitors are used to prepare osteoblast differentiation preparations of odontogenic stem cells, which are then used to treat diseases like periodontitis or peri-implantitis, alter cell epigenetics by applying a medication, and stimulate bone tissue regeneration.	Stomatological Hospital, Tianjin Medical University (Tianjin, China)	Liu D, Gao P, Jia Z, Zhang X, Xiao R, Feng C, Wang Y	11/19/2014
US 9,750,731	Techniques and compounds for adjusting how transcription factors work, particularly those that bind histone-modifying enzymes, or epigenetic regulators, to particular DNA promoters. The myocyte-enhancing factor MEF2, the transcription factor GATA3, and the forkhead/winged helix transcription factor FOXP3 are among the targeted transcription factors. Moreover, therapeutic uses for small molecule modulators of MEF2 and its associated factors, such as p300/CBP, Cabin1, histone deacetylases, among others.	University of Southern California (Los Angeles)	Chen L, Jayathilaka N, Han A, Petasis N	9/5/2017
US 9,797,016	Strategies and biomarkers (such as epigenetic biomarkers) for the identification of bladder cancer in tissue and urine samples and urine sediments that are helpful in differentiating among renal epithelial tumors, bladder cancer, and prostate cancer.	Oslo University Hospital	Lind GE, Lothe RA, Skotheim RI, Jeronimo C, Costa VL, Henrique R, Teixeira MR	10/24/2017
US 9,840,526	Protein 4 that contains a metallodomain is a desirable epigenetic target for cancer treatment. Complexes of iridium (III) are effective as an irreversible BRD4 inhibitor. A formula is given along with a very helpful example.	University of Macau, Hong Kong Baptist University, Kaohsiung Medical University	Leung C-H, Zhong H-J, Ma D-L, Wang H-M	12/12/2017
US 9,857,358	An approach to assessing DNA damage biomarkers in free eukaryotic cell nuclei, as well as transcription factor activation, activity, or expression levels, and/or epigenetic changes to chromatin or chromatin-associated factors. Additionally, methods for integrating nuclear biomarkers into a matrix of endpoints that can clarify the main mechanism through which genotoxicants cause DNA damage	Litron Laboratories (Rochester, NY, USA)	Dertinger SD, Bryce SM, Bemis JC	1/2/2018
US 9,914,981	Techniques and strategies for creating abiotic stress-tolerant plants with higher yields based on the precise detection of a DNA methylation signature in individual plants within a population of plants. The techniques entail identifying and using a plant's epigenetic characteristics that contribute to its remarkable energy efficiency	Bayer CropScience (Diegem, Belgium)	Akpo EI, Laga B, De Block M	3/13/2018
201711378099.9	The application of 11beta-HSD2 and epigenetics as early markers of articular cartilage dysplasia and osteoarthritis susceptibility. Results showed that IUGR neonatal umbilical cord mesenchymal stem cells—or the whole umbilical cord—exhibited less histone acetylation and expression of the 11beta-HSD2 promoter region than in the control group. These stem cells also showed impaired cartilage differentiation ability, making them more prone to glucocorticoid-induced degeneration similar to osteoarthritis. This patent establishes a framework for early detection articular cartilage dysplasia and fetal osteoarthritis in adulthood and introduces new potential strategies for mitigating osteoarthritis.	Wuhan University	Wang Hui	3/23/2018

(Continued)

TABLE 15.1 (*Continued*)
Patents Related to Epigenetics (2013–2023)

Patent No	Description	Assignee	Inventor	Date
US 9,968,580	Rett syndrome, memory loss, chronic inflammation, trinucleotide repeat diseases like fragile X syndrome, precancerous conditions involving cancer stem cells, and disorders involving cancer stem cells that endured cancer treatment can all be treated with FDA-approved dietary supplements and nutraceuticals.	Ott DM	Ott DM	5/15/2018
US 9,982,308	Identifying an epigenetic modification to a minimum of one gene chosen from the NDRG4/NDRG2 subfamily, GATA4, OSMR, GATA5, SFRP1, ADAM23, JPH3, SFRP2, APC, MGMT, TFPI2, BNIP3, FOXE1, SYNE1, SOX17, PHACTR3, and JAM3; detected epigenetic changes suggest a predisposition to, or the incidence of, cancer. Additionally, pharmacogenetic techniques for choosing appropriate cancer patient therapy and care that center on selecting patients in accordance with the invention's techniques	Exact Sciences Development Company (Madison, WI, USA)	Van Engeland M, De Bruine MA, Griffioen A, Louwagie J, Bierau K, Brichard G, Otto G, Penning M	5/29/2018
US 9,988,667	Techniques and instruments for detecting multiple epigenetic markers on chromatin, including its fragments. Through the identification of one or more genetic material characteristics, the techniques and devices allow for analyzing the genetic material streaming through a passageway in a continuous body of fluid.	Cornell University (Ithaca, NY, USA)	Craighead HG, Cipriany BR, Levy S, Soloway P	6/5/2018
US 10,023,592	Compounds that are substituted for heterocyclic derivatives, mixtures that contain said compounds, and applications of said compounds and mixtures for epigenetic regulation through blocking the recognition of acetyllysine regions of proteins like histones by the bromodomain. The treatment of cancer and neoplastic diseases can benefit from the use of said compositions and methods.	Celgene Quanticel Research (San Diego)	Boloor A	7/17/2018
201810443242.6	A micelle drug delivery system for targeted epigenetics modification sites. The system consists of a drug carrier formed by an amphipathic compound and a drug. It interacts with histone, DNA, or RNA epigenetic modification sites, forming host–guest complexes with strong affinity. The system can be designed to load drugs and deliver them to targeted sites, ultimately to regulate and control epigenetic inheritance and cure diseases more effectively.	Guilin Medical University	MO Jingxin	9/21/2018
201811283285.9	Applying recombinant fusion protein in antibody targeting in epigenetics. The protein improves on ChIP-seq technology, simplifying operations, increasing data quality, and expanding the application range, particularly with only a few cells.	Vazyme Biotech Co., Ltd. (Nanjing)	Feng SU	03/01/2019
US 10,508,122	Novel hydroxamic acids exhibiting particular HDAC and/or TTK/Mps1 kinase inhibitor properties, as well as their pharmaceutically appropriate salts, that can be used to modulate the functioning of these enzymes and/or HDAC, as well as pharmaceutical compositions containing these compounds and methods for preparing them.	The Regents of the University of Colorado (Denver)	Liu X, Zhang G, Chan DC-F, Piscopio	12/17/2019

US 10,519,149	Substances, as well as their therapeutically approved salts and compositions, that are helpful in treating illnesses linked to the hindering histone deacetylase (e.g., HDAC2)	Rodin Therapeutics (Boston)	Fuller NO, Lowe JA III	12/31/2019
US 10,533,003	Polyheteroaryl HDAC inhibitors	Karus Therapeutics (Oxfordshire, UK)	Shuttleworth SJ, Cecil AR, Mac Cormick S, Nodes WJ, TomassiCDSilva FA, Fuller NO, Lowe JA III	1/14/2020
US 10,532,053	Investigated the HDAC inhibitory activity of a 4-arylamino quinazoline hydroxamic acid compound, its production, its pharmaceutical layout, and its application in the creation of an HDAC inhibitor medicine	Guangdong Zhongsheng Pharmaceutical (Guangdong, China)	Chen L, Long C, Chen X, Liu Z, Ye H, Xie C	1/14/2020
US 10,538,498	Novel compounds with HDAC6 inhibitory activity, including their stereoisomers and pharmaceutically acceptable salts, as well as their use in preparing therapeutic medicaments. Also covers pharmaceutical compositions containing these compounds, methods for treating diseases using said compositions, and methods for preparing the novel compounds.	Chong Kun Dang Pharmaceutical (Seoul)	Lee J, Han Y, Kim Y, Choi D, Min J, Bae M, Yang II, Kim D	1/21/2020
US 10,537,535	Substances with HDAC inhibitory properties that have an impact on both in vitro and in vivo gene expression. These inhibitors have a wide range of therapeutic applications, including the treatment of infectious conditions, senescence, neurological disorders, neural degenerative diseases, and various cancers.	The Regents of the University of California (Oakland, CA, USA)	Ray A, Yamanaka S	1/21/2020
US 10,544,156	A substance, or a salt or solvate of it that is approved by the FDA, in addition to a pharmaceutical composition that contains it, and the application of such a substance as an inhibitor of DNA methyltransferase for cancer management	Pierre Fabre Medicament (Boulogne-Billancourt, France), Centre National de la Recherche Scientifique (Paris)	Halby L, Menon Y, Kaloun EB, Long C, Arimondo PB	1/28/2020
US 10,532,066	Applying a mixture containing an HDAC inhibitor to a cancer cell to modify programmed death receptor ligand1, enhancing the therapeutic response	H Lee Moffitt Cancer Center and Research Institute (Tampa, FL, USA)	Villagra AV, Sotomayor EM	3/3/2020
PCT/JP2020/017607	A histone demethylase inhibitor comprising a compound represented by formula [I], or a pharmaceutically acceptable salt thereof, or a tautomer thereof. Also relates to a prophylactic and/or therapeutic drug that contains the inhibitor for associated with increased histone demethylase reactivity	Osaka University	ISHII, Hideshi	10/29/2020
202210157876.1	A method for improving the yield of an endophytic fungus fermentation product 4-hydroxyphenethyl alcohol through epigenetics modification	Human Environmental Biological Occupational Technology College	Xie Guangyuan	05/20/2022

(Continued)

TABLE 15.1 (*Continued*)
Patents Related to Epigenetics (2013–2023)

Patent No	Description	Assignee	Inventor	Date
202210349190.2	An epigenetics data analysis cloud platform that consists of a database, a data input module, a data processing module, a storage module, a data backup module, a data extraction module, and a data analysis module and is characterized in that the data input module is linearly connected with the database. The system sorts data in the analysis module, and final effective data is fed back to a user in a chart. Disordered arrangement of the data is avoided, presentation of data rules is guaranteed, the user can conveniently search the data, and the user can conveniently observe and research the data.	Shanghai Jaboinine Biotechnology Limited Company	Wang Chengyang	06/24/2022
PCT/CN2021/123490	A preparation that regulates macrophage epigenetics for use in developing drugs to treat hemophagocytic lymphohistiocytosis, within the field of biology. The preparation regulates macrophage epigenetics through a DNA methyltransferase inhibitor, a histone acetylase inhibitor or a genetic drug that downregulates DNA methyltransferase expression. The preparation can reduce the expression level of inflammatory factors, increase blood cell count, relieve hepatosplenomegaly, and reduce the ratio and number of activated macrophages in spleen cells.	Union Hospital, Tongji Medical College, Huazhong University of Science and Technology	Hu, Yu	03/23/2023

BIBLIOGRAPHY

Abbott. "Can epigenetics help verify the age claims of refugees?" *Nature* 561, no. 7721 (2018a): 15–15.

Abbott. "European scientists seek 'epigenetic clock' to determine age of refugees." *Nature* 561, no. 7721 (2018b): 15–16.

Alavian-Ghavanini, A., Rüegg, J., "Understanding epigenetic effects of endocrine disrupting chemicals: From mechanisms to novel test methods." *Basic Clin Pharmacol Toxicol.* 122 (2018): 38–45.

Alizadeh, Zahra, Mortaz, Esmaeil, Adock, Lan, et al. "Role of epigenetics in the pathogenesis of asthma." *Iran J Allergy Asthma Immunol.* (2017): 82–91.

Allan, R.S., Zueva, E., Cammas, F., et al. "An epigenetic silencing path-way controlling T helper 2 cell lineage commitment." Nature. (2012):487(7406):249–53.

Al-Mahdawi, S., Pinto, R.M., Ismail, O., et al. "The Friedreich ataxia GAA repeat expansion mutation induces comparable epigenetic Changes in human and transgenic mouse brain and heart tissues." *Hum Mol Genet.* 17 (2008): 735–746.

Alpert, J.S., Thygesen, K., Antman, E., et al. "Myocardial infarction redefined—a consensus document of the joint European society of cardiology/American college of cardiology committee for the redefinition of myocardial infarction." *J Am Coll Cardiol Comm.* 36, no. 3 (2000): 959–969.

Angelopoulou, Efthalia, Paudel, Yam Nath, Papageorgiou, Sokratis G., et al. "Environmental impact on the epigenetic mechanisms underlying Parkinson's disease pathogenesis." *Brain Sci.* 12, no. 2 (2022): 175.

Antman, E.M., Tanasijevic, M.J., Thompson, B., et al. "Cardiac-specific troponin I levels to predict the risk of mortality in patients with acute coronary syndromes." *N Engl J Med.* 335 (1996): 1342–1349.

Apple, F.S., Wu, A.H., Mair, J., et al. "Future biomarkers for detection of ischemia and risk stratification in acute coronary syndrome." *Clin Chem.* 51 (2005): 810–824.

Ashoori, H., Ghamarchehreh, M.E., Tavallaei, M., et al. "Evaluation of the epigenetic biomarker bone morphogenic protein 3 for colorectal cancer diagnosis." *J Clin Diagn Res.* 12 (2018): GC07–GC09.

Aune, T.M., Collins, P.L., Chang, S. "Epigenetics and T helper 1 differentiation." *Immunology.* 126, no. 3 (2009): 299–305.

Barrett, M.T., Sanchez, C.A., Prevo, L.J., et al. "Evolution of neoplastic cell lineages in Barrett oesophagus." *Nat Genet.* 22 (1999): 106–109.

Barwell, J., Wootten, D., Simms, J., et al. "RAMPs and CGRP receptors." In Spielman, W.S., Parameswaran, N., Eds., *RAMPs. Advances in Experimental Medicine and Biology*, Springer: New York, 2012, Volume 744, pp. 13–24.

Belinsky, S.A., Nikula, K.J., Palmisano, W.A., et al. "Aberrant methylation of p16(INK4a) is an early event in lung cancer and a potential biomarker for early diagnosis." *Proc Natl Acad Sci USA.* 95 (1998): 11891–11896.

Beltran-Garcia, Jesus, Osca-Verdegal, Rebeca, Roma-Mateo, Carlos, et al. "Epigenetic biomarkers for human sepsis and septic shock: Insights from immunosuppression." *Epigenomics* 12 (2020): 617–646.

Berdasco, Esteller. "Clinical epigenetics: Seizing opportunities for translation." *Nat Rev Genet.* 20, no. 2 (2019): 109–127.

Bergman, Y., Cedar, H. "Epigenetic control of recombination in the immune system." *Semin Immunol.* 22, no. 6 (2010): 323–329.

Best, Carey. "Epigenetic therapies for non-oncology indications." *Drug Discov Today.* 15, no. 1008 (2010): 1010–1011.

Bhan, A., Soleimani, M., Mandal, S.S. "Long noncoding RNA and cancer: A new paradigm." *Cancer Res.* 77 (2017): 3965–3981.

Blüml, S., Zupkovitz, G., Kirchberger, S., et al. "Epigenetic regulation of dendritic cell differentiation and function by oxidized phospholipids." *Blood.* 114, no. 27 (2009): 5481–5489.

Boumber, Issa. "Epigenetics in cancer: What's the future." *Oncology* 25 (2011): 220–226.

Brooks, W.H., Le Dantec, C., Pers, J.O., et al. "Epigenetics and autoimmunity." *J Autoimmun.* 34, no. 3 (2010): J207–J219.

Cai, T.T., Muhali, F.S., Song, R.H., et al. "Genome-wide DNA methylation analysis in graves' disease." *Genomics.* 105, no. 4 (2015): 204–210.

Cao, Y., Zhao, G., Yuan, M., et al. "KCNQ5 and C9orf50 methylation in stool DNA for early detection of colorectal cancer." *Front Oncol.* 10 (2021): 621295.

Cao-Lei, L., Laplante, D.P., King, S. "Prenatal maternal stress and epigenetics: Review of the human research." *Curr Mol Biol Rep.* 2 (2016): 16–25.

Cavalli, Heard. "Advances in epigenetics link genetics to the environment and disease." *Nature* 571, no. 7766 (2019): 489–499.

Cedar, H., Bergman, Y. "Epigenetics of haematopoietic cell development." *Nat Rev Immunol.* 11, no. 7 (2011): 478–488.

Cernuda-Morollon, E., Larrosa, D., Ramon, C., et al. "Interictal increase of CGRP levels in peripheral blood as a biomarker for chronic migraine." *Neurology.* 81 (2013): 1191–1196.

Chaudhry, M.A. "Strategies for detecting DNA methylation: A survey of US patents." *Recent Pat DNA Gene Seq.* 4, no. 2 (2010): 79–85.

Chen, K.L., Wang, S.S., Yang, Y.Y., et al. "The epigenetic eff ects of amyloid-beta(1–40) on global DNA and neprilysin genes in murine cerebral endothelial cells." *Biochem Biophys Res Commun.* 378 (2009): 57–61.

Chen, X., Dai, G., Yu, L., et al. "miR-143-3p inhibits the proliferation, migration and invasion in osteosarcoma by targeting FOSL2." *Sci. Rep.* 8 (2018): 606.

Chen, X., Xu, X., Pan, B., et al. "miR-150-5p suppresses tumor progression by targeting VEGFA in colorectal cancer." *Aging* 10 (2018): 3421–3437.

Chiapperino, L. "Epigenetics: Ethics, politics, biosociality." *Br Med Bull.* 128, no. 1 (2018): 49–60.

Ciechomska, Marzena, Roszkowski, Leszek, et al. "NA methylation as a future therapeutic and diagnostic target in rheumatoid arthritis." *Cells* 8, no. 9 (2019): 953.

Clement, G., Braunschweig, R., Pasquier, N., et al. "Methylation of APC, TIMP3, and TERT: A new predictive marker to distinguish Barrett's oesophagus patients at risk for malignant transformation." *J Pathol.* 208 (2006b): 100–107.

Connolly, M., Mullan, R.H., Cormick, J.M., et al. "Acute-phase serum amyloid A regulates tumor Necrosis factor α and matrix turnover and predicts disease progression In patients with inflammatory arthritis before and after biologic therapy." Arthritis Rheum. 64 (2012): 1035–1045.

Cristina Moran Moguel, Maria, Petarra-del Rio, Stefania, Mayorquin-Galvan, Evangelina, et al. "Rheumatoid arthritis and miRNAs: A critical review through a functional view." J Immunol Res. 2018 (2018): 2474529.

Delcuve, G.P., Rastegar, M., Davie, J.R. "Epigenetic control." *J Cell Physiol.* 219 (2009): 243–250.

Devaraj, S., DuClos, T.W., Jialal, I. "Binding and internalization of C-reactive protein by Fc-gamma receptors on human aortic endothelial cells mediates biological effects." *Arterioscler Thromb Vasc Biol.* 25 (2005): 1359–1363.

Dietrich, D., Meller, S., Uhl, B., et al. "Nucleic acid-based tissue biomarkers of urologic malignancies." *Crit Rev Clin Lab Sci.* 51 (2014): 173–199.

Druse, M.J., Gillespie, R.A., Tajuddin, N.F., et al. "S100B-mediated protection against the pro-apoptotic effects of ethanol on fetal Rhombencephalic neurons." *Brain Res.* 1150 (2007): 46–54.

Dubois, M., Louvel, S., Le Gof, A., et al. "Epigenetics in the public sphere: Interdisciplinary perspectives." *Environ Epigenet.* 5, no. 4 (2019): 019.

Dupras, C., Beauchamp, E., Joly, Y. "Selling direct-to-consumer epigenetic testing: Are we ready?" *Nat Rev Genet.* 22 (2020): 335–336.

Dupras, C., Beck, S., Rothstein, M.A., et al. "Potential (Mis)use of epigenetic age estimators by private companies and public agencies: Human rights law should provide ethical guidance." *Environ Epigenet.* 5, no. 3 (2019): 1–12.

Dupras, C., Bunnik, E.M. "Toward a framework for assessing privacy risks in multi-omic research and databases." *Am J Bioethics.* 21, no. 12 (2021): 46–64.

Dupras, C., Joly, Y., Rial-Sebbag, E. "Human rights in the postgenomic era: Challenges and opportunities arising with epigenetics." *Soc Sci Inf.* 59, no. 1 (2020): 12–34.

Dupras, C., Ravitsky, V. "The ambiguous nature of epigenetic responsibility." *J Med Ethics.* 42, no. 8 (2016): 534–541.

Dupras, C., Ravitsky, V., Williams-Jones, B. "Epigenetics and the environment in bioethics." *Bioethics.* 28, no. 7 (2014): 327–334.

Dupras, C., Saulnier, K.M., Joly, Y. "Epigenetics, ethics, law and society: A multidisciplinary review of descriptive, instrumental, dialectical, reflexive and empirical analyses." *Soc Stud Sci.* 49, no. 5 (2019): 485–810.

Dupras, C., Song, L., Saulnier, K.M., et al. "Epigenetic discrimination: Emerging applications of epigenetics pointing to the limitations of policies against genetic discrimination." *Front Genet.* 8 (2018): 1–6.

Duran-Sanchon, S., Moreno, L., Augé, J.M., et al. "Identification and validation of microRNA profiles in fecal samples for detection of colorectal cancer." *Gastroenterology* 158 (2020): 947–957.

Durham, P.L. "Inhibition of calcitonin gene-related peptide function: A promising strategy for treating migraine." *Headache J Head Face Pain.* 48 (2008): 1269–1275.

Dyke, S.O.M., Cheung, W.A., Joly, Y., et al. "Epigenome data release: A participant-centered approach to privacy protection." *Genome Biol.* 16, no. 1 (2015): 142–154.

Dyke, S.O.M., Ennis, C.A., Joly, Y., et al. "Communicating science: Epigenetics in the spotlight." *Environ Epigenet* 6, no. 1 (2020): dvaa015.

Dyke, S.O.M., Saulnier, K.M., Dupras, C. "Points-to-consider on the return of results in epigenetic research." *Genome Med.* 11, no. 1 (2019): 1–9.

Eads, C.A., Lord, R.V., Wickramasinghe, K., et al. "Epigenetic patterns in the progression of esophageal adenocarcinoma." *Cancer Res.* 61 (2001): 3410–3418.

Edvinsson, L. "Novel migraine therapy with calcitonin gene-regulated peptide receptor antagonists." *Expert Opin Ther Targets* 11 (2007): 1179–1188.

Edvinsson, L. "Role of CGRP in migraine. In calcitonin gene-related peptide (CGRP) mechanisms." Brain, S.D., Geppetti, P., Eds., *Handbook of Experimental Pharmacology*, Springer International Publishing: Cham, Switzerland, 2019, Volume 255, pp. 121–130.

Edvinsson, L., Grell, A.-S., Warfvinge, K. "Expression of the CGRP family of neuropeptides and their receptors in the trigeminal ganglion." *J Mol Neurosci.* 70 (2020): 930–944.

Edvinsson, L., Warfvinge, K. "Recognizing the role of CGRP and CGRP receptors in migraine and its treatment." *Cephalalgia.* 39 (2019): 366–373.

Edris A, den Dekker HT, Melén E, Lahousse L. Epigenome-wide association studies in asthma: A systematic review. Clinical and Experimental Allergy.2019(49):953–68. Available from: doi/pdf/10.1111/cea.13403

Erwin, C. "Ethical issues raised by epigenetic testing for alcohol, tobacco, and cannabis." *Behav Sci Law.* 33, no. 5 (2015): 662–674.

Esteller, K. "Epigenetics in cancer." *N Engl J Med.* 358 (2008): 1148–1154.

Esteller, K. "Cancer epigenetics for the 21st century: What's next?" *Genes Cancer.* 2, no. 604 (2011): 605.

Esteller, K. "DNA methylation and cancer." *Adv Cancer.* 70 (2012): 27–38.

Falcão-Holanda, Renata, Karina Colo Brunialti, Milena, Galvonas Jasiulionis, Miriam, et al. "Epigenetic regulation in sepsis, role in pathophysiology and therapeutic perspective." *Medicine.* 8 (2021): 685333.

Fan, Y., Ma, X., Li, H., et al. "miR-122 promotes metastasis of clear-cell renal cell carcinoma by downregulating Dicer." *Int J Cancer.*142 (2018) 547–560.

Fang, L., Deng, Z., Shatseva, T., et al. "MicroRNA miR–93 promotes tumor growth and angiogenesis by targeting integrin–beta8." *Oncogene.* 30 (2011): 806–821.

Feinberg, A.P. "The key role of epigenetics in human disease prevention and mitigation." *N Engl J Med.* 378, no. 14 (2018): 1323–1334.

Feinberg, A.P., Tycko, B. "The history of cancer epigenetics." *Nat Rev Cancer.* 4, no. 143 (2004): 145–146.

Fenaux, P., Mufti, G.J., et al. "Efficacy of azacitidine compared with that of conventional care regimens in the treatment of higher-risk myelodysplastic Syndromes: A randomised, open-label, phase III study." *Lancet Oncol.* 10 (2009): 223–232.

Ferlay, J., Soerjomataram, I., Ervik, M., et al. "GLOBOCAN 2012 v1.0, cancer incidence and mortality worldwide: IARC CancerBase No. 11 international agency for research on cancer." Lyon, France (2013). http://globocan.iarc.fr

Fernández-Morera, JL., Calvanese, V., Rodríguez-Rodero, S., et al. "Epigenetic regulation of the immune system in health and disease." *Tissue Antigens.* 76, no. 6 (2010): 431–439.

Ferrari, A., Neefs, I., Hoeck, S., et al. "Towards novel non-invasive colorectal cancer screening methods: A comprehensive review." *Cancers.* 13 (2021): 1820.

Finnin, M.S., Donigian, J.R., Cohen, A., et al. "Structures of a histone deacetylase homologue bound to the TSA and SAHA inhibitors." *Nature.* 401 (1999): 188–193.

Fodor, A., Cozma, A., Karnieli, E. "TBC update: Personalized epigenetic management of diabetes." *Per Med.* 14, no. 6 (2017): 531–549.

Fu, B., Yan, P., Zhang, S., et al. "Cell-free circulating methylated SEPT9 for noninvasive diagnosis and monitoring of colorectal cancer." *Dis Markers.* (2018): 6437104.

Galvani, M., Ottani, F., Ferrini, D., et al. "Prognostic influence of elevated values of cardiac troponin I in patients with unstable angina." *Circulation.* 95 (1997): 2053.

García-Giménez, J.L. *Epigenetic Biomarkers and Diagnostics.* Academic Press, 2015.

Garcia-Martinez, L., Zhang, Y., et al. "Epigenetic mechanisms in breast cancer therapy and resistance." *Nat Commun.* 12 (2021): 1786.

Gingell, J.J., Hendrikse, E.R., Hay, D.L. "New insights into the regulation of CGRP-family receptors." *Trends Pharmacol Sci.* 40 (2019): 71–83.

Goyal, A., Sahu, R.K., Kumar, M., et al. "p16 Promoter methylation, expression, and its association with estrogen receptor, Progesterone receptor, and human epidermal growth factor receptor 2 subtype of Breast carcinoma." *J Cancer Res Ther.* 15, no. 5 (2019): 1147–1154. https://doi.org/10.4103/jcrt.JCRT_472_18

Graham, D.Y. "Helicobacter pylori update: Gastric cancer, reliable therapy, and possible benefits." *Gastroenterology* 148 (2015): 719–731.

Greally, J.M. "A user's guide to the ambiguous word-epigenetics." *Nat Rev Mol Cell Biol.* 19, no. 4 (2018): 207–208.

Griffiths, E.A., Garcia-Manero, G., et al. "Oral cedazuridine/decitabine for MDS and CMML: A phase 2 pharmacokinetic/pharmacodynamic randomized." *Blood.* 136 (2020): 674–683.

Gunzburg, M.J., Perugini, M.A., Howlett, G.J. "Structural basis for the recognition and cross-linking of amyloid fibrils by human Apolipoprotein." *E J Biol Chem.* 282 (2007): 35831–35841.

Guo, Shicheng, Xu, Lingxia, Chang, Cen. "Epigenetic regulation mediated by methylation in the pathogenesis and precision medicine of rheumatoid arthritis." *Front Genet.* 11 (2020): 811.

Han, Y.D., Oh, T.J., Chung, T.H., et al. "Early detection of colorectal cancer based on presence of methylated syndecan-2 (SDC2) in stool DNA." *Clin Epigenet.* 11 (2019): 51.

Hellstrom-Lindahl, E., Ravid, R., Nordberg, A. "Age-dependent decline of neprilysin in Alzheimer's disease and normal brain: Inverse correlation with A beta levels." *Neurobiol Aging.* 29 (2008): 210–221.

Henikof, S., Greally, J.M. "Epigenetics, cellular memory and gene regulation." *Curr Biol.* 26, no. 14 (2016): R644–R648.

Heyn, H., Esteller, M. "DNA methylation profiling in the clinic: Applications and challenges." *Nat Rev Genet.* 13 (2012): 679–680.

Hillyar, Christopher, Rallis, S. Kathrine, et al. "Advances in epigenetic cancer therapeutics." *Cureus.* 12, no. 11 (2020): e11725.

Histed, S.N., Lindenberg, M.L., Mena, E., et al. "Review of functional/anatomic imaging in ocology." *Nucl Med Commun.* 33, no. 4 (2015): 349–361. https://doi.org/10.1097/MNM.0b013e32834ec8a5

Ho, Shuk Mei. "Environmental epigenetics of asthma: An update." *J Allergy Clin Immunol.* 126, no. 3 (2010): 453–465.

Houshmand, M., Abbaszadegan, M.R., Kerachian, M.A. "Assessment of bone morphogenetic protein 3 methylation in Iranian patients with colorectal cancer." *Middle East J Dig Dis.* 9 (2017): 158–163.

Huang, J.Y., King, N.B. "Epigenetics changes nothing: What a new scientific field does and does not mean for ethics and social justice." *Public Health Ethics.* 11, no. 1 (2018): 69–81.

Hunter, P. "The second coming of epigenetic drugs: A more strategic and broader research framework could boost the development of new drugs to modify epigenetic factors and gene expression." *EMBO Rep.* 16 (2015): 276–279.

Hwang, Y.G., Balasubramani, G.K., Metes, I.D., et al. "Differential response of serum amyloid A To different therapies in early rheumatoid arthritis and its potential Value as a disease activity biomarker." *Arthritis Res Ther.* 18 (2016): 1–8.

Ihar, Z.D. "Phenotypic personhood: Epigenetics and the biolegality of processing asylum." In *Personhood in the Age of Biolegality*, 2020, pp. 127–147.

Ivey, K.N., Srivastava, D. "microRNAs as developmental regulators." *Cold Spring Harb Perspect Biol.* 7, no. 7 (2015): a008144.

Iwata, N, Tsubuki, S., Takaki, Y., et al. "Identifi cation of the major Abeta1–42-degrading catabolic pathway in brain parenchyma: Suppression leads to biochemical and pathological deposition." *Nat Med.* 6 (2000): 143–150.

Jaberie, H., Hosseini, S.V., Naghibalhossaini, F. "Evaluation of alpha 1-antitrypsin for the early diagnosis of colorectal cancer." *Pathol Oncol Res.* 26 (2020): 1165–1173.

Jaffe, A.S., Ravkilde, J., Roberts, R., et al. "It's time for a change to a troponin standard." *Circulation.* 102 (2000): 1216–1220.

Jakubowski, Jennifer, Labrie, Viviane. "Epigenetic biomarkers for Parkinson's disease: From diagnostics to therapeutics." *J. Parkinson's Dis.* 7, no. 1 (2017): 1–12.

Janson, P.C., Winerdal, M.E., Winqvist, O. "At the crossroads of T helper lineage commitment-Epigenetics points the way." *Biochim Biophys Acta.* 1790, no. 9 (2009): 906–919.

Jeffries, M.A., Sawalha, A.H. "Autoimmune disease in the epigenetic era: How has epigenetics changed our understanding of disease and how can we expect the field to evolve." *Expert Rev Clin Immunol.* (2015): 11.

Jia, Y.J., Liu, Z.B., Wang, W.G., et al. "HDAC6 regulates microRNA27b that suppresses proliferation, promotes apoptosis and target MET in diffuse large B-cell lymphoma." *Leukemia.* 32 (2018): 703–711.

Jin, Z., Cheng, Y., Gu, W., et al. "A multicenter, double-blinded validation study of methylation biomarkers for progression prediction in Barrett's esophagus." *Cancer Res.* 69 (2009): 4112–4115.

Joly, Y., Dupras, C., Pinkesz, M., et al. "Looking beyond GINA: Policy approaches to address genetic discrimination." *Annu Rev Genom Hum Genet.* 21 (2020): 491–507.

Jones, P.L., Veenstra, G.J., Wade, P., et al. "Methylated DNA and MeCP2 recruit histone deacetylase to repress transcription." *Nat Genet.* 19 (1998): 187–191.

Juengst, E.T., Fishman, J.R., McGowan, M.L., et al. "Serving epigenetics before its time." *Trends Genet.* 30, no. 10 (2014): 427–429.

Kabir, T.D., Ganda, C., Brown, R.M., et al. "A microRNA-7/growth arrest specific 6/TYRO3 axis regulates the growth and invasiveness of sorafenib-resistant cells in human hepatocellular carcinoma." *Hepatology.* 67 (2018): 216–231.

Kawakami, K., Brabender, J., Lord, R.V., et al. "Hypermethylated APC DNA in plasma and prognosis of patients with esophageal adenocarcinoma." *J Natl Cancer Inst.* 92 (2000): 1805–1811.

Kersting, M., Friedl, C., Kraus, A., et al. "Differential frequencies of p16(INK4a) promoter hypermethylation, p53 mutation, and K-ras mutation in exfoliative material mark the development of lung cancer in symptomatic chronic smokers." *J Clin Oncol.* 18 (2000): 3221–3229.

Khera, A., McGuire, D.K., Murphy, S.A., et al. "Race and gender differences in C-reactive protein levels." *J Am Coll Cardiol.* 46 (2005): 464–469.

Kim, J.H., Park, S.C. "Syndecan-2 methylation as a new biomarker for early detection of colorectal neoplasm." *Gut Liver.* 12 (2018): 479–480.

Kim, M.S., Lee, J., Sidransky, D. "DNA methylation markers in colorectal cancer." *Cancer Metast Rev.* 29 (2010): 181–189.

Kim, Y.K., Kim, W.-J. "Epigenetic markers as promising prognostic indicators for bladder cancer." *Int J Urol.* 16 (2009): 17–20.

Kingsley, S.M.K., Bhat, B.V. "Role of microRNAs in sepsis." *Inflamm Res.* 66 (2017): 553–569.

Knoppers, T., Beauchamps, E., Dewar, K., et al. "The omics of our lives: Practices and policies of direct-toconsumer epigenetic and microbiomic testing companies." *New Genet Soc.* 40, no. 4 (2021): 541–569.

Koga, S., Aoki, N., Uitti, RJ., et al. "When DLB, PD, and PSP masquerade as MSA: An Autopsy study of 134 patients." *Neurology.* 85, no. 5 (2015): 404–412. https://doi.org/10.1212/WNL.0000000000001807

Kokabu, S., Rosen, V. "BMP3 expression by osteoblast lineage cells is regulated by canonical WNT signaling." *FEBS Open Bio.* 8 (2018): 168–176.

Krauser, D.G., Lloyd-Jones, D.M., Chae, C.U., et al. "Effect of body mass index on natriuretic peptide levels in patients with acute congestive heart failure: A ProBNP Investigation of Dyspnea in the Emergency Department (PRIDE) substudy." *Am Heart J.* 149 (2005): 744–750.

Lal, G., Bromberg, J.S. "Epigenetic mechanisms of regulation of Foxp3 expression." *Blood.* 114, no. 18 (2009): 3727–3735.

Lamb, Y.N., Dhillon, S. "Epi proColon®2.0 CE: A blood-based screening test for colorectal cancer." *Mol Diagn Ther.* 21 (2017): 225–232.

Lappé, M., Landecker, H. "How the genome got a life span." *New Genet Soc.* 34, no. 2 (2015): 152–176.

Lee, E.J., Luo, J., Wilson, J.M., et al. "Analyzing the cancer methylome through targeted bisulfite sequencing." *Cancer Lett.* 340, no. 2 (2012): 171–178.

Lee, G.R., Kim, S.T., Spilianakis, C.G., et al. "T helper cell differentiation: Regulation by cis elements and epigenetics." *Immunity.* 24, no. 4 (2006): 369–379.

Lee, H.J., Li, C.W., Hammerstad, S.S., et al. "Immunogenetics of autoimmune thyroid diseases: A comprehensive review." *J Autoimmun.* 64 (2015): 82–90.

Leung, W.K., Wu, Y., Kakugawa, M.-S., et al. "Screening for gastric cancer in Asia: Current evidence and practice." *Lancet Oncol.* 9 (2008): 279–287.

Lichner, Z., Saleh, C., Subramaniam, V., et al. "miR-17 inhibition enhances the formation of kidney cancer spheres with stem cell/tumor initiating cell properties." *Oncotarget.* 6 (2014): 5567–5581.

Liloglou, Triantafillos, Bediaga, Naiara G., Brown, Benjamin R.B., et al. "Epigenetic biomarkers in lung cancer." *Cancer Lett.* 342 (2014): 200–212.

Limbach, M., Saare, M., Tserel, L., et al. "Epigenetic profiling in CD4+ and CD8+ T cells from graves' disease patients reveals changes in genes associated with T cell receptor signaling." *J Autoimmun.* 67 (2016): 46–56.

Lin, A., Li, C., Xing, Z., et al. "The LINK-A lncRNA activates normoxic HIF1alpha signalling in triple-negative breast cancer." *Nat Cell Biol.* 18 (2016): 213–224.

Liu, A., Zhu, J., Wu, G., Cao, L., Tan, Z., Zhang, S., et al. "Antagonizing miR-455–3p inhibits chemoresistance and aggressiveness in esophageal squamous cell carcinoma." *Mol Cancer.* 16, no. 1 (2017): 106. https://doi.org/10.1186/s12943-017-0669-9

Liu, W.J., Zhao, Y., Chen, X., Miao, M.L., Zhang, R.Q. "Epigenetic modifications in esophageal cancer: An evolving biomarker." *Front Genet.* 13 (2023): 1087479.

Liu, X., Hao, Y., Yu, W., et al. "Long non-coding RNA emergence during renal cell carcinoma tumorigenesis." *Cell Physiol Biochem.* 47 (2018): 735–746.

Liu, X., Xiao, Z.D., Han, L., et al. "LncRNA NBR2 engages a metabolic checkpoint by regulating AMPK under energy stress." *Nat Cell Biol.* 18 (2016): 431–442.

Liu, Y., Zhao, G., Miao, J., et al. "Performance comparison between Plasma and stool methylated SEPT9 tests for detecting colorectal cancer." *Front Genet.* 11 (2020): 324.

Loi, M., Del Savio, L., Stupka, E. "Social epigenetics and equality of opportunity." *Public Health Ethics.* 6, no. 2 (2013): 142–153.

Ma, J., Dong, C., Ji, C. "MicroRNA and dru resistance." *Cancer Gene Ther.* 17 (2010): 523–531.

Ma, Z.Y., Law, W.L., Ng, E.K.O., et al. "Methylated septin 9 and carcinoembryonic antigen for serological diagnosis and monitoring of patients with colorectal cancer after surgery." *Sci Rep.* 9 (2019): 10326.

Maekita, T., Nakazawa, K., Mihara, M., et al. "High levels of aberrant DNA methylation in Helicobacter pylori-infected gastric mucosae and its possible association with gastric cancer risk clin." *Cancer Res.* 12 (2006): 989–995.

Makki, J.J. "Diversity of breast carcinoma: Histological subtypes and clinical relevance." *Clin Med Insights Pathol.* 8 (2015): CPath-S31563.

Malhotra, R. "Understanding migraine: Potential role of neurogenic inflammation." *Ann Indian Acad Neurol.* 19 (2016): 175.

Mancilla, V.J., Peeri, N.C., Silzer, T., et al. "Understanding the interplay between health disparities and epigenomics." *Front Genet.* (2020): 903.

Manolio, T.A., Collins, F.S., Cox, N.J., et al. "Finding the missing heritability of complex diseases." *Nature.* 461 (2009): 747–753.

McAnena, P., Brown, J.A., Kerin, M.J. "Circulating nucleosomes and nucleosome modifications as biomarkers in cancer." *Cancers.* 9 (2017): 5.

McLachlan, S.M., Rapoport, B. "Breaking tolerance to thyroid antigens: Changing concepts in thyroid autoimmunity." *Endocr Rev.* 35, no. 1 (2014): 59–105.

Meda, F., Folci, M., Baccarelli, A., et al. "The epigenetics of autoimmunity." *Cell Mol Immunol.* 8, no. 3 (2011): 226–236.

M'hamdi, H.I., de Beaufort, I., Jack, B., et al. "Responsibility in the age of developmental origins of health and disease (DOHaD) and epigenetics." *J Dev Origins Health Dis.* 9, no. 1 (2018): 58–62.

Miller, M., Freeman, P. "Legal issues relating to epigenetics in life insurance." In *Association of Life Insurance Counsel Meeting*, 2018. Available from https://cdn.ymaws.com/www.alic.cc/resource/collection/85A25C98-739E-46C6-A9A0-BA94A55C001D/Insurance_Section_-_Epigenetics_by_Miller_and_Freeman.pdf

Miller, Michael, Reynolds, Sheila M., Bernard, Brady. "Comprehensive molecular characterization of gastric adenocarcinoma." *Nature.* 513 (2014): 202–209.

Morel, D., Jeffery, D., et al. "Combining epigenetic drugs with other therapies for solid Tumours—Past lessons and future promise." *Nat Rev Clin Oncol.* 17 (2020): 91–107.

Moscarello, M.A., Mastronardi, F.G., Wood, D.D., et al. "The role of Citrullinated proteins suggests a novel mechanism in the Pathogenesis of multiple sclerosis." *Neurochem Res.* 32 (2007): 251–256.

Mowla, Karim, Mohammad, Amin SakiSaki, Mohammad, Taha Jalali, et al. "How to manage rheumatoid arthritis according to classic biomarkers and polymorphisms." *Front Biol.* 12 (2017): 183–191.

Müller, R., Hanson, C., Hanson, M., et al. "The biosocial genome? Interdisciplinary perspectives on environmental epigenetics, health and society." *EMBO Rep.* 18, no. 10 (2017): 1677–1682.

"Myocardial infarction redefined—a consensus document of the Joint European Society of Cardiology/American College of Cardiology Committee for the redefinition of myocardial infarction." *Eur Heart J.* 21 (2000): 1502–1513.

Nell, V.P., Machold, K.P., Stamm, T.A., et al. "Autoantibody profiling as early Diagnostic and prognostic tool for rheumatoid arthritis." *Ann Rheum Dis.* 64, no. 12 (2005): 1731–1736.

Nilsen, T.W. "Mechanisms of microRNA–mediated gene regulation in animal cells." *Trends Genet.* 23 (2007): 243–249.

Niu, F., Wen, J., Fu, X., et al. "Stool DNA test of methylated syndecan-2 for the early detection of colorectal neoplasia." *Cancer Epidemiol Prev Biomark.* 26 (2017): 1411–1419.

Nuovo, G.J., Plaia, T.W., Belinsky, S.A., et al. "In situ detection of the hypermethylation-induced inactivation of the p16 gene as an early event in oncogenesis." *Proc Natl Acad Sci USA.* 96 (1999): 12754–12759.

Oh, E.S., Petronis, A. "Origins of human disease: The chrono-epigenetic perspective." *Nat Rev Genet.* 22, no. 8 (2021): 533–546.

Oh, T., Kim, N., Moon, Y., et al. "Genome-wide identification and validation of a novel methylation biomarker, SDC2, for blood-based detection of colorectal cancer." *J Mol Diagn.* 15 (2013): 498–507.

Oh, T.J., Oh, H.I., Seo, Y.Y., et al. "Feasibility of quantifying SDC2 methylation in stool DNA for early detection of colorectal cancer." *Clin Epigenet.* 9 (2017): 126.

Ohman, E.M., Armstrong, P.W., Christenson, R.H., et al. "Cardiac troponin T levels for risk stratification in acute myocardial ischemia." GUSTO IIA investigators. *N Engl J Med.* 335 (1996): 1333–1341.

Olsson, A.H., Volkov, P., Bacos, K., et al. "Genomewide associations between genetic and epigenetic variation influence mRNA expression and insulin secretion in human pancreatic islets." *PLoS Genet.* 10, no. 11 (2014): e1004735.

Pang, K.C., Dinger, M.E., Mercer, T.R., et al. "Genome-wide identification of long noncoding RNAs in CD8+ T cells." *J Immunol.* 182, no. 12 (2009): 7738–7748.

Park, Y.S., Kim, D.S., Cho, S.W., et al. "Analysis of syndecan-2 methylation in bowel lavage fluid for the detection of colorectal neoplasm." *Gut Liver.* 12 (2018): 508–515.

Pearson, T.A., Mensah, G.A., Alexander, R.W., et al. "Markers of inflammation and cardiovascular disease: Application to clinical and public health practice: A statement for healthcare professionals from the Centers for Disease Control and Prevention and the American Heart Association." *Circulation.* 107 (2003): 499–511.

Peng, W.X., Koirala, P., Mo, Y.Y. "LncRNAmediated regulation of cell signaling in cancer." *Oncogene.* 36 (2017): 5661–5667.

Peplow, Philip V., Martinez, Bridget, Gennarelli, Thomas, et al. "Neurodegenerative diseases biomarkers: Towards translating research to clinical practice." *Springer US.* 173 (2022): 155–180.

Pillai, R.S., Bhattacharyya, S.N., Filipowicz, W. "Repression of protein synthesis by miRNAs: How many mechanisms." *Trends Cell Biol.* 17 (2007): 118–126.

Portela, A., Esteller, M. "Epigenetic modifications and human disease." *Nat Biotechnol.* 28 (2010): 1057–1068.

Probst, A.V., Dunleavy, E., Almouzni, G. "Epigenetic inheritance during the cell cycle." *Nat Rev Mol Cell Biol.* 10 (2009): 192–206.

Qu, L., Ding, J., Chen, C., et al. *Exosome-Transmitted lncARSR Promotes Sunitinib Resistance in Renal Cancer by Acting as a Competing Endogenous RNA.* Cell Press, 2016.

Quddus, J., Johnson, K.J., Gavalchin, J., et al. "Treating activated CD4+ T cells with either of two distinct DNA methyltransferase inhibitors, 5-azacytidine or procainamide, is sufficient to cause a lupus-like disease in syngeneic mice." *J Clin Invest.* 92, no. 1 (1993): 38–53.

Rauch, T., Wang, Z., Zhang, X., et al. "Homeobox gene methylation in lung cancer studied by genome-wide analysis with a microarray-based methylated CpG island recovery assay." *Proc Natl Acad Sci USA.* 104 (2007): 5527–5532.

Rauch, T.A., Zhong, X., Wu, X., et al. "High-resolution mapping of DNA hypermethylation and hypomethylation in lung cancer." *Proc Natl Acad Sci USA.* 105 (2008): 252–257.

Richardson, B. "Effect of an inhibitor of DNA methylation on T cells. II. 5- Azacytidine induces self-reactivity in antigen-specific T4+ cells." *Hum Immunol.* 17, no. 4 (1986): 456–470.

Richardson, B. "Primer: Epigenetics of autoimmunity." *Nat Clin Pract Rheumatol.* 3, no. 9 (2007): 521–527.

Richardson, B.C., Liebling, M.R., Hudson, J.L. "CD4+ cells treated with DNA methylation inhibitors induce autologous B cell differentiation." *Clin Epigenetics.* 11, no. 34 (2019): 12 of 15.

Rifai, N., Ridker, PM. "Proposed cardiovascular risk assessment algorithm using high-sensitivity C-reactive protein and lipid screening." *Clin Chem.* 47 (2001): 28–30.

Rifai, N., Warnick, G.R. "Quality specifications and the assessment of the biochemical risk of atherosclerosis." *Clin Chem Acta.* 346 (2004): 55–64.

Rozek, L.S., Dolinoy, D.C., Sartor, M.A., et al. "Epigenetics: Relevance and implications for public health." *Annu Rev Public Health.* 35 (2014): 105–122.

Rui, J., Deng, S., Lebastchi, J., et al. "Methylation of insulin DNA in response to proinflammatory cytokines during the progression of autoimmune diabetes in NOD mice." *Diabetologia.* 59, no. 5 (2016): 1021–1029.

Russo, A.F. "CGRP as a neuropeptide in migraine: Lessons from mice: CGRP as a neuropeptide in migraine." *Br J Clin Pharmacol.* 80 (2015): 403–414.

Sarnak, M.J., Katz, R., Stehman-Breen, C.O., et al. "Cystatin C concentration as a risk factor for heart failure in older adults." *Ann Intern Med.* 142 (2005): 497–505.

Sawalha, A.H. "Epigenetics and T-cell immunity." *Autoimmunity.* 41, no. 4 (2008): 245–252.

Schaefer, A., Stephan, C., Busch, J., et al. "Diagnostic, prognostic and therapeutic implications of microRNAs in urologic tumors." *Nat Rev Urol.* 7 (2010): 286–297.

Schulmann, K., Sterian, A., Berki, A., et al. "Inactivation of p16, RUNX3, and HPP1 occurs early in Barrett's-associated neoplastic progression and predicts progression risk." *Oncogene.* 24 (2005): 4138–4148.

Seike, M., Gemma, A., Hosoya, Y., et al. "Increase in the frequency of p16INK4 gene inactivation by hypermethylation in lung cancer during the process of metastasis and its relation to the status of p53". *Clin Cancer Res.* 6 (2000): 4307–4313.

Selvakumar, Sushmaa Chandralekha, Preethi, K.A., Tusubira, Deusdedit, et al. MicroRNAs in the epigenetic regulation of disease progression in Parkinson's disease. *Front Cell Neurosci.* 16 (2022): 995997.

Shabani, M., Borry, P., Smeers, I. "Forensic epigenetic age estimation and beyond: Ethical and legal considerations." *Trends Genet.* 34, no. 7 (2018): 489–491.

Shapo, N., Masar, M.S. "Modern regulatory frameworks for the use of genetic and epigenetic underwriting technology in life insurance." *J Insur Regul.* 30, no. 10 (2020):1–23.

Sharifi, M., Moridnia, A. "Apoptosis-inducing and antiproliferative effect by inhibition of miR-182 5p through the regulation of CASP9 expression in human breast cancer." *Cancer Gene Ther.* 24 (2017): 75–82.

Sharma, J., Bhar, S., et al. "A review on Interleukins: The key Manipulators in Rheumatoid Arthritis." *Modern Rheumatol.* 27 (2017): 723–746.

Sharma, S., Kelly, T.K., Jones, P.A. "Epigenetics in cancer." *Carcinogenesis* 31 (2010): 27–33.

Shayan, M., Shariatpanahi, A.M., Seifati, S., et al. "Aberrant SEPT9 methylation in plasma cell-free DNA of CRC patients." *Biomed Res.* 29 (2018): 3654–3659.

Siegmund, K.D., Connor, C.M., Campan, M., et al. "DNA methylation in the human cerebral cortex is dynamically regulated throughout the life span and involves differentiated neurons." *PLoS ONE.* 2 (2007): e895.

Silverman, L.R., Demakos, E.P., et al. "Randomized controlled trial of Azacitidine in patients with the myelodysplastic syndrome: A study of the cancer and leukemia group B." *J Clin Oncol.* 20 (2002): 2429–2440.

Skvortsova, K., Stirzaker, C., et al. "The DNA methylation landscape in cancer." *Essays Biochem.* 63 (2019): 797–811.

Slatkin, M. "Epigenetic inheritance and the missing heritability problem." *Genetics.* 182 (2009): 845–850.

Sole, C., Arnaiz, E., Manterola, L., et al. "The circulating transcriptome as a source of cancer liquid biopsy biomarkers." *Semin Cancer Biol.* 58 (2019): 100–108.

Stirzaker, C., Taberlay, P.C., Statham, A.L., et al. "Mining cancer methylomes: Prospects and challenges." *Trends Genet.* 30 (2014): 75–84.

Strickland, F.M., Richardson, B.C. "Epigenetics in human autoimmunity. Epigenetics in autoimmunity—DNA methylation in systemic lupus erythematosus and beyond." *Autoimmunity.* 41 (2008).

Su, W.C., Kao, W.Y., Chang, T.K., et al. "Stool DNA test targeting methylated syndecan-2 (SDC2) as a noninvasive screening method for colorectal cancer." *Biosci Rep.* 41 (2021): BSR20201930.

Subrahmanyam, R., Sen, R. "Epigenetic features that regulate IgH locus recombination and expression." *Curr Top Microbiol Immunol.* 13 (2012): 1205–212.

Suehiro, Y., Zhang, Y., Hashimoto, S., et al. "Highly sensitive faecal DNA testing of TWIST1 methylation in combination with faecal immunochemical test for haemoglobin is a promising marker for detection of colorectal neoplasia." *Ann Clin Biochem.* 55 (2018): 59–68.

Sun, J., Fei, F., Zhang, M., et al. "The role of mSEPT9 in screening, diagnosis, and recurrence monitoring of colorectal cancer." *BMC Cancer.* 19 (2019): 450.

Sun, J., Zheng, M.Y., Li, Y.W., et al. "Structure and function of Septin 9 and its role in human malignant tumors." *World J Gastrointest Oncol.* 12 (2020): 619–631.

Sung, H., Ferlay, J., Siegel, R.L., Laversanne, M., et al. "Global cancer statistics 2020: GLOBOCANEstimates of incidence and mortality worldwide for 36 cancers in 185 countries." *CA Cancer J Clin.* 71 (2021): 209–249.

Szyf, M. "Epigenetic, DNA methylation, and chromatin modifying drugs." *Annu Rev Pharmacol Toxicol.* 49 (2009): 243–254.

Taki, de Melo-Martin. "Conducting epigenetics research with refugees and asylum seekers: Attending to the ethical challenges." *Clin Epigenet.* 13 (2021): 105.

Tang, S.Y., Zhou, P.J., Meng, Y., Zeng, F.R., Deng, G.T. Gastric cancer: An epigenetic view. *World J Gastrointest Oncol.* 14, no. 1 (2022): 90.

Tendl, K.A., Schulz, S.M., Mechtler, T.P., et al. "DNA methylation pattern of CALCA in preterm neonates with bacterial sepsis as a putative epigenetic biomarker." *Epigenetics* 8 (2013): 1261–1267.

Tessarz, P., Kouzarides, T. "Histone core modifications regulating nucleosome Structure and dynamics." *Nat Rev Mol Cell Bio.* 15 (2014): 703–708.

Thomas, M.L., Marcato, P. "Epigenetic modifications as biomarkers of tumor development, therapy response, and recurrence across the Cancer care Continuum." *Cancers (Basel).* 10, no. 4 (2018).

Thornburg, K.L., Shannon, J., Thuillier, P., Turker, M.S. "In utero life and epigenetic predisposition for disease." *Adv Cancer.* 71, no. 57 (2010): 61–66.

Tomasetti, Marco, Amati, Monica, Neuzil, Jiri, et al. "Circulating epigenetic biomarkers in lung malignancies: From early diagnosis to therapy Author links open overlay panel." *Lung Cancer.* 107 (2017): 65–72.

Turajlic, S., Xu, H., Litchfield, K., et al. "Deterministic evolutionary trajectories influence primary tumor growth: TRACERx renal Cell." *Cell Press.* 173 (2018a): 595–610.

Turajlic, S., Xu, H., Litchfield, K., et al. "Tracking cancer evolution reveals constrained routes to metastases: TRACERx renal Cell." *Cell Press.* 173 (2018b): 581–594.

Urdinguio, Rocio G., Sanchez-Mut, Jose V., Esteller, Manel. "Epigenetic mechanisms in neurological diseases: Genes, syndromes, and therapies." *Lancet Neurol.* 8, no. 11 (2009): 1056–1072.

Ushiku, H., Yamashita, K., Katoh, H., et al. "Promoter DNA methylation of CDO1 gene and its clinical significance in esophageal squamous cell carcinoma." *Dis Esophagus.* 30 (2017): 1–9.

Vaes, N., Lentjes, M.H.F.M., Gijbels, M.J., et al. "NDRG4, an early detection marker for colorectal cancer, is specifically expressed in enteric neurons." *Neurogastroenterol Motil.* 29 (2017): e13095.

Vaes, N., Schonkeren, S.L., Rademakers, G., et al. "Loss of enteric neuronal Ndrg4 promotes colorectal cancer via increased release of Nid1 and Fbln2." *EMBO Rep.* 22 (2021): e51913.

Vanden Berghe, W., Ndlovu, M.N., Hoya-Arias, R., et al. "Keeping up NF-kappaB appearances: Epigenetic control of immunity or inflammation-triggered epigenetics." *Biochem Pharmacol.* 72, no. 9 (2006): 1114–1131.

Vears, D.F., D'Abramo, F. "Health, wealth and behavioural change: An exploration of role responsibilities in the wake of epigenetics." *J Community Genet.* 9, no. 2 (2018): 153–167.

Vidaki, A., Kayser, M. "Recent progress, methods and perspectives in forensic epigenetics." *Forensic Sci Int Genet.* 37 (2018): 180–195.

Wade, Paul, Gegonne, Anne, Jones, P.L., et al. "Mi-2 complex couples DNA methylation to chromatin remodelling and histone deacetylation." *Nat Genet.* 23 (1999): 62–66.

Wang, J., Liu, S., Wang, H., et al. "Robust performance of a novel stool DNA test of methylated SDC2 for colorectal cancer detection: A multicenter clinical study." *Clin Epigenet.* 12 (2020): 162.

Wang, J.S., Guo, M., Montgomery, E.A. et al. "DNA promoter hypermethylation of p16 and APC predicts neoplastic progression in Barrett's esophagus." *Am J Gastroenterol.* 104 (2009): 2153–2160.

Wang, Z., Shang, J., Zhang, G., et al. "Evaluating the clinical performance of a novel dual-target stool DNA test for colorectal cancer detection." *medRxiv* (2021).

Weber, J.A., Baxter, D.H., Zhang, S., et al. "The microRNA spectrum in 12 body fluids." *Clin Chem.* 56 (2010): 1733–1741.

Wen, J., Liu, X., Qi, Y., et al. "BMP3 suppresses colon tumorigenesis via ActRIIB/SMAD2-dependent and TAK1/JNK signaling pathways." *J Exp Clin Cancer Res.* 38 (2019): 428.

Wilson, C.B., Makar, K.W., Shnyreva, M., et al. "DNA methy-lation and the expanding epigenetics of T cell lineage commitment." *Semin Immunol.* 17, no. 2 (2005): 105–119.

Wilson, C.B., Rowell, E., Sekimata, M. "Epigenetic control of T-helper-cell differentiation." *Nat Rev Immunol.* 9, no. 2 (2009): 91–105.

Wu, C.T., Morris, J.R. "Genes, genetics, and epigenetics: A correspondence." *Science* 293, no. 5532 (2001): 1103–1105.

Wu, J., Zhang, X., Nauta, H.J., Lin, Q., et al. "JNK1 reg ulates histone acetylation in trigeminal neurons following chemical stimulation." *Biochem Biophys Res Commun.* 376 (2008): 781–786.

Xiao, W., Lou, N., Ruan, H., et al. "Mir-144–3p promotes cell proliferation, metastasis, sunitinib resistance in clear cell renal cell carcinoma by downregulating ARID1A." *Cell Physiol Biochem Int J Exp Cell Phys Biochem Pharmacol.* 43 (2017): 2420–2433.

Xiao, W., Wang, X., Wang, T., et al. "MiR-223–3p promotes cell proliferation and metastasis by downregulating SLC4A4 in clear cell renal cell carcinoma." *Aging* 11 (2019): 615–633.

Xie, L., Jiang, X., Li, Q., et al. "Diagnostic value of methylated septin9 for colorectal cancer detection." *Front Oncol.* 8 (2018): 247.

Yang, X., Xu, Z.J., Chen, X., et al. "Clinical value of preoperative methylated septin 9 in Chinese colorectal cancer patients." *World J Gastroenterol.* 25 (2019): 2099–2109.

Yuan, J.H., Liu, X.N., Wang, T.T., et al. "The MBNL3 splicing factor promotes hepatocellular carcinoma by increasing PXN expression through the alternative splicing of lncRNA-PXN-AS1." *Nat Cell Biol.* 19 (2017): 820–832.

Yung, R.L., Quddus, J., Chrisp, C.E., et al. "Mechanism of drug-induced lupus. I. Cloned Th2 cells modified with DNA methylation inhibitors in vitro cause autoimmunity in vivo." *J Immunol.* 154, no. 6 (1995): 3025–3035.

Zhao, Yuhai, Jaber, Vivian, Alexandrov, Peter, et al. "Deficiency in the ubiquitin conjugating enzyme UBE2A in Alzheimer's disease (AD) is linked to deficits in a natural circular miRNA-7 sponge (circRNA; ciRS-7)." *Genes* 7, no. 12 (2016): 116.

Zhu, X.T., Yuan, J.H., Zhu, T.T., et al. "Long noncoding RNA glypican 3 (GPC3) antisense transcript 1 promotes hepatocellular carcinoma progression via epigenetically activating GPC3." *FEBS J.* 283 (2016): 3739–3754.

Zochbauer, Muller S., Fong, K.M., Virmani, A.K., et al. "Aberrant promoter methylation of multiple genes in non-small cell lung cancers." *Cancer Res.* 61 (2001): 249–255.

Zou, H., Osborn, N.K., Harrington, J.J., et al. "Frequent methylation of eyes absent 4 gene in Barrett's esophagus and esophageal adenocarcinoma." *Cancer Epidemiol Biomarkers Prev.* 14 (2005): 830–834.

Index

A

A1AT alpha 1 antitrypsin, 214
ACS acute coronary syndrome, 219, 220
AI (Artificial Intelligence), 196
Akt Pathways, 66, 95, 108
alkaloids, 146, 187, 190
allicin, 73
AML (acute myeloid lymphoma), 8, 64
amyloid precursor protein (APP), 101, 103, 107
angiogenesis, 7, 162, 185, 211, 213
ankylosing spondylitis, 92, 95
apoptosis, 3, 6, 17, 20, 48, 50–52, 59–61, 71–74, 76, 94, 108, 185, 197, 211
arginine methyltransferases, 66, 89, 212
Artemisia annua, 146
artemisinin, 136
asthma, 208, 209, 223, 224
autoantibodies, 91, 92, 217
autoimmune diseases, 226
azacitidine, 37, 38, 63, 64, 65, 67, 68, 107
azumamides, 188
Aβ amyloid beta, 102, 103

B

BACE1, 19, 103
bacillus, 165
bacterial β-glucuronidases, 167, 168
Bacteroides, 163, 165
barberry, 73, 75
Barrett's esophagus, 232
belinostat, 38, 39, 60, 61, 64, 65
berberine, 38, 39, 60, 61, 64, 65
betaine, 24
betanin, 74
betulinic acid, 73
big data analytics, 202
bioinformatics, 189
biological processes, 15, 48, 90, 102, 115, 153, 155, 165, 189, 190, 202, 203, 209, 210
biomarkers, 202, 203, 209, 210
biosynthesis, 146, 147
bisulfite conversion, 3, 9
black cumin, 51
bladder cancer, 227
blood-brain barrier, 108, 162, 199
BMP3 (bone morphogenetic protein 3), 41, 213, 214
BNP (brain-type natriuretic peptide), 219, 220
bone, 41, 94, 213, 214
brain tumors, 51
breakthroughs, 2, 5, 201
breast cancer, 20, 25, 39, 42, 50, 53, 59, 63–67, 73, 76, 123, 156, 199, 214, 215
bromodomain, 51, 187, 215, 228
butyrate, 160, 166, 169, 171, 172

C

calebin-A, 137, 146
campotothecin, 74, 167, 168
cancer, 1–3, 6, 182, 185–190, 197–200, 203, 205, 208–216, 225–229
cancer stem cells (CSCs), 63, 193, 228
cannabidiol, 75, 137
cardiovascular diseases (CVDs), 21, 24–26, 28, 36
carmustine, 43
carnosol, 74
Cas9, 172
celastrol, 75
cell differentiation, 14
cell-free nucleosomes, 42
cell-free tumor DNA (ccfDNA), 215
central nervous system, 64, 170, 173
cervical cancer, 19, 42
challenges, 189, 199, 224, 225
chemotherapy, 27, 40, 42, 156, 173, 210, 212
ChIP chromatin immunoprecipitation, 4
cholangiocarcinoma, 5
chromatin modifications, 4, 69
chromatin remodeling, 104
chronic inflammatory diseases, 86, 91
chronic myeloid leukemia, 188
chronic myelomonocytic leukemia (CMML), 64
circadian rhythms, 8
citrate, 220
citrus fruits, 51, 73–75, 140, 146
cladribine, 64
clinical triats, 221
CLOCK, 8, 89
cognitive function, 25, 170
colitis, 58, 62, 72, 82, 163, 168, 172
Cologuard®, 41, 213
colorectal cancer (CRC), 41, 42, 166–168, 173, 213
computational methods, 116, 119, 123
Coptis chinensis Franch, 147
coronary artery disease, 19, 116
CpG, 143, 146, 162, 210, 212, 218
C-reactive protein (CRP), 94, 219, 221
CREB binding protein (CBP), 65, 69, 88, 227
CRISPR technology, 11
curcumin, 25, 50, 51, 71
cutaneous T-cell lymphoma (CTCL), 64
cyclin-dependent kinase inhibitor 2A (CDKN2A), 212
cyclostellettamines, 188
cytosine-phosphodiester-guanine (CpG), 9, 197, 210, 212
cytoskeleton, 213
cytotoxic T-lymphocyte–associated antigen 4 (CTLA 4), 218

D

dacinostat, 38, 39
decitabine, 37, 63, 64

dementia, 99–103
demethylase, 106, 117, 123
detoxification, 159, 170, 185
developmental disorders, 48
diabetes, 49, 134, 159
diagnostic tools, 2, 11
diet, 167
differentiation, 160, 184, 185, 190
digestive system tumor, 146
DNA hypomethylation, 7, 8, 218
DNA methylation, 11, 19, 30, 38, 47, 53, 80, 114, 123, 125, 130–132, 134, 142, 143, 196, 229
DNA repair, 53, 114, 125, 144, 146, 148, 149
DNMT DNA methyltransferase, 47, 53, 131, 132, 134, 135
docking, 115, 116, 120, 121, 131–133
DOT1L, 61
drug delivery systems (DDSs), 197, 199, 201
drug design research, 116
drug discovery, 116, 117
drug repurposing, 117
dysregulation, 69, 103, 196

E

early detection, 42, 211
efficiency, 69, 78, 117, 157, 190, 217
efflux mechanisms, 59, 200
EGCG epigallocatechin gallate, 7, 72, 73, 122, 126
ELISA, 42, 98
embryonic development, 17, 87, 180, 190
ENCODE, 22, 124, 170
environmental factors, 11, 14, 21, 22, 103, 115, 180, 189
enzalutamide, 66
enzyme-linked immunosorbent assay, 9
enzymes, 9, 22, 23, 26, 88, 160, 210
epigenetic biomarkers, 41
epigenetic healing, 47, 53, 172, 174
epigenetic mechanisms, 2, 16, 100, 209
epigenetics, 209, 215–217, 220
epigenome editing, 11, 29, 33
epigenome Project, 2
epithelioid sarcoma, 43
esophageal cancer, 212
ethical considerations, 204
eukaryotes, 48, 90
EZH2 inhibitor, 43, 80

F

fatty acids, 50, 55, 165, 168, 220
FDA approval, 213
fecal hemoglobin, 41, 213
fetal alcohol exposure, 9
firmicutes, 163, 165, 167
flaxseeds, 25
fludarabine, 64
folate, 7, 8, 24, 30, 162, 184
FOXP3, 50, 227
fragile X syndrome, 20, 185, 228
functional components, 124, 155
fungus PRMT, 119
future directions, 205

G

gastric cancer, 65, 168, 211, 212
gene activity, 1, 22, 25, 26, 36, 85, 86, 184
gene expression, 159, 160, 162, 163, 167, 170–174, 180, 182, 226, 229
gene regulation, 2, 14, 22, 24, 91, 215
genetic code, 1, 14, 25, 48
genetic discrimination, 224, 225
genetic markers, 136
genistein, 7, 51, 73
genome-wide association studies, 93
genomics, 117, 118, 136, 147, 153, 203
Ginkgo biloba, 146
ginsenosides, 146
glioblastoma multiforme (GBM), 42, 43
glucocorticoid receptor gene promoter, 8
Glycyrrhiza uralensis, 147
GPR109a, 160
Graves' disease (GD), 218
green tea, 7, 51, 71, 73, 75
guadecitabine, 38, 68
gut microbiota, 159, 163, 173
gynecological cancer, 43

H

HAT (histone acetyltransferase), 7, 23, 66, 69, 75, 116, 166
HD Huntington's disease, 99, 102, 105
HDAC histone deacetylase, 116, 169
hematological malignancies, 215
HER2-negative, 42
herbal medicine, 50, 51, 134
hesperidin, 74
high-throughput screening, 186, 190
hippocampus, 8, 103, 169
histone acetylation, 4, 9, 18, 24, 25, 28, 40, 44, 61, 173
histone acetyltransferases (HATs), 23, 66, 75, 88, 116, 166, 169
histone deacetylation, 39, 60, 209
histone demethylase, 116, 229
histone methyltransferase (HMT), 61, 75, 116
histone modification, 2, 15, 16, 18, 20, 26, 27, 30, 59, 65, 69, 79, 90, 104, 203, 221
HLA (human leucocyte antigen), 62, 92–95, 107
homology modelling, 119, 121
Human Epigenome Project, 2
huntingtin gene (HTT), 105, 106
hydroxamic acid, 196
hypermethylation, 18, 60, 208, 235
hypertension, 20, 21, 53, 219
hypomethylation, 103, 104, 107, 196, 208, 212–214

I

IBD, 62, 71, 76, 92, 93, 160, 171, 172
immune checkpoint inhibitors, 2
immune system, 8, 55, 61, 62, 76, 91, 95, 170, 217, 218
immunomodulatory, 51, 75
immunosuppression, 61, 209, 221
in silico tools, 117
in vitro diagnostic (IVD) test, 41–43, 45
insulin, 50, 51, 90, 95, 163, 168, 172, 218, 220
integrative health, 46

intellectual property rights, 224
interleukin 6 (IL6R), 93
introns, 99, 104, 226
iridium (III), 227
ischemia, 51, 220
isoforms, 119, 122, 123, 188

J

JNK Jun N-terminal kinase, 168, 216

K

kaempferol, 142, 168
KAT family, 88
KDM histone demethylase, 116
kidney cancer, 210
kinase inhibitors, 43
KMT histone methyltransferase, 49

L

lactic-co-glycolic acid, 197
lactobacillus, 163, 165, 172, 173
lead optimization, 117, 118
legal experts, 224
leukemia, 37, 39, 64, 73, 93, 95, 188, 199, 209
Lewy bodies, 104, 223
lifestyle, 1, 2, 7, 14, 51, 172, 213
lipid metabolism, 6, 21, 25, 103, 163, 226
liquid biopsies, 11, 29, 214
lncRNAs, 17, 90, 91, 102, 211
lung cancer, 26, 43, 65, 73, 214–216, 222
lupus, 61, 71, 72, 75, 91, 226
luteolin, 142, 146
lysine, 184, 187, 188
lysine demethylases (KDMs), 39, 212

M

macrophage, 63, 160, 223, 230
malaria, 146
mammals, 18, 48, 60, 89
marine sources, 185, 190
mass spectrometry, 156, 186
medicinal plants, 135, 136, 147
meditation, 46, 49
memory, 5, 6, 19, 103
messenger RNA (mRNA), 5, 94, 153
metabolic disorders, 20, 28, 165
metabolite profiles, 136
metabolites, 51, 136, 147, 162, 173, 174
methionine, 86, 89
methylated cytosine, 2
methylated DNA immunoprecipitation (MeDIP), 9
methylation-sensitive restriction enzymes, 9
MGMT promoter methylation, 42
microRNAs (miRNAs), 5, 90, 197
microtiter plates, 186
migraine, 209, 216
mitochondrial oxidative stress, 51
molecular docking, 116, 123
molecular dynamics (MD), 119
multi-omics integration, 28, 157
multiple myeloma, 8, 38, 53
multiple sclerosis (MS), 62, 75
mycoplasma, 165
myeloperoxidase, 219

N

nano inhibitors, 66
nanoparticles, 197, 199
nanopore sequencing, 10, 28
natural compounds, 185, 186
natural products, 37, 38, 46
n-butylidenephthalide, 142
ncRNAs noncoding RNAs, 5, 66, 71, 90
neurodegenerative diseases, 100, 190, 222
neurons, 22, 87, 106, 107
Nigella sativa, 51
nimbolide, 74, 143, 146
non-hematological cancers, 39
non-invasive biomarkers, 211
N-terminal pro b-type natriuretic peptide (NT-proBNP), 219
nucleosomes, 16, 36, 42, 102, 115
nutraceuticals, 50, 228
nutrition, 11, 190

O

obesity, 6, 9, 20, 23, 184, 210
off-target effects, 6, 10, 203
oleanolic acid, 143, 146
oligofructose, 167, 168
omega-3 fatty acids, 25
omics technologies, 190
oncogene activation, 19
Ophiorrhiza pumila, 147
oral cancers, 42
osteoarthritis, 226, 227
ovarian cancer, 65–67
oxidative stress, 6, 7, 8, 21, 24, 26, 27, 51, 108, 226

P

paclitaxel, 67, 136, 199
pancreatic cancer, 43, 44
panobinostat, 39, 65, 107
peripheral blood leukocytes, 49
peripheral T-cell lymphoma (PTCL), 64
P-glycoproteins, 200
phenotypes, 17, 36, 47, 85, 92, 180, 186, 189, 190, 218
phenyl butyric acid, 39
phytochemicals, 51, 135
Pivanex, 39
placental growth factor, 220
plumbagin, 74
polymerase chain reaction (PCR), 208
polyphenols, 25, 51
post-translational modification, 37, 89, 197
prebiotics, 165, 167
precision medicine, 40, 41, 155, 190
pregnancy-associated plasma protein-A, 220
prion disease, 102, 106
probiotics, 171–173
prostate cancer, 43, 65, 68, 74, 227

proteins, 181, 182, 186, 222, 224, 228
proteomics, 190, 203
psoriasis, 92
psychological disorders, 185
pterostilbene, 73

Q

QSAR, 124, 128, 129
quercetin, 25, 51, 71, 73, 75, 76
quisinostat, 39

R

reactive oxygen species (ROS), 26, 65
recombinant DNA technologies, 153
reduced representation bisulfite sequencing, 9
renal cell carcinoma (RCC), 210, 211
resveratrol, 25, 71–73, 75, 76, 122
RGQ-PCR kit, 42
rheumatoid arthritis (RA), 20, 58, 75
romidepsin, 37–39, 60, 63–65
ROS1-rearranged NLSC, 43
rutin, 75

S

S-adenosyl methionine (SAM), 119
Salvia miltiorrhiza Bunge, 147
selenium, 7, 25, 31
senescence, 211, 229
sennoside A, 144
septin 9 (SEPT9), 213
SETDB1, 66
silibinin, 145, 146
single-cell bisulfite sequencing, 10
single-cell epigenomics, 11, 28
single-molecule real-time sequencing, 10
sirtuin (SIRT) antagonists, 119
sirtuin 1 (SIRT1), 51
skin diseases, 44
SMAD4, 43
small molecule inhibitors, 39
SOCS1, 9, 62
Sophora flavescens, 75
sophoridine, 75
SORBS3, 222
sperm quality, 170
stress response gene, 170
suberic bishydroxamic acid, 37
sulforaphane, 25
synaptic plasticity, 21, 102, 103, 168
synovial sarcoma, 64
systemic autoimmune rheumatic diseases (SARDs), 217
systemic lupus erythematosus (SLE), 6, 49, 61–64, 91, 217
systemic sclerosis (SSc), 6, 217

T

tazemetostat, 39, 67
T-cell lymphoma, 38, 44, 60, 64
tefinostat, 39
TET proteins, 87
tetramethylpyrazine, 138, 146
theranostic test, 42
therapeutic interventions, 2, 29, 40, 62, 147, 172, 190
thiol, 38
thymoquinone, 51
thyroid autoimmune diseases (AITDs), 209
TNF tumor necrosis factor, 50
traditional medicine, 13, 137, 139
transcription factors, 4, 212
transcriptome, 10, 100, 153, 155, 168, 189
trapoxin, 37, 188
trichostatin A (TSA), 188
triple negative breast cancer, 53, 64
trithorax groups, 186
tumor heterogeneity, 200
tumor protein p53 (TP53), 43
tumor suppressor gene (TSG), 25
tumors, 20, 39, 43, 44, 199, 200, 211, 212, 214, 215, 227
tyrosine kinase inhibitors, 43

U

ubiquitin, 89, 223
Unani systems, 51
unknown cancers, 43
UQCRC1 gene, 107
urea cycle disorders, 39
urinary tract infection, 172
urine, 208, 211, 227
US Food and Drug Administration (FDA), 2, 65, 220
UTX demethylase, 106

V

valeric acid, 139, 146
valproic acid, 39, 118, 119, 196
vesicular transport deficit, 105
vinblastine, 146
vincristine, 146
virtual screening, 117, 118, 129
vitamin B12, 7, 24, 108, 162
vitamins, 7, 24, 31, 72, 103, 107, 108
vorinostat, 37–39, 60, 61, 63, 64, 67, 76, 168

W

walnuts, 25
whole genome bisulfite sequencing, 9
Withaferin A, 52
Withania somnifera, 52, 75

X

X chromosome inactivation, 3, 5, 18, 90, 197
Xestospongia, 188
X-ray crystallography, 187

Y

yoga, 46, 47, 49, 50, 53

Z

zebularine, 10, 38
zinc, 37–39, 42, 88, 160, 188
ZNF582, 42
Zolinza, 37, 64

Made in the USA
Las Vegas, NV
15 September 2026

239c74a5-1de6-42e5-a3bb-c1595c1b7552R02